*IP-Based Next-Generation
Wireless Networks*

IP-Based Next-Generation Wireless Networks

Systems, Architectures, and Protocols

Jyh-Cheng Chen

National Tsing Hua University

Tao Zhang

Telcordia Technologies

A John Wiley & Sons, Inc., Publication

For general information on our other products and services please contact our Customer Care Department within the U.S. at 877-762-2974, outside the U.S. at 317-572-3993 or fax 317-572-4002.

Wiley also publishes its books in a variety of electronic formats. Some content that appears in print, however, may not be available in electronic format.

Library of Congress Cataloging-in-Publication Data

Chen, Jyh-Cheng
  IP-based next-generation wireless networks : systems, architectures, and protocols / Jyh-Cheng Chen and Tao Zhang.
    p.   cm.
Includes bibliographical references and index.
  ISBN 0-471-23526-1 (Cloth)
1. Wireless LANs. I. Zhang, Tao II. Title.
  TK5105.78.C49 2004
  004.6′8—dc21

                                        2003012945

Printed in the United States of America

10 9 8 7 6 5 4 3 2

Contents

Foreword

Two technologies that have profoundly impacted people on this planet recently are cellular telephony and the Internet. The former, with its tremendous advantages of tetherless and ubiquitous communication capabilities, was accepted worldwide. It met the expectations of a success story for wealthy nations. On the other hand, its reach into the developing and the not-so-prosperous parts of the world was even more profound. These parts of the world did not have the infrastructure for providing PSTN services for the vast majority of the population, for the obvious reason that tremendous investment was needed. At the end of the twentieth century, the demographics of the most populous nations of the world changed, with a tilt towards a large middle-class population that could afford the luxury of a telephone in every household. This need was a big impetus for the growth of the cellular telephony worldwide.

The second most important technology with a global appeal is the Internet. Personal Computers (PCs), laptops, personal digital assistants, and even cellular phones can be connected to the Internet. The Internet has touched almost every segment of the population on the face of this planet with applications (besides worldwide email) in business, education, healthcare, and manufacturing, to name a few.

Cellular telephone networks could be either circuit switched or packet switched. The former could be viewed as wireless versions of the traditional PSTN with voice telephony being the primary application. The latter are wireless extensions to the Internet and hence are suitable for mobile data networking. Such cellular networks adopt the well-known Internet Protocol (IP) for networking and can be exploited for providing mobile multimedia services.

This book, *IP-Based Next-Generation Wireless Networks*, by Jyh-Cheng Chen and Tao Zhang, deals with wireless IP networking architectures, protocols, and techniques at the IP layer and above. It is a valuable reference for academicians, engineers, and business personnel. It comes at an appropriate time.

Prathima Agrawal, Ph.D.

Assistant Vice President, Network Systems Research Laboratory, and
Executive Director, Mobile Networking Research Department,
Telcordia Technologies
Email: pagrawal@research.telcordia.com
August 2003

Preface

Wireless networks are evolving into wireless IP (Internet Protocol) networks to overcome the limitations of traditional circuit-switched wireless networks. Wireless IP networks are more suitable for supporting the rapidly growing mobile data and multimedia applications. IP technologies bring the globally successful Internet service creation and offering paradigm to wireless networks, bringing the vast array of Internet services to mobile users and providing a successful platform for fostering future mobile services. IP-based protocols, which are independent of the underlying radio technologies, are also better suited for supporting seamless services over heterogeneous radio technologies and for achieving global roaming.

Wireless networks are evolving on two major fronts. First, radio access systems are evolving to third and fourth generation systems that can support significantly higher system capacity and per-user data rates with enhanced quality-of-service (QoS) support capabilities. Second, wireless IP networking technologies are profoundly changing the overall wireless network architectures and protocols.

Many books are available on radio access systems, examining the physical, link, and network layers specific to each radio system. Few books, however, have been designed to systemically address the wireless IP networking aspect, i.e., architectures, protocols, and techniques at the IP layer and above of a wireless IP network. This book seeks to provide a systematic description and comparison of next-generation wireless IP network architectures, systems, and protocols, with a focus on the IP layer and above.

Several major efforts have emerged to define global standards for wireless IP networks. The two most influential standards bodies are 3GPP (Third Generation Partnership Project) and 3GPP2 (Third Generation Partnership Project 2). Different standards efforts have been taking significantly different approaches, which lead to different architectures and different migration paths toward future wireless IP networks. This book provides insights into critical issues in wireless IP networking, thoroughly illustrates the standards and network architectures defined by leading standards bodies such as 3GPP and 3GPP2, and discusses in detail protocols and techniques in four major technical areas: signaling, mobility, security, and QoS.

To provide the necessary background, the book starts by presenting a historical overview of the evolution of wireless networks in Chapter 1, Introduction. Chapter 1 then reviews the evolution of public mobile services by examining the first, second, and current waves of mobile data services. It continues on to discuss the motivations for IP-based wireless networks and provides an overview of related standards activities.

Chapter 2 details the network architectures defined by 3GPP and 3GPP2. To help readers quickly get a sense of the solutions proposed by 3GPP and 3GPP2 and to easily identify their fundamental differences, Chapter 2 presents the most important aspects of the architectures proposed by 3GPP and 3GPP2 in a consistent format and highlights their major differences. In addition, the all-IP mobile network architecture proposed by the Mobile Wireless Internet Forum (MWIF) is also discussed.

Chapters 3 to 6 address systematically four of the most critical topic areas in next-generation wireless networks: signaling, mobility management, security, and QoS. Because Chapter 2 discusses network-layer signaling and control necessary for the operations of the networks, Chapter 3 focuses on application-level signaling and session control needed to support real-time and multimedia applications in IP networks and in the IP Multimedia Subsystems (IMS) defined by 3GPP and 3GPP2. Chapters 4, 5, and 6 discuss issues and solutions related to mobility management, network security, and QoS, respectively. Each chapter looks first at the subject in IP networks, then at the architectures and protocols defined by 3GPP and 3GPP2. The MWIF specifications are discussed in some chapters if related issues in MWIF are also addressed.

The book is designed primarily for researchers, engineers, technical managers, and graduate and undergraduate students. People entering the field of wireless IP networking will also find this book a helpful reference. The book emphasizes the principles underlying each major architecture and illustrates these principles with abundant technical details. It provides the audience with perspectives that are difficult to obtain from reading the standards specifications directly.

We are grateful to the ITSUMO (Internet Technologies Supporting Universal Mobile Operation) team from Telcordia Technologies, Inc. and Toshiba America Research, Inc. (TARI). Our work on the ITSUMO project and discussions with the ITSUMO members contributed to the book. Special thanks are due to Dr. Prathima Agrawal of Telcordia and Dr. Toshikazu Kodama of TARI for their continuous support and invaluable advice throughout the writing of the book. We thank Mr. Chi-Chen Lee, Mr. Jui-Hung Yeh, and Mr. Chih-Hsing Lin of the National Tsing Hua University for preparing many of the figures, tables, and references in the book. Jyh-Cheng Chen would also like to acknowledge the project members of the "Program for Promoting Academic Excellence of Universities" for many insightful discussions. Jyh-Cheng Chen's work was supported in part by the Ministry of Education, Taiwan, National Science Council (NSC), and Industrial Technology Research Institute (ITRI).

Jyh-Cheng Chen
Tao Zhang

August 2003

For instructor and student resources, please visit the companion web page of the book at http://www.cs.nthu.edu.tw/~jcchen/book.html

Acronyms

1G	First Generation
2G	Second Generation
3G	Third Generation
3GPP	Third-Generation Partnership Project
3GPP2	Third-Generation Partnership Project 2
AAA	Authentication, Authorization, Accounting
AAAF	AAA Foreign
AAAH	AAA Home
AAAL	AAA Local
AAL	ATM Adaptation Layer
AC	Authentication Center
AES	Advanced Encryption Standard
AF	Assured Forwarding
AGW	Access Gateway
AH	Authentication Header
AHAG	Ad Hoc Authentication Group
AK	Anonymity Key
AKA	Authentication and Key Agreement
AL	Air Link
AMA	AA-Mobile-Node-Answer
AMF	Authentication Management Field
AMPS	Advanced Mobile Phone Systems
AMR	AA-Mobile-Node-Request
ANSI	American National Standards Institute
API	Application Programming Interface
APN	Access Point Name

ARIB	Association of Radio Industries and Business
ARP	Address Resolution Protocol
ARPU	Average Revenue Per User
ATM	Asynchronous Transfer Mode
AuC	Authentication Center
AUTN	Authentication Token
AV	Authentication Vector
AVP	Attribute Value Pair
BA	Behavior Aggregate
	Binding Acknowledgment
BER	Bit Error Ratio
BGCF	Breakout Gateway Control Function
BGP	Border Gateway Protocol
BITS	Bump In The Stack
BITW	Bump In The Wire
BR	Border Router
BRAN	Broadband Radio Access Network
BS	Base Station
	Bearer Service
BSC	Base Station Controller
BSS	Base Station Subsystem
BSSAP+	Base Station System Application Part+
BTS	Base Transceiver Station
	Base Transceiver System
BU	Binding Update
CA	Certification Authority
CAMEL	Customized Applications for Mobile Enhanced Logic
CAP	CAMEL Application Part
CAVE	Cellular Authentication and Voice Encryption
CDMA	Code Division Multiple Access
CDR	Call Detail Record
CFN	Connection Frame Number
CH	Correspondent Host
CHAP	Challenge Handshake Authentication Protocol
CK	Cipher Key
CKSN	Cipher Key Sequence Number
CM	Connection Management
CMEA	Cellular Message Encryption Algorithm
CMS	Cryptographic Message Syntax
CN	Core Network
	Correspondent Node
CoA	Care-of Address
COPS	Common Open Policy Service
CQM	Core QoS Manager
CS	Circuit Switched

CSCF	Call State Control Function
	Call Session Control Function
CSE	CAMEL Service Environment
CSM	Communication Session Manager
CS-MGW	Circuit Switched Media Gateway
CT2	Cordless Telephone, Second Generation
CVSE	Critical Vendor/Organization Specific Extension
CWTS	China Wireless Telecommunication Standard
DB	Database
DECT	Digital European Cordless Telecommunications
DES	Data Encryption Standard
DH	Diffie-Hellman
DHCP	Dynamic Host Configuration Protocol
Diff-Serv	Differentiated Service
DNS	Domain Name System
DoS	Denial of Service
DP	Data Privacy
DRS	Data Ready to Send
DS	Differentiated Service
DS-CDMA	Direct Sequence Code Division Multiple Access
DSCP	Differentiated Service Code Point
DSI	Dynamic Subscriber Information
DSNP	Dynamic SLS Negotiation Protocol
DSS	Digital Signature Standard
DSSS	Direct Sequence Spread Spectrum
ECMEA	Enhanced Cellular Message Encryption Algorithm
EDGE	Enhanced Data Rates for Global GSM Evolution
EF	Expedited Forwarding
EIR	Equipment Identity Register
ESA	Enhanced Subscriber Authentication
ESN	Electronic Serial Number
ESP	Encapsulating Security Payload
	Enhanced Subscriber Privacy
ETSI	European Telecommunications Standards Institute
FA	Foreign Agent
FDD	Frequency Division Duplex
FHSS	Frequency Hopping Spread Spectrum
FQDN	Fully Qualified Domain Name
GEA	GPRS Encryption Algorithm
GERAN	GSM EDGE Radio Access Network
GFA	Gateway Foreign Agent
GGSN	Gateway GPRS Support Node
GHDM	General Handoff Direction Message
GLM	Geographical Location Manager
GMM	GPRS Mobility Management

GMSC	Gateway MSC
GNS	Global Name Server
GPRS	General Packet Radio Service
GRE	Generic Routing Encapsulation
GSCF	GPRS Service Control Function
GSM	Global System for Mobile Communications
GSN	GPRS Support Node
GTP	GPRS Tunneling Protocol
HA	Home Agent
HAA	Home-Agent-MIP-Answer
HAAA	Home AAA
HAR	Home-Agent-MIP-Request
HAWAII	Handoff-Aware Wireless Access Internet Infrastructure
HDB	Home Database
HDR	High Data Rate
HE	Home Environment
HFN	Hyper Frame Number
HLR	Home Location Registrar
HMM	Home Mobility Manager
HSS	Home Subscriber Server
HTTP	Hypertext Transfer Protocol
IAB	Internet Architecture Board
IAPP	Inter Access Point Protocol
ICMP	Internet Control Message Protocol
I-CSCF	Interrogating Call State Control Function
ICV	Integrity Check Value
IESG	Internet Engineering Steering Group
IETF	Internet Engineering Task Force
IK	Integrity Key
IKE	Internet Key Exchange
IM	Instant Message
IMEI	International Mobile Station Equipment Identity
IM-MGW	IP Multimedia Media Gateway
IMS	IP Multimedia Subsystem
IMSI	International Mobile Subscriber Identity
IM-SSF	IP Multimedia Service Switching Function
IN	Intelligent Network
Int-Serv	Integrated Service
IPHC	IP Header Compression
IPsec	IP Security
IPv4	Internet Protocol version 4
IPv6	Internet Protocol version 6
ISAKMP	Internet Security Association and Key Management Protocol
ISC	IMS Service Control
ISDN	Integrated Services Digital Network

ISM	Industrial, Scientific, and Medical
ISP	Internet Service Provider
ISUP	ISDN User Part
ITU	International Telecommunication Union
ITU-T	ITU Telecommunication Standardization Sector
KAC	Key Administration Center
KDC	Key Distribution Center
KSI	Key Set Identifier
L2TP	Layer-2 Tunneling Protocol
LA	Location Area
LAC	L2TP Access Concentrator
	Link Access Control
	Location Area Code
LAI	Location Area Identifier
LAN	Local Area Network
LDAP	Lightweight Directory Access Protocol
LEC	Local Exchange Carrier
LLC	Logical Link Control
LNS	L2TP Network Server
MA	Mobile Attendant
MAC	Medium Access Control
	Message Authentication Code
MAP	Mobile Application Part
MAPsec	MAP Security
MC	Message Center
MCC	Mobile Country Code
MC-CDMA	Multi-Carrier Code Division Multiple Access
MD5	Message Digest 5
ME	Mobile Equipment
MEK	MAP Encryption Key
MF	Multi Field
MG	Media Gateway
MGC	Media Gateway Controller
MGCF	Media Gateway Control Function
MIDCOM	Middlebox Communications
MIK	MAP Integrity Key
MIN	Mobile Identification Number
MIP	Mobile IP
MM	Mobility Management
MMD	Multimedia Domain
MMS	Multimedia Messaging Service
MN	Mobile Node
MNC	Mobile Network Code
MRC	Multimedia Resource Controller
MRF	Multimedia Resource Function

MRFC	Multimedia Resource Function Controller
MRFP	Multimedia Resource Function Processor
MS	Mobile Station
MSC	Mobile-services Switching Center
	Mobile Switching Center
MSIN	Mobile Subscriber Identification Number
MSISDN	Mobile Subscriber ISDN Number
MT	Mobile Termination
	Mobile Terminal
MTP	Message Transfer Part
MWIF	Mobile Wireless Internet Forum
NAI	Network Access Identifier
NANP	North American Numbering Plan
NAS	Network Access Server
NAT	Network Address Translator
NID	Network ID
NIST	National Institute of Standards and Technology
NMSI	National Mobile Subscriber Identity
NMT	Nordic Mobile Telephone
NPDB	Number Portability Database
NSAPI	Network-Layer Service Access Point Identifier
NTP	Network Time Protocol
NVSE	Normal Vendor/Organization Specific Extension
OAM&P	Operation, Administration, Maintenance, and Provisioning
OMA	Open Mobile Alliance
OSA	Open Service Access
OSPF	Open Shortest Path Protocol
OTASP	Over-The-Air Service Provisioning
PACS	Personal Access Communications System
PAN	Personal Area Network
PAP	Password Authentication Protocol
PBX	Private Branch Exchange
PCF	Packet Control Function
	Policy Control Function
P-CSCF	Proxy Call State Control Function
PDC	Personal Digital Cellular
PDCP	Packet Data Convergence Protocol
PDE	Position Determining Entity
PDF	Policy Decision Function
PDP	Packet Data Protocol
	Policy Decision Point
PDS	Packet Data Subsystem
PDSN	Packet Data Serving Node
PDU	Packet Data Unit
PEP	Policy Enforcement Point

PHB	Per-Hop Behavior
PHS	Personal Handyphone System
PKC	Public Key Certificate
PKI	Public Key Infrastructure
PLCM	Private Long Code Mask
PLMN	Public Land Mobile Network
P-MIP	Paging in Mobile IP
PMM	Packet Mobility Management
PPP	Point-to-Point Protocol
PS	Packet Switched
PSTN	Public Switched Telephone Network
P-TMSI	Packet TMSI
PZID	Packet Zone ID
QoS	Quality of Service
RA	Routing Area
RAB	Radio Access Bearer
RAC	Routing Area Code
RADIUS	Remote Authentication Dial In User Service
RAI	Routing Area Identifier
RAN	Radio Access Network
RANAP	Radio Access Network Application Part
RAU	Routing Area Update
RB	Radio Bearer
RED	Random Early Detection
RFC	Request For Comments
RLC	Radio Link Control
RN	Radio Network
RNC	Radio Network Controller
RNS	Radio Network Subsystem
ROHC	Robust Header Compression
RRC	Radio Resource Control
RSA	Rivest, Shamir, Adleman
RSVP	Resource Reservation Protocol
RTP	Real-Time Transport Protocol
RTT	Radio Transmission Technology
SA	Security Association
SAD	Security Association Database
SBLP	Service Based Local Policy
SCCP	Signaling Connection Control Part
SCP	Service Control Point
SCS	Service Capability Server
S-CSCF	Serving Call State Control Function
SCTP	Stream Control Transmission Protocol
SDO	Standards Development Organization
SDP	Session Description Protocol

SDU	Selection and Distribution Unit
	Service Data Unit
SGSN	Serving GPRS Support Node
SHA	Secure Hash Algorithm
SID	Session ID
	System ID
SIM	Subscriber Identity Module
SIP	Session Initiation Protocol
SLA	Service Level Agreement
SLP	Service Location Protocol
SLS	Service Level Specification
SM	Session Management
SME	Signaling Message Encryption
SMEKEY	Signaling Message Encryption Key
SMS	Short Message Service
SN	Service Node
	Serving Network
SNMP	Simple Network Management Protocol
SPD	Security Policy Database
SPI	Security Parameter Index
SQM	Subscription QoS Manager
SQN	Sequence Number
SRNS	Serving Radio Network Subsystem
SS7	Signaling System No. 7
SSD	Shared Secret Data
TA	Terminal Adapter
TACS	Total Access Communications Services
TCA	Traffic Conditioning Agreement
TCAP	Transaction Capabilities Application Part
TCP	Transmission Control Protocol
TCS	Traffic Conditioning Specification
TDD	Time Division Duplex
TDMA	Time Division Multiple Access
TE	Terminal Equipment
TEID	Tunnel Endpoint Identifier
TIA	Telecommunications Industry Association
TLS	Transport Layer Security
TMSI	Temporary Mobile Subscriber Identity
TOS	Type of Service
TRIP	Telephony Routing over IP Protocol
TTA	Telecommunications Technology Association
TTC	Telecommunications Technology Committee
TTL	Time to Live
UA	User Agent
UAC	User Agent Client

UAK	UIM Authentication Key
UAS	User Agent Server
UDP	User Datagram Protocol
UE	User Equipment
UEA	UMTS Encryption Algorithm
UHDM	Universal Handoff Direction Message
UIA	UMTS Integrity Algorithm
UICC	UMTS IC Card
UIM	User Identity Module
UMTS	Universal Mobile Telecommunications System
URA	UTRAN Registration Area
URI	Uniform Resource Identifier
USIM	UMTS Subscriber Identity Module
	Universal Subscriber Identity Module
UTRAN	UMTS Terrestrial Radio Access Network
	Universal Terrestrial Radio Access Network
VAAA	Visited AAA
VAS	Value-Added Service
VDB	Visited Database
VLR	Visitor Location Register
VMS	Voice Message System
VoIP	Voice over IP
VP	Voice Privacy
VPMASK	Voice Privacy Mask
VPN	Virtual Private Network
WCDMA	Wideband Code Division Multiple Access
WLAN	Wireless Local Area Network
WWW	World Wide Web

1

Introduction

Wireless networks are increasingly based on IP (Internet Protocol) technologies. An IP-based wireless network, or wireless IP network, uses IP-based protocols to support one or more key aspects of network operations. These may include network-layer routing and transport of user packets, mobility management at the network or higher protocol layers, signaling and control of real-time voice and multimedia services, and support for network security and quality of service. An all-IP wireless network would use IP-based protocols to support all or most aspects of network operations at the network layer or above in the core networks or even in the radio access networks.

IP-based wireless networks offer a range of advantages over traditional circuit-switched wireless networks. For example, IP-based networks are more suitable for supporting the rapidly growing mobile data and multimedia applications. IP-based wireless networks bring the globally successful Internet service creation and offering paradigm into wireless networks. This not only makes Internet services available to mobile users but also provides a proven successful platform for fostering future mobile services. Furthermore, IP-based protocols are independent of the underlying radio technologies and therefore are better suited for supporting services seamlessly over different radio technologies and for achieving global roaming.

Realizing IP-based wireless networks introduces many new technical challenges, especially in the areas of network architecture, signaling and control, mobility management, network security, and quality of services (QoS). These areas are therefore the focus of this book. For each of these technical areas, we will discuss the

IP-Based Next-Generation Wireless Networks: Systems, Architectures, and Protocols,
By Jyh-Cheng Chen and Tao Zhang. ISBN 0-471-23526-1 © 2004 John Wiley & Sons, Inc.

key technical challenges and examine the technologies and solutions that are available or being developed to address these challenges.

To provide a background for detailed discussions, this chapter reviews the evolutions of wireless networks and mobile services, discusses business and technology drivers for the migration toward IP-based wireless networks, and outlines the content of the rest of the book.

1.1 EVOLUTION OF WIRELESS NETWORKS

Due to the nature of radio propagation, radio systems covering small geographical areas typically could provide higher data rates and require lower levels of radio transmission power than radio systems covering larger geographical areas. Therefore, wireless networks are usually optimized to fit different coverage areas and communications needs. Based on radio coverage ranges, wireless networks can be categorized into wireless Personal Area Networks (PANs), wireless Local Area Networks (WLANs), low-tier wireless systems, public wide-area (high-tier) cellular radio systems, and mobile satellite systems. The coverage area sizes versus bit rates for various types of radio systems are illustrated in Figure 1.1.

PANs use short-range low-power radios to allow a person or device to communicate with other people or devices nearby. For example, Bluetooth radios [22] could support three power classes, which provide radio coverage ranges up to approximately 10 m, 50 m, and 100 m, respectively. Bluetooth can support bit rates up to about 720 Kbps. Other radio technologies for PANs include HomeRF [27] and

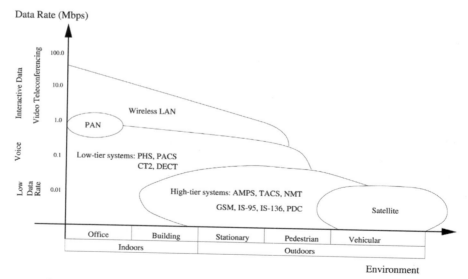

Fig. 1.1 Wireless systems: bit rates vs. coverage areas. Copyright 2003 IEEE.

IEEE 802.15 [29]. IEEE 802.15 is defining a short-range radio system to support data rates over 20 Mbps. PANs may have many applications. For example, they allow a person to communicate wirelessly with devices inside a vehicle or a room. People with Personal Digital Assistants (PDAs) or laptop (notebook) computers may walk into a meeting room and form an ad-hoc network among themselves dynamically. A service discovery protocol may be used over a PAN to help individuals to locate devices or services (e.g., a printer, a viewgraph projector) that are nearby as they move about.

Low-tier wireless systems use radio to connect a telephone handset to a base station that is connected via a wireline network to a telephone company. They are designed mainly to serve users with pedestrian-moving speeds. Typically, the coverage ranges of such low-tier base stations are less than 500 m outdoors and less than 30 m indoors. Some of the well-known low-tier standards include Cordless Telephone, Second Generation (CT2), Digital European Cordless Telecommunications (DECT), Personal Access Communications Systems (PACS), and Personal Handyphone System (PHS) [25, 32]. CT2 and DECT primarily are used as wireless extensions of residential or office telephones. PACS and PHS, on the other hand, operate in public areas and provide public services.

- *Cordless Telephone, Second Generation (CT2)*: CT2, designed in the United Kingdom in 1989, uses digital radio technologies to make a telephone cordless. CT2-based cordless telephones are designed for use in homes, offices, or public telephone booths. CT2 supports only circuit-switched voice services.
- *Digital European Cordless Telecommunications (DECT)*: DECT was defined by the European Telecommunications Standards Institute (ETSI) in 1992 to be the cordless telephone standard for Europe. DECT was designed primarily for use in an office environment. It provides short-range wireless connectivity (i.e., up to a few hundred meters) to Private Branch Exchanges (PBXs), which are telephone switches used in an enterprise telephone network. DECT supports circuit-switched voice and data services.
- *Personal Access Communications Systems (PACS)*: PACS was designed by Telcordia (then, Bellcore) in the United States in 1992 to provide wireless access to local exchange carriers (LECs). It was designed to provide radio coverage within a 500-m range and to support voice, data, and video for use in both indoor and outdoor microcells.
- *Personal Handyphone System (PHS)*: PHS was designed by the Telecommunications Technical Committee of Japan to support both voice and data services. It was designed to support a channel rate of 384 Kbps, which is significantly higher than the data rates supported by most other low-tier standards, on both the forward (from network to mobile terminals) and the reverse directions.

Next, we focus on WLANs and wide-area wireless networks.

1.1.1   Wireless Local Area Networks

A WLAN provides a shared radio media for users to communicate with each other and to access an IP network (e.g., the Internet, an enterprise network, an Internet Service Provider, or an Internet Application Provider).

WLANs typically use the unlicensed Industrial, Scientific, and Medical (ISM) radio frequency bands. In the United States, the ISM bands include the 900-MHz band (902–928 MHz), 2.4-GHz band (2400–2483.5 MHz), and the 5.7-GHz band (5725–5850 MHz).

The most widely adopted WLAN standard around the world is IEEE 802.11 [28] today. IEEE 802.11 consists of a family of standards that defines the physical layers (PHY) and the Medium Access Control (MAC) layer of a WLAN, WLAN network architectures, how a WLAN interacts with an IP core network, and the frameworks and means for supporting security and quality of service over a WLAN. The IEEE 802.11 standards family includes the following key standards:

- *IEEE 802.11*: Defines the MAC and different physical layers based on radio frequency (RF) and Infrared (IR). Direct Sequence Spread Spectrum (DSSS) and Frequency Hopping Spread Spectrum (FHSS) operating in the 2.4-GHz ISM band are specified for the RF physical layer. The DSSS PHY provides 2 Mbps of peak rate and optional 1 Mbps in extremely noisy environments. On the other hand, the FHSS PHY operates at 1 Mbps with optional 2 Mbps in very clean environments. The IR PHY supports both 1 Mbps and 2 Mbps for receiving, and 1 Mbps with an optional 2 Mbps bit rate for transmitting.
- *IEEE 802.11b*: Defines a physical layer that provides data rates up to 11 Mbps in the 2.4-GHz ISM radio frequency band. IEEE 802.11b is the most widely deployed WLAN today.
- *IEEE 802.11a*: Defines a physical layer that supports data rates up to 54 Mbps using the 5.7-GHz ISM radio frequency band.
- *IEEE 802.11g*: Defines an extended rate physical layer to support data rates up to 54 Mbps using the 2.4-GHz ISM radio frequency band.
- *IEEE 802.11i*: Defines a framework and means for supporting security over IEEE 802.11 WLANs.
- *IEEE 802.11e*: Defines a framework for supporting QoS for delay-sensitive applications (e.g., real-time voice and video) over IEEE 802.11 WLANs.
- *IEEE 802.11f*: Defines the Inter Access Point Protocol (IAPP) to assure interoperability of multi-vendor access points.

The ETSI HIPERLAN/2 [26] and the Multimedia Mobile Access Communication (MMAC) [33] are also developing wireless LAN standards for supporting bit rates as high as 50 Mbps.

WLANs are being used to support an increasingly broader range of mobile applications:

- *Enterprise WLANs*: WLANs were first adopted and are now widely used in enterprise networks to provide wireless data services inside buildings and over campuses or building complexes.
- *Commercial Public WLANs*: Today, WLANs are being deployed rapidly around the world to provide public wireless services. Public WLANs were first deployed in airports, café shops, and hotels. Today, public WLANs are being deployed in train stations, gas stations, shopping malls, parks, along streets, highways, or even on trains and airplanes. Public WLANs are being used to provide mobile Internet services to business travelers and consumers. They are also used to provide customized telematics services to people inside moving vehicles and to in-vehicle computers that monitor or control the vehicles.
- *Wireless Home Networks*: WLANs started to be used in private homes to replace wired home networks.

The growing worldwide deployment of public WLANs is of special significance because it is creating a growing impact on what public wireless networks will look like and how public mobile services will be provided in the near future.

- Public WLANs that are available today can provide significantly higher data rates than cellular networks that are expected to be available in the near future.
- By seamlessly integrating services over public WLANs with services over wide-area wireless networks, mobile network and service providers could take full advantage of both WLAN and wide-area radio technologies to create new services and reduce networking costs. For example, public WLANs can be used for high-speed mobile data and multimedia services in limited geographical areas while wide-area wireless networks can provide continuous service coverage at lower speeds over large geographical areas.
- Public WLANs are the first wave of all-IP radio access networks that have emerged in public wireless networks, making public wireless networks one step forward on their migrations to IP-based wireless networks.
- Broad range of alternatives to how public WLANs will be owned and operated and how services will be offered over public WLANs calls for new and innovative business models for providing public mobile services.

Today, mobile network operators worldwide have deployed commercial public WLANs. Substantial growth of public WLANs are expected to continue. Figure 1.2 shows the worldwide WLAN equipment sales projected by Forrester Research for public, enterprise, and home WLANs. The expected compound annual growth rate (CAGR) of public WLAN equipment sales between 2002 and 2006 is 65.5%, which is over twice the CAGR for home WLAN equipment sales (30.4%) and about five times the CAGR for enterprise WLAN equipment sales (13.4%) during the same period.

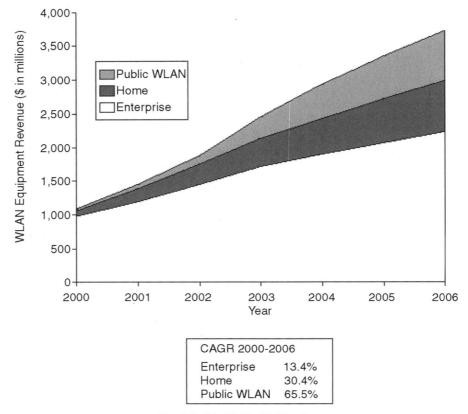

CAGR 2000-2006	
Enterprise	13.4%
Home	30.4%
Public WLAN	65.5%

Fig. 1.2 *Worldwide WLAN sales*

1.1.2 Public Wide-Area Wireless Networks

Public (commercial) wide-area wireless networks provide public mobile services over large geographical areas to users moving on both pedestrian and vehicular speeds. A commercial wide-area wireless network typically consists of the following:

- *Radio Access Networks (RAN) or Radio Systems*: A RAN provides radio resources (e.g., radio channels) for mobile users to access a core network. A RAN consists of wireless base stations, each providing radio coverage to a geographical area called a *radio cell* or *cell*. These radio cells may be significantly larger than the coverage area of a WLAN. For example, a radio cell in a wide-area network may exceed 10 km in diameter. Multiple cells may be deployed to provide continuous radio coverage over an entire country or beyond. As the radio cells are typically arranged in a cellular formation to increase radio frequency reusability, wide-area radio systems are commonly referred to as *cellular systems*.

- *Core Network*: A Core Network is typically a wireline network used to interconnect RANs and to connect the RANs to other networks such as the Public Switched Telephone Network (PSTN) and the Internet.

Wide-area radio systems are classified into *generations* based on the technologies they use and the networking capabilities they provide.

1.1.2.1 1G, 2G, and 2.5G Wireless Networks

First-generation (1G) wide-area radio systems [25, 32], which first became commercially available in the early 1980s, use analog radio technologies and circuit-switched transmission and networking technologies. The main mobile services provided by 1G radio systems are circuit-switched voice services. There were three main 1G radio system standards in the world:

- *Advanced Mobile Phone Systems (AMPS)* in North America.
- *Total Access Communications Services (TACS)* in the United Kingdom. Variants of TACS include ETACS, JTACS, and NTACS.
- *Nordic Mobile Telephone (NMT)* in Nordic countries.

First-generation radio systems lack the ability to support roaming between different network operators. For example, AMPS specifies only the air interface between mobile terminals and wireless base stations. Each network operator therefore operates its proprietary core network. Consequently, automatic roaming between different network operators' AMPS networks was infeasible. When roaming into a new network provider's AMPS network, a user had to manually register with the new network by calling a human operator to request for the registration. Furthermore, as incompatible 1G standards were used in different countries, it was impossible for a user to roam from one country to another.

Second-generation (2G) radio systems [25, 32] began to emerge in the early 1990s. They brought a number of significant advancements over 1G wireless networks:

- Digital signal processing and transmission technologies were introduced to replace the analog signal processing technologies used in 1G radio systems. Digital technologies increased radio system capacity and radio spectrum utilization efficiency, enhanced voice quality, and reduced power consumption on mobile terminals.
- Standards for core networks were introduced to support roaming between network operators and between countries.
- In addition to circuit-switched voice services, 2G wireless networks enabled the very first waves of mobile data and mobile Internet services that were widely adopted by customers (Section 1.2).

Several significantly different 2G standards were developed and used around the world:

- *In North America*: Two major standards for 2G radio systems inaugurated in the 1990s:
    - IS-136 for Time Division Multiple Access (TDMA) radio systems.
    - IS-95 for Code Division Multiple Access (CDMA) radio systems. CDMA uses spread spectrum technologies. A user's traffic is not transported over a single frequency, but it is instead spread over a frequency band that is much broader than the spectrum occupied by the user traffic originally.

  When IS-136 and IS-95 replaced AMPS in the RAN, a new standard IS-41 for 2G core networks was also introduced to support roaming between different network operators.

  Today, IS-136 is primarily used in the United States. IS-95 is primarily used in the United States and South Korea. By the end of 2001, IS-136 and IS-95 were supporting around 200 million subscribers.
- *In Europe*: European countries decided to jointly develop a single set of universal standards for 2G radio system *and* 2G core network to replace the different 1G radio systems used in Europe. This resulted in GSM (Global System for Mobile communications). GSM operates at 900-MHz and 1800-MHz radio frequencies in Europe and at 800 MHz and 1900 MHz in the United States.

  GSM enables users to roam between mobile network providers and even between countries. GSM also enables a user to use the same wireless handset when the user changes network providers. In addition to circuit-switched voice services, GSM offers a 9.6 Kbps circuit-switched symmetric channel for a mobile terminal to use as a data connection to access the Internet.

  Commercial GSM services were launched first in 1991 in Finland. GSM quickly became a spectacular success throughout Europe and soon became the most widely used 2G wireless network standards in the world. By the end of 2001, GSM was used in over 170 countries and was serving around 677 million subscribers.
- *In Japan*: Mobile network operator NTT DoCoMo developed its own 2G radio system—the Personal Digital Cellular (PDC) network. PDC supports both circuit-switched voice and data services over 9.6-Kbps radio channels. By the end of 2001, NTT DoCoMo's PDC networks were supporting around 50 million subscribers.

As mobile data services grow, 2G wireless networks have been enhanced into what are commonly referred to as 2.5G wireless networks to meet the demands for higher data rates to support the growing mobile data services. 2.5G wireless networks provide significantly higher radio system capabilities and per-user data rates than do 2G systems, but they do not yet achieve all the capabilities promised by

3G systems. For example, GSM has been enhanced into the following 2.5G networks:

- *General Packet Radio Services (GPRS)*: GPRS provides a packet-switched core network as an extension to GSM core networks in order to better support packet services over GSM radio systems.
- *Enhanced Data Rates for Global GSM Evolution (EDGE)*: EDGE provides advanced modulation and channel coding techniques to significantly increase the data rates of GSM radio systems. It promises to support data rates up to 384 Kbps. Due to its high speed, some people also regard EDGE as a 3G system.

1.1.2.2 3G Wireless Networks

In the late 1990s, standardization efforts for third-generation (3G) wireless networks began. 3G wireless networks have been designed to

- *Significantly increase radio system capacities and per-user data rates over 2G systems*: 3G radio systems promise to support data rates up to 144 Kbps to users moving up to vehicular speeds, up to 384 Kbps to users moving at pedestrian speeds, and up to 2 Mbps to stationary users.
- *Support IP-based data, voice, and multimedia services*: 3G systems are designed to support a broader range of IP-based mobile services than are 2G systems. The objective is to achieve seamless integration between 3G wireless networks and the Internet so that mobile users can access the vastly available resources and applications on the Internet.
- *Enhance QoS support*: 3G systems seek to provide better QoS support than 2G systems. 3G systems are designed to support multiple classes of services, including, for example, real-time voice, best-effort data, streaming video, and non-real-time video.
- *Improve interoperability*: An important goal of 3G systems is to achieve greater degree of interoperability than 2G systems to support roaming among different network providers, different radio technologies, and different countries.

Two leading international partnerships are taking different approaches to define 3G wireless network standards:

- *Third-Generation Partnership Project (3GPP)*: 3GPP (Section 1.4) seeks to produce globally applicable standards for a third-generation mobile system *based on evolved GSM core networks* and the *radio access technologies* these evolved GSM core networks support. More specifically,
  - 3G core networks will evolve the GSM core network platform to support circuit-switched mobile services and to evolve the GPRS core network platform to support packet-switched services.

- 3G radio access technologies will be based on the Universal Terrestrial Radio Access Networks (UTRANs) that use Wideband-CDMA (WCDMA) radio technologies.

- *Third-Generation Partnership Project 2 (3GPP2)*: 3GPP2 (Section 1.4) seeks to produce globally applicable standards for a third-generation mobile system *based on evolved IS-41 core networks*. In particular,

  - 3G core networks will evolve the IS-41 core network to support circuit-switched mobile services and define a new packet core network architecture that leverages capabilities provided by the IS-41 core network to support IP services.
  - 3G radio access technologies will be based on cdma2000 radio technologies.

WCDMA and cdma2000 are two incompatible CDMA technologies designed for 3G radio systems. Table 1.1 summarizes some of the differences between WCDMA and cdma2000.

A key difference between WCDMA and cdma2000 is that they use different multiple access schemes. WCDMA uses two modes of Direct Sequence CDMA (DS-CDMA): Frequency Division Duplex (FDD) DS-CDMA and Time Division Duplex (TDD) DS-CDMA. Direct Sequence refers to a specific spread spectrum technology used in a CDMA system. With DS-CDMA, each user's traffic is spread by a unique pseudo-random (PN) code into pseudo noises over the same radio frequency band. The receiver uses the exact pseudo-random code to unscramble the pseudo noise to extract the user traffic. FDD and TDD refer to the methods for separating uplink traffic (from mobile to network) from downlink traffic (from

TABLE 1.1 WCDMA vs. cdma2000

	WCDMA	cdma2000
Multiple Access Scheme	Frequency Division Duplex Direct-Sequency CDMA (FDD DS-CDMA) and Time Division Duplex Direct-Sequence CDMA (TDD DS-CDMA)	Frequency Division Duplex Multicarrier CDMA (FDD MC-CDMA)
Spreading Chip Rate	3.84 Mcps	1.2288 Mcps for 1xRTT 3×1.2288 Mcps for 3xRTT
Base Station Synchronization	Asynchronous	Synchronous
Network Signaling	GSM-MAP	IS-41, GSM-MAP
Frame Size	10 ms for physical layer frames 10, 20, 40, and 80 ms for transport layer frames	5 (for signaling), 20, 40 and 80 ms for physical layer frames

network to mobile). FDD uses different frequency bands to transmit uplink and downlink traffic (2110–2170 MHz for downlink and 1920–1980 MHz for uplink). TDD uses the same frequency band for both uplink and downlink transmissions, but it schedules uplink and downlink transmissions in different time slots. On the other hand, cdma2000 uses Frequency Division Duplex (FDD) Multicarrier CDMA (MC-CDMA). A single carrier in cdma2000 uses a Radio Transmission Technology (RTT) that provides data rates up to 144 Kbps. A cdma2000 system that uses a single carrier is referred to as cdma2000 1xRTT. Three carriers may be used together to provide data rates up to 384 Kbps. A cdma2000 system using three carriers is commonly referred to as cdma2000 3xRTT. Details on the WCDMA and cdma2000 can be found in [30] and [31].

The first commercial 3G systems and services were launched worldwide in 2001. Today, commercial cdma2000 1xRTT networks and services are operating in Asia (China, Japan, South Korea, Taiwan), Europe (Romania, United Kingdom), North America (Canada, US, Puerto Rico), and South America (Brazil). The worldwide subscribers of cdma2000 services reached 24 million in September 2002 (CDMA Development Group [24]). The adoption of 3G technologies is especially impressive in Asia. For example, Japanese mobile network operator KDDI launched its cdma2000 1xRTT network in April 2002. KDDI's cdma2000 promises up to 144-Kbps data rates. In addition to voice services, KDDI's cdma2000 network offers a range of high-speed mobile Internet services, including supporting camera phones. The subscribers of KDDI's cdma2000 1xRTT services grew over 1 million within the first three months of service launch and reached over 4 million in December 2002. Subscribers of KDDI's cdma2000 services reached 10 million in September 2003.

Although the approaches taken by 3GPP and 3GPP2 are different, they share the following fundamental principles:

- 3G core networks will be based on IP technologies.
- Evolutionary, rather than revolutionary, approaches are used to migrate wireless networks to full IP-based mobile networks, and the evolution starts in the core networks.

The Internet Engineering Task Force (IETF) has been developing IP-based protocols for enabling the mobile Internet. These protocols are designed to work over any radio system.

The Mobile Wireless Internet Forum (MWIF), formed in January 2000, was among the first international industrial forums that sought to develop and promote an all-IP wireless network architecture independent of radio access technologies. In 2002, MWIF merged with the Open Mobile Alliance (OMA), a global organization that develops open standards and specifications for mobile applications and services.

The evolution of standards for public wide-area wireless networks is illustrated in Figure 1.3.

The evolution of technologies for public wide-area wireless networks is illustrated in Figure 1.4.

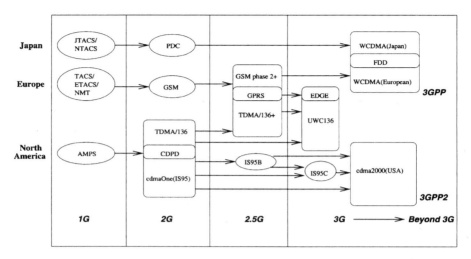

Fig. 1.3 *Evolution of standards for wide-area radio systems. Copyright 2003 IEEE.*

The different paths taken by different standards organizations and industry forums for the migration from second-generation to third-generation wireless networks are converging to a similar target IP-based wireless network illustrated in Figure 1.5. This conceptual architecture has several important characteristics:

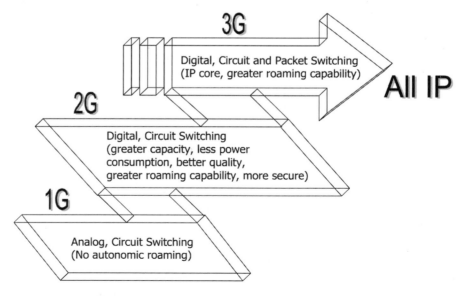

Fig. 1.4 *Evolution of network technologies from 1G to 3G*

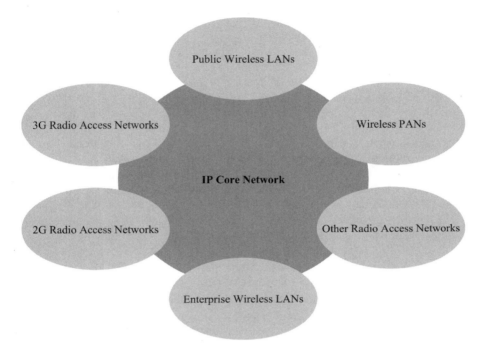

Fig. 1.5 *Wireless IP network supporting heterogeneous radio technologies*

- The core network will be based on IP technologies.
- A common IP core network will support multiple types of radio access networks.
- A broad range of mobile voice, data, and multimedia services will be provided over IP technologies to mobile users.
- IP-based protocols will be used to support mobility between different radio systems.
- All-IP radio access networks will increase over time. The first all-IP radio access networks that have emerged in public wireless networks are public WLANs.

1.2 EVOLUTION OF PUBLIC MOBILE SERVICES

The dominant services over public wide-area wireless networks have been circuit-switched mobile voice services. Today, many mobile operators still generate most of their revenues from circuit-switched mobile voice services.

However, a fundamental shift has been occurring rapidly throughout the world since 2G wireless networks became commercially available. The spectacular growth of 2G wireless networks over the past 10 years helped to push mobile voice services to reach saturation in many regions of the world. For example, the penetrations of mobile voice services (i.e., subscribers as percentage of the population in a given region) have reached 70% or higher in many western European and Asian countries.

Mobile data and multimedia services have become the main drivers for the future growth of mobile services throughout the world. Mobile data and multimedia services have been growing more rapidly than mobile voice services worldwide. Mobile data and multimedia services have matured significantly, evolving rapidly from text-based instant messaging services to low-speed mobile Internet services based on proprietary technologies, to higher speed and broader range of mobile Internet services based on open and standard Internet protocols, and to high-speed and multimedia Internet services. Mobile data and multimedia services are poised to become the dominant mobile services in the near future.

Given the importance of mobile data and multimedia services, we next take a closer look at how these services have been evolving.

1.2.1 First Wave of Mobile Data Services: Text-Based Instant Messaging

The first globally successful mobile data service is SMS (Short Message Services), which was first introduced in Europe over GSM networks. SMS allows a mobile user to send and receive short text messages (up to 160 text characters) *instantly.*

Supporting SMS does not require a packet core network. Instead, SMS messages are delivered using the signaling protocol—Mobile Application Part (MAP)—that was originally designed to support mobility in GSM networks. This allowed SMS services to be provided over the completely circuit-switched 2G GSM networks long before packet core networks were introduced into wireless networks.

SMS services grew rapidly first in Europe. Today, SMS services are booming throughout the world.

- In the United Kingdom, the number of transmitted SMS messages more than doubled in the two-year period from 2001 to 2002. Based on statistics from the Mobile Data Association, an average of 52 million SMS messages were transmitted every day in the United Kingdom in December 2002, which translates into about 2.2 million messages per hour on the average. Figure 1.6 shows the SMS subscriber growth in the United Kingdom from 1998 to June 2003.

- In Europe, 186 billion SMS messages were transmitted in 2002.

- In China, the revenues of SMS and value-added services (VAS) totaled 750 million in 2002. During an eight-day period around the Chinese New Year in 2003 (early February), approximately 7 billion SMS messages were transmitted.

- In the United States, SMS is experiencing an explosive growth, although its usage levels have not reached the levels in Western Europe and the Asia

TEXT MESSAGING GROWTH (SMS): UK GSM
NETWORK OPERATOR TOTALS
June 1998 – June 2003

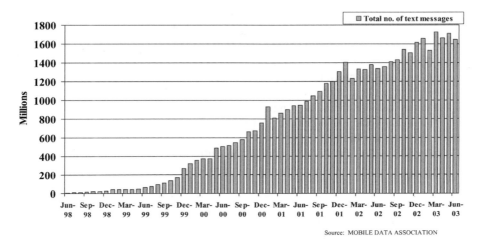

Source: MOBILE DATA ASSOCIATION

Fig. 1.6 *Growth of SMS message transmissions in the United Kingdom*

Pacific region. InphoMatch, a company that handles intercarrier SMS services for AT&T Wireless, Verizon, and T-Mobile, delivered 100 million SMS messages in August 2002 and saw the intercarrier SMS traffic doubling every two months.

In addition to providing a highly valuable data service to mobile users, SMS allowed mobile users to become familiar and comfortable with mobile data services and to appreciate the value of mobile data services. This helped pave the road for mobile users to adopt the more advanced data services that arrived next.

1.2.2 Second Wave of Mobile Data Services: Low-Speed Mobile Internet Services

Interactive and information-based mobile Internet services emerged as the second wave of widespread mobile data services. An example of successful mobile Internet applications is *i-Mode*, which was launched by NTT DoCoMo over its PDC radio systems in Japan in February 1999. The i-Mode services include:

- Sending and receiving emails and instant messages.
- Commercial transactions, e.g., banking, ticket reservation, credit card billing inquiry, and stock trading.

- Directory services, e.g., dictionary, restaurant guides, and phone directory.
- Daily information, e.g., news, weather reports, road conditions, and traffic information.
- Entertainment, e.g., Karaoke, network games, and horoscope.

The i-Mode services has experienced a rapid and steady growth ever since its launch. Figure 1.7 shows the growth of i-Mode subscribers for the three-year period from 2001 to 2003 (Source: NTT DoCoMo). By the summer of 2003, i-Mode was supporting over 38 million subscribers.

i-Mode represents a significant milestone in the evolution of mobile services. It was the first major success in bringing Internet-based services to a large population of mobile subscribers. It demonstrated the values and the potentials of the mobile Internet to the world.

Today, however, the i-Mode services are suffering from two major limitations:

- i-Mode services are limited by the low data rate of the PDC radio networks.
- i-Mode users rely on proprietary protocols developed by NTT DoCoMo, rather than on standard IP-based protocols, to access i-Mode services. The i-Mode services are provided by World-Wide Web (WWW) sites specifically designed for mobile users. Mobile devices use a set of proprietary protocols developed by NTT DoCoMo to communicate with these WWW sites via a gateway. The gateway converts between the protocols over the radio access network and the

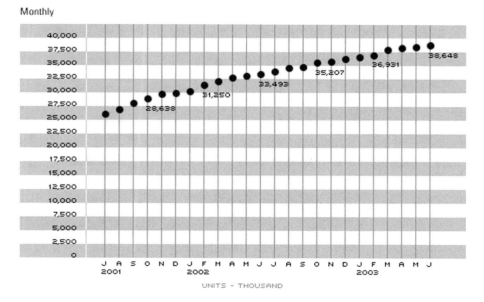

Fig. 1.7 *Growth of i-Mode subscribers*

protocols used by the WWW sites. The proprietary protocols make it difficult for i-Mode to be adopted by other countries.

Today, 2.5G and the early forms of 3G wireless networks allow mobile users to access the Internet via standard IP-based protocols. This is enabling mobile Internet services similar to those provided by i-Mode to be provided to users around the world and over radio channels of higher data rates than 2G radio systems.

1.2.3 Current Wave of Mobile Data Services: High-Speed and Multimedia Mobile Internet Services

The higher system capacities and data rates provided by 2.5G and 3G wireless networks plus the closer integration of 2.5G and 3G wireless networks with the Internet enabled by the IP technologies used in these wireless networks are enabling many new ways for people to communicate. People are not only talking over mobile phones and using instant text-based messages. They can now use their mobile phones to take pictures, record videos, and send the pictures and videos to other people or devices; receive location-dependent services; and play sophisticated games in real time with remote users.

Examples of advanced mobile data and multimedia applications include the following:

- *Camera phones*: Mobile phones with integrated cameras that allow a user to take still pictures, record short videos with sound, and send the photos and videos as multimedia messages or email to other users.
- *Multimedia Messaging Services (MMS)*: Send and receive messages with multimedia contents (data, voice, still pictures, videos, etc.).
- *Networked gaming*: People may download games to their mobile handsets and play the games locally. They may also use their mobile handsets to play games with remote users in real time.
- *Location-based services*: People may use their mobile devices to receive real-time navigation services, local maps, and information on local points of interest (e.g., restaurants, tourist locations, cinemas, gas stations, shopping malls, hospitals, and vehicle repair shops).
- *Streaming videos to mobile devices*: People may use their mobile devices to view real-time and non-real-time videos, for example, short videos received from friends' camera phones, watch TV.
- *Vehicle information systems*: People on moving vehicles (e.g., cars, trains, boats, airplanes) may access the Internet or their enterprise networks the same way as when they are at their offices or homes. They may be able to surf the Internet, access their corporate networks, download games from the network, play games with remote users, obtain tour guidance information, obtain real-time traffic and route conditions information, etc.

Many of the services described above are already available over 2.5G or early forms of 3G wireless networks and are changing the ways people communicate in a fundamental manner. Camera phones, for example, enabled the first multimedia mobile applications that are truly useful to and accepted by large populations of mobile users worldwide.

The world's first commercial camera phones that allow users to take pictures and videos and allow them to send the pictures and videos over wireless networks to other users emerged in late 2001. Ever since their introduction, camera phones have been experiencing a spectacular growth. According to Strategy Analytics (January 2003), 10 million camera phones were sold worldwide in the first nine months of 2002. The number of camera phones sold in 2002 was estimated to be 16 million, which is approximately the same level of worldwide sales of PDAs in 2002. Within merely around a year of camera phones' existence, their annual worldwide sales in 2002 were already close to the approximately 22 million digital cameras sold in the same year (Strategy Analytics).

Today, camera phones are improving rapidly. For example, picture display screens are larger, picture resolutions are higher, and user applications for handling pictures and videos are richer (e.g., applications for editing pictures and videos, taking snapshot pictures from videos, and sending pictures and videos as instant messages or email attachments over wireless networks to other users). Although the functionalities of camera phones are improving, their prices are declining due to intensive competition as all major mobile phone manufacturers around the world are competing in the camera phone market now. For example, camera phones priced at

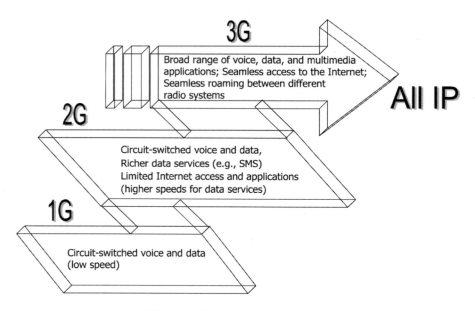

Fig. 1.8 *Evolution of mobile services*

around US $100 became available in late 2002, whereas the average camera phone prices were around US $300 in early 2002. The improving functionalities and the falling prices of the camera phones, with the higher data rates of 2.5G and 3G wireless networks, make camera phones more useful and affordable to the consumers and will therefore further accelerate the growth of camera phones. Strategy Analytics expects that camera phones will outsell digital cameras by 2004.

The widespread use of camera phones is not only changing the ways people communicate, but also it is changing the mix of mobile services and the nature of the network traffic generated by mobile services. In particular, camera phones help increase multimedia traffic over the wireless networks significantly.

The evolution of mobile services is illustrated in Figure 1.8.

1.3 MOTIVATIONS FOR IP-BASED WIRELESS NETWORKS

Wireless networks are evolving into IP-based mobile networks. Is this worldwide march toward IP-based wireless networks a short-term phenomenon or a long-term trend? Several fundamental reasons suggest that IP-based wireless networks are more promising choices than circuit-switched wireless networks for the future.

IP-based wireless networks are better suited for supporting the rapidly growing mobile data and multimedia services. As mobile data and multimedia services continue to grow more rapidly than mobile voice services, they will overtake mobile voice services to become the dominant mobile services in the near future. First, we will see traffic volume of mobile data services surpass that of mobile voice services. As mobile data services become increasingly important to consumers, the revenues generated by network operators from mobile data services will also surpass the revenues from mobile voice services.

Figure 1.9 illustrates the estimated and the forecast (by Analysis Research Limited) Average Revenue Per User (ARPU) for mobile voice and non-voice services in Western Europe, 2000–2007. It shows that non-voice mobile services will grow significantly and steadily over the next few years to account for over 35% of the total revenue from mobile services in 2007. In the US, the penetration of mobile Internet services is expected to be even higher than in Western Europe, suggesting that non-voice mobile services could account for an even higher percentage of mobile operators' total revenue.

Figure 1.10 shows the growth of mobile services projected by the Universal Mobile Telecommunications System (UMTS) Forum. Here, non-voice mobile services are projected to grow even faster than the projection for Western Europe shown in Figure 1.9.

Wireless networks should evolve to support predominately mobile data and multimedia services and traffic, rather than circuit-switched voice services and traffic. IP technologies, which are already universal over wireline data networks, are the most promising solutions available today for supporting data and multimedia applications over wireless networks.

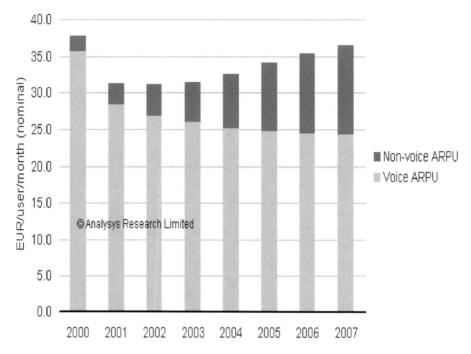

Fig. 1.9 *Growth of mobile voice and non-voice services*

IP-based wireless networks bring the successful Internet service paradigm to mobile providers and users. Perhaps the most important factor to the success of any type of future wireless networks is whether they can provide valuable services to the mass mobile users in ways that can be easily adopted by the users. IP technologies provide a proven and globally successful open infrastructure that fosters innovations of network services and facilitates the creation and offering of these services. A key reason for the success of the Internet is that the IP-based Internet paradigm enables everyone in the world to create and offer services over the Internet anytime and anywhere, as long as they have a computer connected to the Internet. This paradigm has led to the rich and rapidly growing information content, applications, and services over the Internet. This is significantly different from the circuit-switched PSTN or wireless networks, where only the network operators and their partners or suppliers could create and offer services. An IP-based wireless network would bring the service innovation potentials of the Internet paradigm to future wireless networks.

IP-based wireless networks can integrate seamlessly with the Internet. Radio systems need to be connected to the Internet to allow mobile users to access the information, applications, and services available over the Internet. Connecting an IP-based wireless network to the Internet is easier and more cost-effective than connecting a circuit-switched wireless network to the Internet.

Worldwide Revenues - All Services

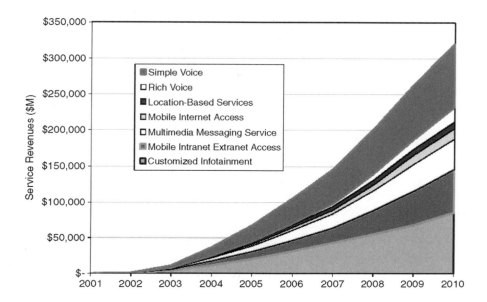

Fig. 1.10 *Growth of mobile voice and non-voice services*

Many mobile network operators also operate wireline networks. They have already built out IP core networks to support wireline IP services or as a backbone network for transporting circuit-switched voice traffic. Mobile network operators could leverage their existing IP core networks to support radio access networks and provide services to mobile users.

IP-based radio access systems are becoming important components of public wireless networks. IP-based radio access systems, e.g., IEEE 802.11 WLANs, are becoming increasingly important parts of public wireless networks worldwide. WLANs, which generally assume IP as the network-layer protocol for supporting user applications, are best supported by IP-based core networks rather than circuit-switched core networks.

Public WLANs could become "pico-cells" used to provide high system capacities and data rates to target geographical areas. Before public WLANs became available, pico-cells in public wireless networks are implemented using cellular radio technologies. Such a pico-cell is implemented using a pico-cellular radio base station to cover a small area. Alternatively, a wireless base station may use smart antennas to implement a pico-cell by shaping one of its radio beams to cover a small geographical area. Implementing a large number of pico-cells using cellular radio technologies are typically expensive—a key reason that pico-cells are not widely available today. Public WLANs offer a new way to provide such pico-cells at much lower costs.

IP technologies provide a better solution for making different radio technologies transparently to users. Different radio technologies will continue to coexist in public wireless networks. These radio technologies include not only different wide-area radio technologies but also the fast growing IP-based public WLANs. One radio technology (e.g., public WLANs) may meet communications needs other radio technologies (e.g., cellular radio systems) may not be able to meet easily. Therefore, heterogeneous radio systems are expected to coexist in the long run.

Mobile users typically do not want to be bothered with the specifics of each radio technology. They want to receive services not technologies. They want the technologies to be made transparent to them.

Therefore, there is a long-term need to interconnect radio systems that use different radio technologies, to support roaming between different radio systems, to provide mobile services over different radio systems in a seamless manner, and to support global roaming between different mobile providers and different countries. IP-based protocols, which are independent of the underlying radio technologies, are better suited than circuit-switched network technologies for achieving these goals.

With IP as the common network-layer protocol, a terminal with multiple radio interfaces (or a single radio interface capable of accessing different types of radio systems) could roam between different radio systems. IP-based network services and applications could be provided to all users in a seamless manner, regardless of which specific radio systems or mobile devices (e.g., PDAs, laptops, phones, or any other special-purpose devices) they are using.

1.4 3GPP, 3GPP2, AND IETF

In this section, we briefly describe the three main international organizations—3GPP, 3GPP2, and IETF—that are defining standards for wireless IP networks.

1.4.1 3GPP

The 3GPP is a partnership or collaboration formed in 1998 to produce international specifications for third-generation wireless networks. 3GPP specifications include all GSM (including GPRS and EDGE) and 3G specifications.

3GPP members are classified into the following categories:

- *Organizational Partners*: An Organizational Partner may be any Standards Development Organization (SDO) in any geographical location of the world. An SDO is an organization that is responsible for defining standards. 3GPP was formed initially by five SDOs: the Association of Radio Industries and Business (ARIB) in Japan, the European Telecommunication Standards

Institute (ETSI), T1 in North America, Telecommunications Technology Association (TTA) in Korea, and the Telecommunications Technology Committee (TTC) in Japan. Today, 3GPP also includes a new Organizational Partner—the China Wireless Telecommunication Standard (CWTS) group of China.

The Organizational Partners are responsible for producing the 3GPP specifications or standards. The 3GPP specifications are published as *3GPP Technical Specifications (TS)*, and *Technical Reports (TR)*.

- *Market Representation Partners*: A Market Representation Partner can be any organization in the world. It will provide advice to 3GPP on market requirements (e.g., services, features, and functionality). A Market Representation Partner does not have the authority to define, modify, or set standards within the scope of the 3GPP.

- *Individual Members*: Members of any Organizational Partner may become an individual member of 3GPP. An Individual Member can contribute, technically or otherwise, to 3GPP specifications.

- *Observers*: Any organization that may be qualified to become a future 3GPP partner may become an Observer. Representatives of an Observer may participate in 3GPP meetings and make contributions to 3GPP, but they will not have authority to make any decision within 3GPP.

The 3GPP Technical Specifications and Technical Reports are prepared, approved, and maintained by Technical Specification Groups (TSGs). Each TSG may have Working Groups to focus on different technical areas within the scope of the TSG. A project Coordination Group (PCG) coordinates the work among different TSGs. Currently, 3GPP has five TSGs:

- *TSG CN (Core Network)*: TSG CN is responsible for the specifications of the core network part of 3GPP systems, which is based on GSM and GPRS core networks. More specifically, TSG CN is responsible primarily for specifications of the layer-3 radio protocols (Call Control, Session Management, Mobility Management) between the user equipment and the core network, signaling between the core network nodes, interconnection with external networks, core network aspects of the interface between a radio access network and the core network, management of the core network, and matters related to supporting packet services (e.g., mapping of QoS).

- *TSG GERAN (GSM EDGE Radio Access Network)*: TSG GERAN is responsible for the specification of the radio access part of GSM/EDGE. This includes the RF layer; layer 1, 2, and 3 for the GERAN; interfaces internal to the GERAN, interfaces between a GERAN and the core network, conformance test specifications for all aspects of GERAN base stations and terminals, and GERAN-specific network management specifications for the nodes in the GERAN.

- *TSG RAN (Radio Access Network)*: TSG RAN is responsible for the definition of the functions, requirements, and interfaces of the UTRAN. This includes radio performance; layer 1, 2, and 3 specifications in UTRAN; specifications of the UTRAN internal interfaces and the interface between UTRAN and core networks; definition of the network management requirements in UTRAN and conformance testing for base stations.

- *TSG SA (Service and System Aspects)*: TSG SA is responsible for the overall architecture and service capabilities of systems based on 3GPP specifications. This includes the definition and maintenance of the overall system architecture, definition of required bearers and services, development of service capabilities and a service architecture, as well as charging, security, and network management aspects of 3GPP system.

- *TSG T (Terminal)*: TSG T is responsible for specifying terminal interfaces (logical and physical), terminal capabilities (such as execution environments), and terminal performance/testing.

3GPP specifications produced in different time periods are published as *Releases*. Each Release contains a set of Technical Specifications and Technical Reports. A Release is said to be *frozen* at a specific date if its content can only be revised in case a correction is needed after that date. Initially, 3GPP planned to standardize a new release each year. The first release therefore is named as Release 99 (frozen in March 2000). Release 99 (R99 in short) mainly focuses on a new RAN based on WCDMA. It also emphasizes the interworking and backward compatibility with GSM. Due to a variety of modifications proposed, Release 00 (R00) was scheduled into two different releases, which are named as Release 4 (R4) and Release 5 (R5). Release 4, frozen in March 2001, is a minor release with some enhancements to R99. IP transport was also introduced into the core network. Release 5 was frozen in June 2002. It comprises major changes in the core network based on IP protocols. More specifically, phase 1 of the IP Multimedia Subsystem (IMS) was defined. In addition, IP transport in the UTRAN was specified. Release 6 is expected to be frozen in March 2004. It will focus on IMS phase 2, harmonization of the IMS in 3GPP and 3GPP2, interoperability of UMTS and WLAN, and multimedia broadcast and multicast.

1.4.2 3GPP2

The 3GPP2, like 3GPP, is also an international collaboration to produce global standards for third-generation wireless networks. 3GPP2 was formed soon after 3GPP when the American National Standards Institute (ANSI) failed to convince 3GPP to include "non-GSM" technologies in 3G standards. 3GPP2 members are also classified into Organizational Partners and Market Representation Partners. Today, 3GPP2 has five Organizational Partners: ARIB (Japan), CWTS (China), TIA

(Telecommunications Industry Association) in North America, TTA (Korea), and TTC (Japan).

Standards produced by 3GPP2 are published as 3GPP2 Technical Specifications. Technical Specification Groups (TSGs) are responsible for producing Technical Specifications. A Steering Committee coordinates the works among different TSGs. Currently, 3GPP2 has the following TSGs:

- *TSG-A (Access Network Interfaces)*: TSG-A is responsible for the specifications of interfaces between the radio access network and core network, as well as within the access network. Specifically, it has a responsibility for the specifications of the following aspects of radio access network interfaces: physical links, transports and signaling, support for access network mobility, 3G capability (e.g., high-speed data support), interfaces inside the radio access network, and interoperability specification.

- *TSG-C (cdma2000)*: TSG-C is responsible for the radio access part, including its internal structure, of systems based on 3GPP2 specifications. Specifically, it has a responsibility for the requirements, functions, and interfaces for the cdma2000 radio infrastructure and user terminal equipment. These include specifications of radio layers 1–3, radio link protocol, support for enhanced privacy, authentication and encryption, digital speech codecs, video codec selection and specification of related video services, data and other ancillary services support, conformance test plans, and location-based services support.

- *TSG-S (Service and System Aspects)*: TSG-S is responsible for the development of service capability requirements for systems based on 3GPP2 specifications. It is also responsible for high-level architectural issues, as required to coordinate service development across the various TSGs. Some specific responsibilities include
  - Definition of services, network management, and system requirements.
  - Development and maintenance of network architecture and associated system requirements and reference models.
  - Management, technical coordination, as well as architectural and requirements development associated with all end-to-end features, services, and system capabilities, including, but not limited to, security and QoS.
  - Requirements for international roaming.

- *TSG-X (Intersystem Operations)*: TSG-X is responsible for the specifications of the core network part of systems, based on 3GPP2 specifications. Specifically, it has a responsibility for:
  - Core network internal interfaces for call associated and noncall associated signaling.
  - IP technology to support wireless packet data services, including voice and other multimedia services.
  - Core network internal interfaces for bearer transport.

- Charging, accounting, and billing specifications.
- Validation and verification of specification text it develops.
- Evolution of core network to support interoperability and intersystem operations, and international roaming.
- Network support for enhanced privacy, authentication, data integrity, and other security aspects.
- Wireless IP services.

Although 3GPP2 specifies standards for both core network and radio access network, *revisions* of 3GPP2 specifications are primary based on the cdma2000 radio access network. As shown in Figure 1.11, there are three revisions in cdma2000 1x and 3x. They are specified by 3GPP2 C.S001-0005 Revision 0 [2, 3, 4, 5, 6], C.S001-0005 Revision A [1, 10, 13, 16, 19], and C.S001-0005 Revision B [7, 11, 14, 17, 20]. The specifications are based on the TIA IS-2000 series [35]. There are two evolutions (EV) of cdma2000 1x. The cdma2000 1x EV-DO, specified by IS-856 [34]/3GPP2 C.S0024 [9], defined the enhancement of cdma2000 1x for data only (DO). It is based on the HDR developed by QUALCOMM for direct Internet access. The specifications of 3GPP2 C.S001-0005 Revision C [8, 12, 15, 18, 21] specify cdma2000 1x EV-DV, the evolution of cdma2000 1x for both data and voice (DV) enhancement. In addition to conventional circuit-switching network, packet-switching network based on IP is also incorporated.

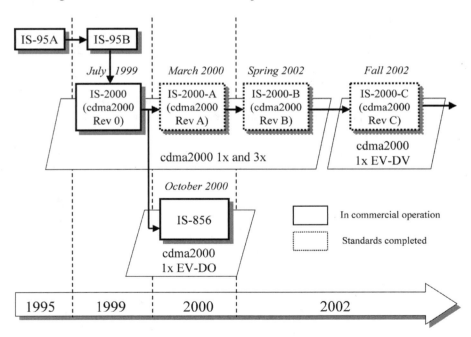

Fig. 1.11 *cdma2000 family. Reproduced with the kind permission of ITU.*

1.4.3 IETF

The Internet Engineering Task Force (IETF) is a large open international community of network designers, operators, vendors, and researchers who are concerned with the evolution of the Internet architecture and smooth operation of the Internet. The Internet is a loosely organized international collaboration of autonomous and interconnected networks that supports host-to-host communication through voluntary adherence to open protocols and procedures defined by *Internet Standards*. Internet Standards are produced by the IETF and specify protocols, procedures, and conventions that are used in or by the Internet. An Internet Standard is in general a specification that is stable, well understood, technically competent, has multiple, independent, and interoperable implementations with substantial operational experience; enjoys significant public support; and is recognizably useful in some or all parts of the Internet.

Internet Standards are archived and published by the IETF as *Request for Comments (RFC)*. RFCs are classified into Standards-Track and Non-Standards-Track RFCs (e.g., Informational, Best Current Practices, etc.). Only Standards-Track RFCs can become Internet Standards. Non-Standards-Track RFCs are used primarily to document best current practices, experiment experiences, historical, or other information.

Standards-Track RFCs are further classified, based on their maturity levels, into the following categories [23]:

- *Proposed Standard*: The entry-level maturity for a Standards-Track RFC is a Proposed Standard. A Proposed Standard specification is generally stable, has resolved known design choices, is believed to be well understood, has received significant community review, and appears to enjoy enough community interest to be considered valuable. However, further experience might result in a change or even retraction of the specification before it advances to the next maturity level of Standards-Track RFC.

   Usually, neither implementation nor operational experience is required for the designation of a specification as a Proposed Standard. However, such experience is highly desirable and will usually represent a strong argument in favor of a Proposed Standard designation.

   A Proposed Standard RFC remains valid for at least six months, but only up to a maximum of 2 years. Then, it is either deprecated or elevated to the next higher level of maturity level: Draft Standard.

- *Draft Standard*: A Draft Standard RFC documents a complete specification from which at least two independent and interoperable implementations have been implemented on different software code bases, and sufficient successful operational experience has been obtained. Here, the term "interoperable" means functionally equivalent or interchangeable system components.

   A Draft Standard RFC remains valid for at least four months but not longer than two years. It may be elevated to the next higher level of maturity (i.e., Internet Standard), returned to Proposed Standard, or deprecated.

- *Internet Standard*: An Internet Standard RFC documents a specification for which significant implementation and successful operational experience have been obtained. An Internet Standard is characterized by a high degree of technical maturity and by a generally held belief that the specified protocol or service provides significant benefit to the Internet community.

The work in progress to produce the potential RFC will be documented and published by the IETF as *Internet Drafts*. Internet Drafts expire six months after their publication. To keep an Internet Draft valid, it needs to be updated before its expiration date.

The IETF operates in ways significantly different from other standardization organizations such as 3GPP and 3GPP2. The IETF is open to any individual. It does not require any membership. The technical work is performed in *Working Groups*. The Working Groups produce RFCs. Anyone can participate in the discussions of any Working Group, contribute Internet Drafts to present ideas for further discussions, and make contributions in any other way to the creation of a RFC. Technical discussions in each Working Group are carried out mostly on mailing lists. The IETF holds face-to-face meetings three times a year.

The Working Groups are organized by technical topics into *Areas*. Areas are managed by Area Directors. The Area Directors form an Internet Engineering Steering Group (IESG) to coordinate the works in different Areas. An Internet Architecture Board (IAB) provides architectural oversight. Currently, the active Areas include Applications Area, General Area, Internet Area, Operations and Management Area, Routing Area, Security Area, Sub-IP Area, and Transport Area.

Decision-making in the Working Groups (e.g., what should be included or excluded in a RFC) is based on the following key principles:

- *Rough consensus*: The principle of "rough consensus" suggests that no formal voting takes place in order to make a decision. Decisions are made if there is a rough consensus among all the individuals who participate in Working Group discussions. For example, a Working Group may submit an Internet Draft to the Area Director and the IESG for approval to become an RFC when there is a rough consensus among the Working Group participants that the Internet Draft is ready to become an RFC. Once approved by the Area Director and the IESG, an Internet Draft will become an RFC.
- *Running code*: The principle of "running code" suggests that the ideas and specifications need to be backed up by actual implementations to demonstrate their feasibility, stability, performance, etc. Implementations and experiences from the implementations are important criteria for an idea to be adopted by a Working Group, for an Internet Draft to be elevated to an RFC, and for an RFC to finally reach the Internet Standard level.

Any individual could propose the creation of a Working Group. To create a Working Group, one must first propose a BOF or Birds of a Feather. A BOF is essentially a group of people who are interested in discussing whether a new Working Group should be created in a specific topic area. A BOF is used to define the goals and milestones of a proposed Working Group and to gauge whether there is enough interest from the IETF participants to create the new Working Group. If there is a rough consensus among the participants that a new Working Group should be created, the chairperson of the BOF will present the results to the Area Director for approval. A new Working Group will be then created if it is approved by the Area Director, the IESG, and the IAB.

1.5 ORGANIZATION OF THE BOOK

This book focuses on network architecture, signaling and control, mobility management, network security, and QoS specified by 3GPP and 3GPP2. The MWIF specifications are discussed in some chapters if related issues are also defined in MWIF. The rest of the book is organized as follows:

- *Chapter 2: "Wireless IP Network Architectures"*: Describes the 3G wireless network architectures defined by 3GPP, 3GPP2, and the all-IP wireless network architecture defined by MWIF. Signaling and session control for network connectivity are also specified.
- *Chapter 3: "IP Multimedia Subsystems and Application-Level Signaling"*: Discusses the IP Multimedia Subsystem (IMS) defined by 3GPP and 3GPP2. It also discusses issues and solutions related to signaling and session control in IP networks and the IMS defined by 3GPP and 3GPP2.
- *Chapter 4: "Mobility Management"*: Discusses issues and solutions for mobility management in IP networks and IP-based wireless networks defined by 3GPP, 3GPP2, and MWIF.
- *Chapter 5: "Security"*: Discusses issues and solutions for network security in IP networks and IP-based wireless networks defined by 3GPP and 3GPP2.
- *Chapter 6: "Quality of Service"*: Discusses issues and solutions for supporting quality of service in IP networks and IP-based wireless networks defined by 3GPP and 3GPP2.

REFERENCES

1. 3rd Generation Partnership Project 2 (3GPP2). cdma2000—introduction. 3GPP2 C.S0001-A, Version 5.0, Release A, July 2001.
2. 3rd Generation Partnership Project 2 (3GPP2). Introduction to cdma2000 spread spectrum systems. 3GPP2 C.S0001-0, Version 3.0, Release 0, July 2001.
3. 3rd Generation Partnership Project 2 (3GPP2). Medium access control (MAC) standard for cdma2000 spread spectrum systems. 3GPP2 C.S0003-0, Version 3.0, Release 0, July 2001.

4. 3rd Generation Partnership Project 2 (3GPP2). Physical layer standard for cdma2000 spread spectrum systems. 3GPP2 C.S0002-0, Version 3.0, Release 0, July 2001.

5. 3rd Generation Partnership Project 2 (3GPP2). Signaling link access control (LAC) standard for cdma2000 spread spectrum systems. 3GPP2 C.S0004-0, Version 3.0, Release 0, July 2001.

6. 3rd Generation Partnership Project 2 (3GPP2). Upper layer (layer 3) signaling standard for cdma2000 spread spectrum systems. 3GPP2 C.S0005-0, Version 3.0, Release 0, July 2001.

7. 3rd Generation Partnership Project 2 (3GPP2). cdma2000—introduction. 3GPP2 C.S0001-B, Version 1.0, Release B, April 2002.

8. 3rd Generation Partnership Project 2 (3GPP2). cdma2000—introduction. 3GPP2 C.S0001-C, Version 1.0, Release C, May 2002.

9. 3rd Generation Partnership Project 2 (3GPP2). cdma2000 high rate packet data air interface specification. 3GPP2 C.S0024-0, Version 4.0, October 2002.

10. 3rd Generation Partnership Project 2 (3GPP2). Medium access control (MAC) standard for cdma2000 spread spectrum systems. 3GPP2 C.S0003-A, Version 6.0, Release A, February 2002.

11. 3rd Generation Partnership Project 2 (3GPP2). Medium access control (MAC) standard for cdma2000 spread spectrum systems. 3GPP2 C.S0003-B, Version 1.0, Release B, April 2002.

12. 3rd Generation Partnership Project 2 (3GPP2). Medium access control (MAC) standard for cdma2000 spread spectrum systems. 3GPP2 C.S0003-C, Version 1.0, Release C, May 2002.

13. 3rd Generation Partnership Project 2 (3GPP2). Physical layer standard for cdma2000 spread spectrum systems. 3GPP2 C.S0002-A, Version 6.0, Release A, February 2002.

14. 3rd Generation Partnership Project 2 (3GPP2). Physical layer standard for cdma2000 spread spectrum systems. 3GPP2 C.S0002-B, Version 1.0, Release B, April 2002.

15. 3rd Generation Partnership Project 2 (3GPP2). Physical layer standard for cdma2000 spread spectrum systems. 3GPP2 C.S0002-C, Version 1.0, Release C, May 2002.

16. 3rd Generation Partnership Project 2 (3GPP2). Signaling link access control (LAC) standard for cdma2000 spread spectrum systems. 3GPP2 C.S0004-A, Version 6.0, Release A, February 2002.

17. 3rd Generation Partnership Project 2 (3GPP2). Signaling link access control (LAC) standard for cdma2000 spread spectrum systems. 3GPP2 C.S0004-B, Version 1.0, Release B, April 2002.

18. 3rd Generation Partnership Project 2 (3GPP2). Signaling link access control (LAC) standard for cdma2000 spread spectrum systems. 3GPP2 C.S0004-C, Version 1.0, Release C, May 2002.

19. 3rd Generation Partnership Project 2 (3GPP2). Upper layer (layer 3) signaling standard for cdma2000 spread spectrum systems. 3GPP2 C.S0005-A, Version 6.0, Release A, February 2002.

20. 3rd Generation Partnership Project 2 (3GPP2). Upper layer (layer 3) signaling standard for cdma2000 spread spectrum systems. 3GPP2 C.S0005-B, Version 1.0, Release B, April 2002.

21. 3rd Generation Partnership Project 2 (3GPP2). Upper layer (layer 3) signaling standard for cdma2000 spread spectrum systems. 3GPP2 C.S0005-C, Version 1.0, Release C, May 2002.

22. Bluetooth SIG, Inc. http://www.bluetooth.org.

23. S. Bradner. The Internet standards process—revision 3. IETF RFC 2026, October 1996.

24. CDMA Development Group. http://www.cdg.org.

25. D. Goodman. *Wireless Personal Communications Systems*. Addison-Wesley Publishing Company, Reading, MA, 1997.

26. ETSI HIPERLAN/2 standard. http://portal.etsi.org/bran/kta/Hiperlan/hiperlan2.asp.

27. HomeRF. http://www.homerf.org/.

28. IEEE P802.11, the working group for wireless LANs. http://grouper.ieee.org/groups/802/11/index.html.

29. IEEE 802.15 working group for WPANs. http://grouper.ieee.org/groups/802/15/.

30. The ultimate IMT-2000 gateway on the World-Wide Web. http://www.imt-2000.org/.

31. Code Division Multiple Access II. http://www.cdg.org.

32. Y.-B. Lin and I. Chlamtac. *Wireless and Mobile Network Architectures*. Wiley, New York, 2001.

33. Multimedia mobile access communication systems. http://www.arib.or.jp/mmac/e/.

34. TIA/EIA TR45.4. cdma2000 high rate packet data air interface specification. November 2000.

35. TIA/EIA TR45.5. CDMA 2000 series, revision A. March 2000.

2

Wireless IP Network Architectures

This chapter examines the wireless IP network architectures defined by 3GPP and 3GPP2, respectively, and the all-IP wireless network architecture defined by MWIF.

2.1 3GPP PACKET DATA NETWORKS

This section discusses the 3GPP packet network architecture based on Release 5 of the 3GPP Technical Specifications. Release 5, completed in June 2002, was the latest release during the writing of this book. We will describe

- *3GPP network architecture* (Section 2.1.1)
- *Protocol reference model* (Section 2.1.2)
- *Traffic and signaling bearers and connections for supporting packet-switched services* (Section 2.1.3)
- *Packet Data Protocol (PDP) context* (Section 2.1.4)
- *Steps for a mobile to access packet data network and services* (Sections 2.1.5)
- *User packet routing and transport* (Section 2.1.6)
- *How a mobile acquires IP addresses for accessing 3GPP packet data services* (Section 2.1.7)
- *Key procedures used in the packet data network* (Sections 2.1.8 through 2.1.10)

IP-Based Next-Generation Wireless Networks: Systems, Architectures, and Protocols,
By Jyh-Cheng Chen and Tao Zhang. ISBN 0-471-23526-1 © 2004 John Wiley & Sons, Inc.

Key procedures used in the packet data network (Sections 2.1.8 through 2.1.10)
Protocol stacks for packet data network (Section 2.1.11)
How to use a 3GPP packet network to access other IP networks (Section 2.1.2)

2.1.1   Network Architecture

A public network administrated by a single network operator for providing land mobile services is referred to as a Public Land Mobile Network (PLMN). The conceptual architecture of a 3GPP PLMN is illustrated in Figure 2.1. It consists of one or more Radio Access Networks (RANs) interconnected via a Core Network (CN).

A RAN provides radio resources (e.g., radio channels, bandwidth) for users to access the CN. Release 5 currently supports GSM/EDGE RAN (GERAN) and UMTS Terrestrial RAN (UTRAN). Work is underway on 3GPP to specify how to

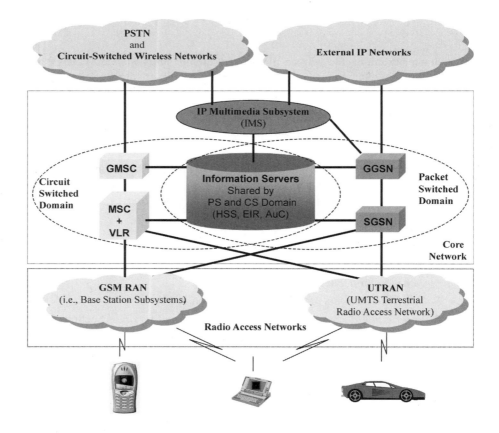

Fig. 2.1   *3GPP conceptual network architecture (Release 5)*

support Broadband Radio Access Networks (BRANs), such as IEEE 802.11 [25], [35].

A GERAN is divided into Base Station Subsystems (BSS). Each BSS consists of one or multiple Base Transceiver Stations (BTSs) and Base Station Controllers (BSCs). A BTS maintains the air interface. It handles signaling and speech processing over the air interface. A BSC controls the radio connections toward the mobile terminals as well as the wireline connections toward the CN. Each BSC can control one or more BTSs.

A UTRAN is divided into Radio Network Subsystems (RNS). Each RNS consists of one or multiple Node Bs controlled by a Radio Network Controller (RNC). A Node B is a wireless base station, which is analogous to a BTS in GSM, and it provides the air interface with mobile terminals. An RNC, which is analogous to a BSC in GSM, controls the radio connections toward the mobile terminals and the wireline interfaces with the CN.

The CN implements the capabilities for supporting both circuit-switched and packet-switched communication services to mobile users. These communication services include both *basic services* and *advanced services*. Basic circuit-switched services include switching of circuit-switched voice and data calls and call control functions for supporting basic point-to-point circuit-switched calls. Basic packet-switched services include the routing and transport of user IP packets. Advanced services, commonly referred to as supplementary services or value-added services, include any service that provides added value beyond the basic services. Examples of advanced circuit-switched services include prepaid calls, toll-free calls, call forwarding (e.g., forwarding a voice phone call to another phone or to an E-mail box), multiparty communications, and pay-per-view. Advanced packet-switched services may include all services or applications over IP networks beyond simple packet transport. Some examples are e-mail, World-Wide Web, location-based services, multimedia messaging services, networked gaming, and e-commerce.

The CN is divided into the following functional building blocks [21], [23]:

- *Circuit-Switched (CS) Domain*
- *Packet-Switched (PS) Domain*
- *IP Multimedia Subsystem (IMS)*
- *Information Servers*

Each RAN routes circuit-switched traffic to the CS CN domain and routes packet-switched traffic to the PS CN domain.

2.1.1.1 *Mobile Devices, Subscribers, and Their Identifiers* A mobile device in GSM is called Mobile Station (MS), which is synonymous with User Equipment (UE) in UMTS.[1] Figure 2.2 illustrates the functional architecture of a UE, which consists of Mobile Equipment (ME) and UMTS Subscriber Identity

---

[1]In this book, MS and UE are used interchangeably.

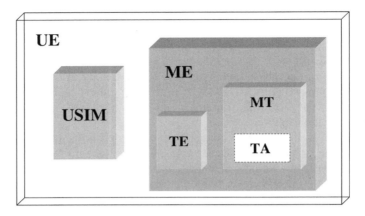

Fig. 2.2   *Functional architecture of a user equipment (UE)*

Module (USIM) [13]. USIM is developed based on the *Subscriber Identity Module (SIM)* used in GSM systems. A ME, consisting of Mobile Termination (MT) and Terminal Equipment (TE), is the device a user uses to access the network services. TE provides functions for the operations of the access protocols. MT, on the other hand, supports radio transmission and channel management. Depending on applications, an MT may have a combination of different Terminal Adapters (TA). In realization, MT could also be a mobile handset and TE could be a laptop computer. It is also possible to integrate MT and TE in the same device.

Each MT is identified by a globally unique International Mobile Station Equipment Identity (IMEI) [15].

A mobile station may be configured to access the PS domain only, the CS domain only, or both the CS and the PS domains.

Each subscriber to 3GPP network services is assigned a globally unique *International Mobile Subscriber Identity (IMSI)* as its permanent identifier. A subscriber uses its IMSI as its common identifier for accessing PS services, CS services, or both PS and CS services at the same time.

A subscriber's IMSI is stored on a USIM on a mobile station. A subscriber can move its USIM from one mobile station to another so that the subscriber can use different mobile stations to access the network while being identified by the network as the same subscriber.

The network uses the IMSI to identify a subscriber and to identify the network services and resources used by a subscriber for billing purpose. A mobile's IMSI may be used as the mobile's identifier at multiple protocol layers in 3GPP, e.g., at the physical layer, link layer, and the network layer.

An IMSI can consist of only numerical characters 0 through 9. It contains three parts as shown in Figure 2.3 [15]:

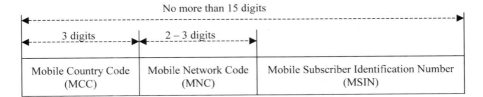

Fig. 2.3 *Structure of International Mobile Subscriber Identity (IMSI)*

- *Mobile Country Code (MCC)*: The MCC uniquely identifies a mobile subscriber's home country.
- *Mobile Network Code (MNC)*: The MNC uniquely identifies a mobile subscriber's home PLMN in the mobile subscriber's home country. The MNC can be two or three digits in length, depending on the value of the MCC.
- *Mobile Subscriber Identification Number (MSIN)*: The MSIN uniquely identifies a mobile subscriber within one PLMN.

Allocation of MCCs is administrated by the ITU-T according to ITU-T Blue Book Recommendation E.212. MNC + MSIN is commonly referred to as the National Mobile Subscriber Identity (NMSI). The NMSIs are allocated by the numbering administrations in each country. When more than one PLMN exists in a country, a unique MNC is assigned to each of these PLMNs.

To reduce the need to transmit IMSI, which uniquely identifies a mobile subscriber, over the air, 3GPP uses a Temporary Mobile Subscriber Identity (TMSI) to identify a mobile whenever possible. A TMSI is a four-octet number assigned to a mobile temporarily by an MSC/VLR for circuit-switched services or by an SGSN for packet-switched services. The two most significant bits in a TMSI indicates whether the TMSI is for packet-switched services. A TMSI for packet-switched services is referred to as a Packet TMSI or P-TMSI. An MSC or SGSN uses a TMSI to uniquely identify a mobile. The TMSI will only be allocated in ciphered form. Furthermore, measures will be taken to ensure that the mapping between a mobile's IMSI and TMSI is known only by the mobile and the network node (MSC or SGSN) that assigned the TMSI (Section 2.1.8). A mobile's TMSI, instead of its IMSI, will then be used as the mobile's identity whenever possible in signaling messages transmitted over the air. As only the mobile and the MSC or SGSN that assigned the TMSI to the mobile know the mapping between the mobile's IMSI and TMSI and that a TMSI is valid only when the user is served by the MSC or SGSN that assigned the TMSI, the security impact of transmitting unencrypted TMSI over the air is lower than transmitting unencrypted IMSI.

To send and receive IP packets over the PS CN, a mobile also needs to be configured with at least one IP address. The mobile may use multiple IP addresses simultaneously. However, a mobile is not required to have a valid IP address at all times while it is attached to the PS domain. Instead, a mobile may acquire an IP address only when it needs to activate packet data services over the PS CN.

2.1.1.2 Circuit-Switched Domain in Core Network

The CS domain consists of all the CN entities for providing circuit-switched voice and data services to mobile users. The CS CN domain is built on the GSM core network technologies. Its main network entities are:

- *Mobile-services Switching Center (MSC)*
- *Gateway MSC*
- *Visitor Location Register (VLR)*
- *Home Subscriber Server (HSS), Equipment Identity Register (EIR), and Authentication Center (AuC)*

The MSC performs switching and call control functions needed to provide basic circuit-switched services to mobile terminals. In addition, it also performs mobility management functions, including location registration and handoff functions for mobile terminals. The MSC interconnects RANs to the CS CN domain. One MSC may interface with multiple GSM BSSs or UTRAN RNSs.

In 3GPP Release 5, the CS CN made a significant improvement over the previous releases: It allows the switching and call control functions of an MSC to be separated and implemented on separate network entities:

- *MSC Server* for handling call control and mobility management.
- *CS Media Gateway (CS-MGW)* for handling circuit switching, media conversion, and payload processing (e.g., echo canceller, codec) and payload transport over the circuits.

Separation of switching and call control allows switching and call control technologies to evolve independently. It also helps increase network scalability. For example, one MSC Server can support multiple CS-MGWs and new MSC Servers and/or CS-MGWs can be added to increase call control and/or switching capabilities.

A dedicated MSC called Gateway MSC (GMSC) may be used to interface with external circuit-switched networks. A GMSC is responsible for routing a circuit-switched call to its final destination in external networks. The switching and call control functions of a GMSC can also be separated and implemented on separate network entities: CS-MGW for switching and media control and a GMSC Server for call control.

A VLR maintains location and service subscription information for visiting mobiles temporarily while they are inside the part of a network controlled by the VLR. It tracks a visiting mobile's location and informs the visiting mobile's HLR of the mobile's current location. It retrieves a visiting mobile's service subscription information from the mobile's HLR, maintains a copy of the information while the visiting mobile is inside the part of the network controlled by the VLR, and uses the information to provide service control for the visiting mobile.

A VLR is typically integrated with each MSC because no open standard interface has been defined between an MSC and a VLR. The Mobile Application Part (MAP) [12] protocol is used for signaling between a VLR and an HLR.

The other information servers HSS, EIR, and AuC are shared by the CS and the PS domains and will be discussed in Section 2.1.1.5.

2.1.1.3 Packet-Switched Domain in the Core Network

The PS CN domain provides the following main functions for supporting packet-switched services:

- *Network access control*: Determines which mobiles are allowed to use the PS domain. These functions include registration, authentication and authorization, admission control, message filtering, and usage data collection.
- *Packet routing and transport*: Route user packets toward their destinations either inside the same PLMN or in external networks.
- *Mobility management*: Provides network-layer mobility management functions. These functions include tracking the locations of mobile terminals, initiating paging to determine an idle mobile's precise location when the network has data to send to the mobile, and maintaining up-to-date CN routes to mobiles as they move.

The PS domain is built on the GPRS network platform. As in GPRS, the 3GPP PS CN domain consists of two main types of network nodes:

- *Serving GPRS Support Node (SGSN)*
- *Gateway GPRS Support Node (GGSN)*

An SGSN interconnects one or more RANs to a PS CN. A SGSN performs the following specific functions:

- *Access control*: The SGSN is responsible for the first line of control over users' access to the PS CN domain. The GGSN provides an additional line of control for access to the PS CN domain.
- *Location management*: The SGSN tracks the locations of mobiles that use packet-switched services. It may report the location information to the HLR so that the location information may be used, for example, by the GGSN to perform network-initiated procedures to set up connections to mobiles.
- *Route management*: The SGSN is responsible for maintaining a route to a GGSN for each mobile and to relay user traffic between the mobile and the GGSN.

- *Paging*: The SGSN is responsible for initiating paging operations upon receiving user data destined to idle mobiles.
- *Interface with service control platforms*: The SGSN is the contact point with CAMEL functions for GPRS and IP-based services. CAMEL (Customized Applications for Mobile Enhanced Logic) [11] is a set of procedures and protocols that allow a network operator to provide operator-specific services to its subscribers even when the subscribers are currently in foreign networks. For example, CAMEL can be used to provide prepaid services, such as prepaid calls and prepaid Short Message Services (SMS).

A GGSN serves as the interface between the PS CN domain and any other packet network (e.g., the Internet, an intranet, the 3GPP IP Multimedia Subsystem). One GGSN can be used to support both GERANs and UTRANs. A GGSN provides the following specific functions:

- *Packet routing and forwarding center*: A GGSN acts as a packet routing and forwarding center for user packets. All user packets to and from a mobile in a PLMN will be sent first to a GGSN, which we will refer to as the mobile's *serving GGSN*. The mobile's serving GGSN will then forward these user packets toward their final destinations.
- *Route and mobility management*: A mobile's serving GGSN tracks the SGSN that is currently serving each mobile (which we will refer to as the mobile's serving SGSN). The GGSN maintains a route to the mobile's serving SGSN and uses the route to exchange the user traffic with the SGSN.

IP is used as the basic protocol for transporting traffic between SGSNs and between an SGSN and a GGSN. IP is also the routing protocol between GGSNs and between a GGSN and any other IP network.

Private IP addresses may be used to address the SGSNs and GGSNs inside a PLMN. When a PLMN uses private IP addresses, Network Address Translators (NATs) will be needed to translate between the private IP addresses used inside a PLMN and the public IP addresses used over the public network so that mobiles inside the PLMN can communicate with terminals outside the PLMN. Each PLMN may have multiple logically separated IP networks referred to as *IP Addressing Domains*. Each IP Addressing Domain may also use private IP addresses internally. Gateways and firewalls may be used to interconnect IP Addressing Domains.

SGSNs and GGSNs are also identified by *SGSN Numbers* and *GGSN Numbers*, respectively. SGSN Numbers and GGSN Numbers are used primarily with non-IP protocols, e.g., MAP or other SS7 (Signaling System 7)-based protocols. SGSNs and GGSNs may need to use such non-IP protocols to communicate, for example, with the HSS.

2.1.1.4 *IP Multimedia Subsystem*   3GPP Release 5 introduced the IP Multimedia Subsystem (IMS) that provides core network entities for supporting

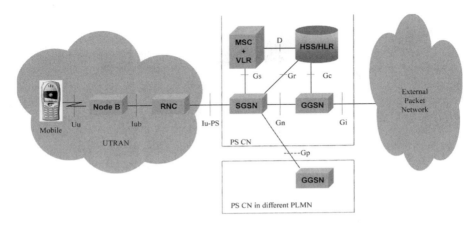

Fig. 2.5 *Protocol reference model for 3GPP PS domain*

is using services handled by the MSC. This is a critical capability that allows close integration of the networking capabilities provided by the PS and the CS domains. Signaling over Interface G_s uses SS7 connectionless SCCP (Signaling Connection Control Part) protocol.

The protocol reference model for the PS domain is illustrated in Figure 2.5, where the RAN is assumed to be UTRAN for illustration purpose. The main interfaces for supporting packet-switched services are the G_n, G_p, G_i, G_s, G_c, and G_r interfaces inside the PS CN domain, the I_u interface connecting a RAN with the PS CN domain, and interfaces inside a RAN (e.g., the I_{ub} and U_u interfaces inside a UTRAN).

2.1.3 Packet Data Protocols, Bearers, and Connections for Packet Services

A mobile uses a *Packet Data Protocol (PDP)* to exchange user packets over a 3GPP PS CN domain with other mobiles either inside the same 3GPP network or in other IP networks. In the rest of this book, we will assume that the PDP is IP unless stated otherwise explicitly.

The PDP Packet Data Units (PDUs) (i.e., user packets) are transported inside a 3GPP network over *traffic bearers*. A traffic bearer is a set of network resources and data transport functions used to deliver user traffic between two network entities. A traffic bearer can be a path, a logical connection, or a physical connection between two network nodes.

The structure of traffic bearers for supporting packet-switched services is illustrated in Figure 2.6. For a mobile to send or receive user packets over a 3GPP PS CN, a dedicated path needs to be maintained between the mobile and its serving

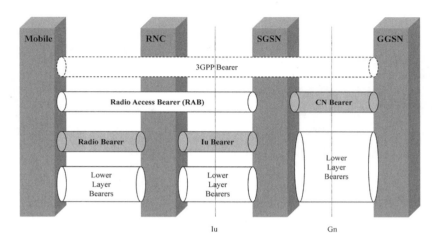

Fig. 2.6 *3GPP bearers (connections) for supporting packet-switched services*

GGSN. We shall refer to this path as the *3GPP Bearer* for the mobile. For example, a 3GPP Bearer in a UMTS network would be a UMTS Bearer.

A 3GPP Bearer is constructed by concatenating a *Radio Access Bearer (RAB)* that connects a mobile over a RAN to the edge of the CN (i.e., a SGSN) and a *CN Bearer* that carries user traffic between the edge of the CN and a GGSN.

- A RAB is a logical connection that is constructed by concatenating a *Traffic Radio Bearer* and an I_u *Traffic Bearer*.

  A Traffic Radio Bearer is a logical connection provided by the protocol layer immediately below the PDP layer for transporting user packets between a mobile and an RNC (or a BSC). An I_u Traffic Bearer is a logical connection provided by the protocol layer immediately below the PDP layer for transporting user packets between the RAN (i.e., an RNC or a BSC) and an SGSN.

- A CN Bearer is a logical connection provided by the protocol layer immediately below the PDP layer for transporting user packets between an SGSN and a GGSN.

A Traffic Radio Bearer, an I_u Traffic Bearer, and a CN Bearer collectively form a logical link immediately below the PDP layer for transporting user packets between the mobile and its serving GGSN.

The Traffic Radio Bearers, I_u Traffic Bearers, Radio Access Bearers, and CN bearers are managed by different protocols and procedures. This creates several benefits.

- The resource management requirements differ significantly in different parts of a 3GPP network, such as inside a RAN, a CN, and between a RAN and a CN. Separating the bearers in these parts of the network allows different protocols and procedures to be used to address the unique resource management needs in each part of the network. It also allows the technologies used in each part of the network to evolve with less dependency on each other.

- Separating the bearers in these parts of the network facilitates mobility management. For example, when a mobile moves inside the service area of the same SGSN, new CN Bearer will not be needed for the mobile. This reduces the time it takes for a mobile to handoff from one RNC to another under the same SGSN. It also means that the GGSNs do not need to be aware of how mobility is managed inside each RAN. Instead, a GGSN only needs to keep track of which SGSN is currently serving a mobile.

The Traffic Radio Bearers, I_u Traffic Bearers, and CN Bearers are supported by bearers provided by lower-layer protocols. The full protocol stacks for supporting Traffic Radio Bearers, I_u Traffic Bearers, and CN Bearers will be described in Section 2.1.11.

Before network resources in a RAN or the PS CN can be allocated to provide packet-switched services to a mobile, a dedicated logical *Signaling Connection* needs to be established between the mobile and an SGSN. This signaling connection is used, for example, for the mobile to register with the PS CN domain, for the mobile to request the SGSN to establish CN Bearers, and for the establishment of Radio Access Bearers.

As illustrated in Figure 2.7, the signaling connection between a mobile and an SGSN is constructed by concatenating a Signaling Radio Bearer between the mobile and the RAN (e.g., the RNC in a UTRAN) and an I_u Signaling Bearer between the RAN and the SGSN.

The Signaling Radio Bearers and Traffic Radio Bearers for the same mobile are collectively referred to as a *RRC connection*. This is because the Radio Resource Control (RRC) protocol [16], [18] is used to establish, maintain, and release the Radio Bearers. A mobile will use a common RRC connection to carry the signaling and user traffic for both PS and CS services.

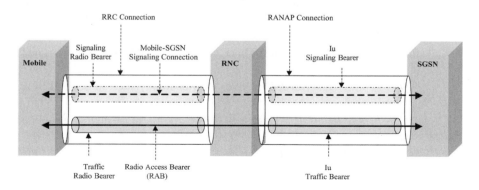

Fig. 2.7 *Signaling and traffic connections between mobile and SGSN*

The I_u Signaling Bearers and the I_u Traffic Bearers for the same mobile are collectively referred to as a *RANAP connection*. This is because all these I_u Bearers are established, maintained, modified, changed, and released by the Radio Access Network Application Part (RANAP) protocol [20].

The RNC is responsible for interconnecting the user's RRC connection with the user's RANAP connection to construct a signaling connection or a RAB between the mobile and the SGSN. In other words, the RNC acts as a protocol converter and converts between protocols used in the RAN and the CN.

2.1.4   Packet Data Protocol (PDP) Context

To send or receive user packets, a mobile needs to acquire and configure itself with a *PDP Address* (i.e., an IP address when the PDP is IP). A mobile may use multiple PDP addresses simultaneously. Before user packets destined to or originated from a PDP address can be transported over a 3GPP PS CN, a *PDP Context* for this PDP address has to be established and activated in the PS CN domain (i.e., on an SGSN and a GGSN) and on the mobile.

A PDP Context maintained by a network node contains a set of information that the network uses to determine how to forward user packets destined to and originated from a particular PDP address. The PDP context maintained by a mobile, an SGSN, and a GGSN link together a Radio Access Bearer and a CN Bearer to form a 3GPP Bearer for the mobile.

PDP contexts maintained on the SGSN and the GGSN contain the following main information [22]:

- *PDP Address*: The PDP address used by the mobile to send and receive PDP packets.
- *Routing Information*: Information used by the network node to determine where to forward a user packet. Such information includes the identifiers of the tunnels established between SGSN and GGSN for this PDP context, and an *Access Point Name (APN)*.

   An APN is a logical name used by an SGSN to determine which GGSN should be used for the mobile and by a GGSN to determine the services requested by the user or the address of an access point in an external packet network to which user packets should be forwarded. The use of APN will be discussed further in Section 2.1.9.
- *Quality of Service (QoS) Profiles*: Three categories of QoS profiles are defined [22]:
   - *QoS Profile Subscribed*: Describes QoS characteristics subscribed by a mobile user.
   - *QoS Profile Requested*: Describes QoS currently requested by a mobile user.
   - *QoS Profile Negotiated*: Describes the QoS actually provided by the network to the mobile at the current time.

The SGSN maintains all three types of QoS profiles. But the GGSN maintains only QoS Profile Negotiated.

A PDP context can be in either ACTIVE or INACTIVE state:

- *ACTIVE state*: A PDP context in ACTIVE state contains up-to-date information for forwarding PDP packets between the mobile and the GGSN. However, the fact that a PDP context is in ACTIVE state does not necessarily mean that the Radio Access Bearers (RABs) required to transport user packets over the RAN are established. Instead, RABs may be established only when the mobile has user packets to send to the network or the network has user packets to the mobile.

- *INACTIVE state*: A PDP context in INACTIVE state may contain a valid PDP address, but it will not contain valid routing and mapping information needed to determine how to process PDP packets. No user data can be transferred between the mobile and the network. A changing location of a mobile user will not cause an update for the PDP context.

  If a GGSN has user packets to send to a mobile but the PDP context for the destination PDP address is in INACTIVE state, the GGSN may use Network-requested PDP Context Activation procedure (Section 2.1.9.2) to change the PDP context of the destination mobile into ACTIVE state. The GGSN may also discard packets destined to a mobile if the corresponding PDP context is in INACTIVE state.

Figure 2.8 illustrates the actions that can be performed on a PDP context and how these actions cause the PDP context to transition between ACTIVE and INACTIVE states.

- *PDP Context Activation*: PDP Context Activation creates and activates a PDP Context. Successful PDP Context Activation moves the PDP context from INACTIVE state to ACTIVE state. The mobile or the GGSN may initiate the PDP Context Activation procedure. But the GGSN can only initiate PDP Context Activation under some strict limitations (Section 2.1.9.2).

- *PDP Context Modification*: PDP Context Modification changes the characteristics of an active PDP Context. For example, it can be used to modify the PDP Address or the attributes of the QoS profile to be supported by the network. The mobile terminal, the RNC in a UTRAN, the SGSN, or the GGSN may initiate the PDP Context Modification procedure. However, Release 5 of the 3GPP Specifications only allows the GGSN-initiated PDP Context Modification procedure (Section 2.1.9.3) to modify the PDP Address.

- *PDP Context Deactivation*: PDP Context Deactivation removes an existing PDP Context. This procedure moves a PDP context from ACTIVE state to INACTIVE state. The state of a PDP context can also be transitioned from

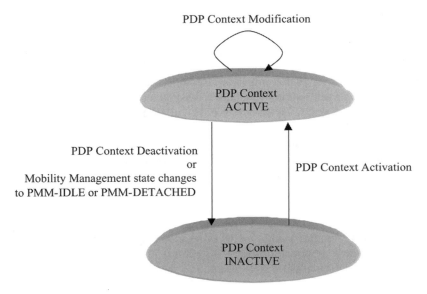

Fig. 2.8 *3GPP PDP context state transitions*

ACTIVE to INACTIVE when mobile's Packet Mobility Management (PMM) state changes into PMM-IDLE or PMM-DETACHED (Chapter 4, "Mobility Management").

2.1.5   Steps for a Mobile to Access 3GPP Packet-Switched Services

A three-phased process is used to activate a mobile user's access to 3GPP packet-switched network and services. These three phases, as illustrated in Figure 2.9, are as follows:

- *GPRS Attach*: A mobile uses the GPRS Attach procedure to register with the PS CN domain or, more precisely, with an SGSN. When a mobile performs GPRS Attach with an SGSN, we say that the mobile is attaching to the SGSN. During GPRS Attach, a mobile provides its identity and service requirements to the SGSN and will be authenticated and authorized by the SGSN.

    In addition to registering the mobile with an SGSN, a successful GPRS Attach also accomplishes the following:
    - Establish a Mobility Management Context on the mobile, in the RAN (e.g., on the RNC in a UTRAN), and on the SGSN (Chapter 4, "Mobility Management"). This allows the RAN and the SGSN to track the mobile's location.
    - Establish a signaling connection (Section 2.1.3) between the mobile and the SGSN. The mobile and the SGSN use this signaling connection to

(a) Phase 1: Mobile registers with PS CN via GPRS Attach

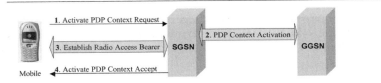

(b) Phase 2: Activate PDP Context and establish Radio Access Bearer.

(c) Phase 3: Registers with the IMS (only if the mobile wishes to use services provided by IMS).

Fig. 2.9 *Three-phased access to 3GPP packet-switched network and services*

exchange signaling and control messages needed to perform the GPRS Attach procedure. After GPRS Attach is completed, the mobile can continue to use this signaling connection to exchange signaling messages with the SGSN, for example, to perform PDP context activation.

– Allow the mobile to access some services provided by the SGSN. Such services may include, for example, sending and receiving SMS messages and being paged by the SGSN. SMS messages are delivered over signaling connections using the signaling protocol MAP.

GPRS Attach procedure alone does not establish any Radio Access Bearer or CN Bearer for the mobile. Therefore, GPRS Attach alone is not sufficient to enable a mobile to send or receive user packets over the PS CN domain.

GPRS Attach procedure is described in more detail in Section 2.1.8.

• *PDP Context Activation and RAB Establishment*: Recall that before a mobile can use a PDP address to send or receive user packets over a 3GPP PS CN, a PDP context for the PDP address has to be established and activated on the mobile, and in the PS CN. A mobile can request the network to establish and activate a PDP context for its PDP address after the mobile has performed GPRS Attach successfully. A successful PDP Context Activation will also trigger the PS CN domain to establish the CN Bearer and the Radio Access Bearer (i.e., the I_u Bearer and the Radio Bearer) needed to transport user packets to and from the mobile. Therefore, after a successful PDP context activation, a mobile will be able to send and receive user packets over the PS CN domain.

- *Register with the IMS*: When a mobile wishes to use the IP-based real-time voice or multimedia services provided by the IMS, the mobile needs to perform registration with the IMS. As SIP is the signaling protocol for the real-time services provided by the IMS, the SIP registration procedure is used for a user to register with the IMS. Procedure for registration with the IMS is described in detail in Chapter 3.

2.1.6   User Packet Routing and Transport

Inside the PS CN domain, IP is the main protocol for transporting user packets between network nods (e.g., between RNC and SGSN, between SGSN and GGSN, between GGSNs, and between SGSNs). IP is also used for routing between GGSNs. However, user packets are not directly routed using the IP routing protocol between SGSN and GGSN. Instead, routing of user packets between SGSN and GGSN is based on GPRS-specific protocols and procedures. In particular,

- *GGSN acts as a central point for routing of all user packets*. As discussed in Section 2.1.1.3, all user packets to and from any mobile will be forwarded first to the mobile's serving GGSN which will then forward the packets toward their final destinations. This is illustrated in Figure 2.10. The figure shows that the source and the destination mobiles are connected to different GGSNs. Even when the source and the destination mobiles are connected to the same GGSN, user packets will still have to go to the GGSN first, which will then forward the packets to the destination.

- *User packets are tunneled between RNC and SGSN, between SGSN and GGSN, and between two GGSNs*. Tunneling a packet is to put the packet inside another packet (called a *encapsulating packet*) and then route the encapsulating packet based on the information in the header of the encapsulating packet. Tunneling in 3GPP packet domain is done using the GPRS Tunneling Protocol (GTP) (Section 2.1.11.1).

  The GTP tunnel used to transport a mobile's packets between an SGSN and the user's serving GGSN forms a CN Bearer for the mobile.

  Tunneling achieves two main objectives here:

  - Tunneling between SGSN and GGSN enables GPRS protocols, rather than the IP routing protocol and IP mobility management protocols, to be used inside the PS domain for routing and mobility management.

  - Tunneling makes the transport of user packets inside a PS CN domain independent of the protocols used in external packet networks. This will avoid the need for the 3GPP network to understand external packet data

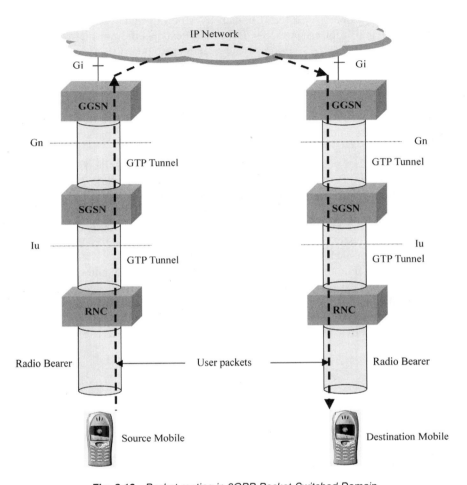

Fig. 2.10 *Packet routing in 3GPP Packet-Switched Domain*

protocols and therefore allow the 3GPP protocols to evolve independently of the protocols used by the mobiles and the external packet networks.

- *Host-specific routes are used to forward user packets between a mobile and a GGSN.* The SGSN and the GGSN maintain an individual routing entry as part of a PDP context (Sections 2.1.4, 2.1.5) for every mobile terminal (host) that has an active PDP context—hence the term host-specific routes. In particular, for each mobile with an active PDP context, the routing entry maintained by a GGSN identifies the SGSN that is currently serving the mobile. When a mobile moves from one SGSN to another, the GPRS mobility management procedures are used to ensure that these routing entries on the SGSN and the

GGSN are properly updated (Chapter 4, "Mobility Management").

Regular IP routing protocols do not use host-specific routing, but instead, they use *prefix-based routing*. A router running standard IP routing protocols will maintain a route to a subnetwork identified by the network prefix assigned to that subnetwork.

Figure 2.11 illustrates how the GTP tunnels, the Radio Access Bearers, and the Radio Bearers are addressed for the purpose of routing user packets between a mobile and its serving GGSN.

On a mobile, user packets originated from a PDP address on the mobile are sent to a Service Access Point provided by the protocol layer immediately below the IP layer for transmission to the PS CN. Packets addressed to the PDP address are also delivered by the lower protocol to the IP layer through the Service Access Point. Such a Service Access Point on a mobile for transporting user IP packets is identified by a Network-layer Service Access Point Identifier (NSAPI). A unique NSAPI is used for each IP address configured on a mobile.

A mobile will send the NSAPI for an IP address to the SGSN and GGSN during the activation process of PDP context. The SGSN and the GGSN will then use the NSAPI to identify the PDP context corresponding to the IP address.

The receiving end side of a GTP tunnel locally assigns a Tunnel Endpoint Identifier (TEID) that the transmitting side of the tunnel has to use to send user

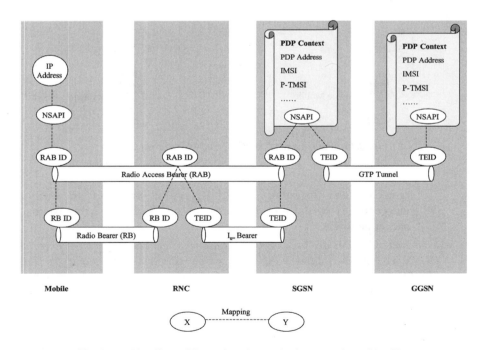

Fig. 2.11 *Identifiers of Bearers and mapping between these identifiers*

packets over the tunnel to the receiving side. A TEID is generated based on the mobile's IMSI and the NSAPI associated with the PDP context for which the GTP tunnel is established. The TEIDs for a GTP tunnel are exchanged between the receiving and the transmitting sides of the tunnel during tunnel setup process.

Each RAB is identified by a RAB Identifier (RAB ID) on the mobile, the RNC, and the SGSN. The Radio Bearer (RB) that makes up the mobile-RNC portion of a RAB is identified by a RB ID on the mobile and the RNC.

There is a one-to-one mapping between NSAPI, PDP context, and RAB. In fact, the RAB ID used by a mobile and an SGSN to communicate with an RNC will be identical to the NSAPI. Furthermore, for PS services, each PDP context can use only one RB. Hence, there is also a one-to-one mapping between NSAPI, PDP context, RAB, and RB.

2.1.7 Configuring PDP Addresses on Mobile Stations

A mobile has the following ways to acquire a PDP address (i.e., IP address for our discussions) for use to send or receive user packets over a 3GPP network:

- *Use a static PDP address assigned by the visited 3GPP network*, i.e., the 3GPP network to which the mobile is attached currently.
- *Use a static PDP address assigned by an external IP network.* Here, an external IP network can be any IP network that is external to the visited PS domain. For example, an external IP network can be any IP network belonging to the operator of the visited PS domain, an Internet Service Provider (ISP) network, or the mobile's home IP network.
- *Acquire a PDP address dynamically from the visited 3GPP network.*
- *Acquire a PDP address dynamically from an external IP network.*

When a mobile uses a static PDP address assigned by the visited 3GPP network, the visited 3GPP network has to ensure that user packets addressed to the mobile's static PDP address can always be routed to the mobile's serving GGSN (which will in turn forward the packets to the mobile). When a mobile uses a static PDP address assigned by an external IP network, the mobile's PDP address may not be part of the PDP addressing space of the visited 3GPP network. The visited 3GPP network and the external IP network will have to collectively ensure that user packets addressed to the mobile's static PDP address can be routed to the mobile's serving GGSN.

If a mobile wants to use a static PDP address, it will inform the visited 3GPP network of its static PDP address during PDP context activation (Section 2.1.9). This static PDP address will then be used by the SGSN and the GGSN in the mobile's PDP context for forwarding packets destined to and originated from the PDP address.

A mobile may acquire a PDP address dynamically from a visited 3GPP network. In this case, the mobile's serving GGSN in the visited 3GPP network will be responsible for allocating a PDP address from its PDP address space to the mobile during the PDP context activation procedure.

If a mobile wishes to acquire a PDP address dynamically from an external network, it will inform the visited network of this decision during PDP context activation. This will trigger the visited PS domain to first activate a PDP context without a PDP address for the mobile. This PDP context will allow the mobile to send requests to external networks to acquire PDP addresses and receive replies from the PDP address servers in the external network. Besides these messages used by the mobile to contact IP address servers in an external IP network, the visited PS CN will not forward other user packets to or from the mobile before a valid PDP address is added to the mobile's PDP context. The mobile's serving GGSN in the visited network will have to learn the PDP address assigned to the mobile by the external network in order to be able to forward packets destined to the mobile's PDP address. More details on how a mobile acquires an PDP address dynamically from an external IP network are discussed in Section 2.1.12.3.

Similar to the case of using a static PDP address assigned by an external network, when a mobile uses a dynamic PDP address assigned by an external network, the mobile's PDP address may not be part of the PDP addressing space of the visited 3GPP network. The visited 3GPP network and the external IP network will have to collectively ensure that user packets addressed to the mobile's PDP address can be routed to the mobile's serving GGSN.

When a mobile uses a dynamically assigned PDP address, the mobile's PDP address may change when the mobile moves. As a result, correspondent hosts may not be able to know which PDP address a destination mobile is using currently and, therefore, may not be able to initiate communication with a destination mobile. Instead, it may have to wait to be contacted first by the mobile before it can send user packets to the mobile.

2.1.8   GPRS Attach Procedure

A mobile uses the GPRS Attach procedure to attach to the PS domain but uses the IMSI Attach procedure to attach to the CS domain. The GPRS Attach procedure and the IMSI Attach procedure may be combined to allow a mobile to attach to the PS and the CS domain simultaneously.

This section describes the GPRS Attach procedure which is illustrated in Figure 2.12. We assume that mobiles use the I_u mode network access, the RAN is a UTRAN, the GPRS Attach is not combined with IMSI Attach, and the mobile is not currently IMSI-attached.

Only a mobile can initiate GPRS Attach. The mobile or the network (i.e., the SGSN or the HLR) may initiate the GPRS Detach procedure to remove the states established by the GPRS Attach procedure on the mobile and network nodes.

The SGSN that the mobile is attempting to attach to is referred to as the new SGSN. The SGSN (if any) used by the mobile before it attaches to the new SGSN is called the old SGSN.

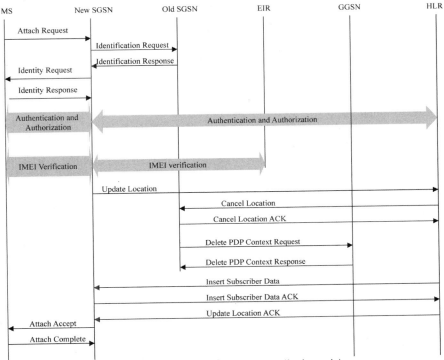

Fig. 2.12   *GPRS Attach procedure (for I$_u$ mode)*

A mobile initiates GPRS Attach by sending an Attach Request message to the SGSN. The Attach Request message provides the following key information (among other information) to the SGSN:

- *Identifiers of the mobile*: The mobile includes its P-TMSI or its IMSI (only when it does not have a P-TMSI), but not both IMSI and P-TMSI, as its identity in the GPRS Attach Request message. This ensures that if someone intercepts the GPRS Attach Request message, it will not be able to learn the mapping between the mobile's P-TMSI and IMSI from this message alone.

- *P-TMSI Signature*: A P-TMSI Signature is a three-octet number assigned to the mobile by the SGSN that assigned the P-TMSI. The P-TMSI Signature is used by the SGSNs to authenticate a P-TMSI. It can also be used by the mobile to authenticate the network node that is assigning the P-TMSI.

- *Attach Type*: The Attach Type indicates whether the Attach Request is for GPRS Attach only, GPRS Attach while already IMSI attached, or combined GPRS/IMSI Attach.

- *Location information*: The SGSN will start to track the mobile's location after a successful GPRS Attach. Therefore, the Attach Request message provides the Routing Area Identity (RAI) of the mobile's current Routing Area. A Routing Area consists of one or multiple RNCs connected to the same SGSN. Routing Areas are used by the SGSN to track mobiles' locations. Location and mobility management in a 3GPP packet network will be discussed in more detail in Chapter 4, "Mobility Management."

Upon receiving an Attach Request message from a mobile, the SGSN performs two main functions:

- Authenticate the mobile user.
- Perform location update when necessary.

The SGSN uses a mobile user's IMSI as a user's permanent identity to authenticate the user. However, due to security considerations, the Attach Request message will often carry only the mobile's P-TMSI as long as a valid P-TMSI exists (Section 2.1.1.1). If only the P-TMSI is provided in the Attach Request received from the mobile, the SGSN may have to use other means to determine the mobile's IMSI. Two cases exist:

- *Case 1*: The mobile's P-TMSI was assigned by this new SGSN. In this case, the SGSN should also know the mobile's IMSI. This is because the SGSN needed to know a mobile's IMSI to assign a P-TMSI to the mobile in the first place.
- *Case 2*: The mobile's P-TMSI was assigned by the old SGSN.

In Case 2 above, the SGSN has two basic ways to get the mobile's IMSI: Ask the mobile directly, or ask the old SGSN. The SGSN will first attempt to retrieve the mobile's IMSI from the old SGSN. As the SGSNs are typically connected via a high-speed wireline network, it could be much faster for an SGSN to communicate with another SGSN over this wireline network than to communicate with the mobile over the bandwidth-limited radio access network. Retrieving information from the old SGSN also reduces over-the-air signaling overhead. To ask the old SGSN for the mobile's IMSI, the SGSN sends an Identification Request message to the old SGSN. The old SGSN responds with an Identification Response message containing the mobile's IMSI if it knows the mobile's IMSI. In case the old SGSN does not know the mobile's IMSI, the new SGSN will then request the mobile to provide its IMSI. The new SGSN does so by sending an Identity Request message to the mobile. The mobile responds with an Identity Response message carrying its own IMSI.

Once the SGSN obtains the mobile's IMSI, the SGSN will initiate the security procedures (Chapter 5, "Security") to authenticate and authorize the user. Although authentication is based primarily on the IMSI, the SGSN may optionally check the mobile's IMEI that identifies the equipment currently used by the user. The SGSN does so by sending an Identity Request message to the mobile. The mobile will respond with an Identity Response message that carries the mobile's IMEI.

After positive authentication and authorization of the mobile, the SGSN updates the mobile's location with the mobile's HLR in any of the following cases:

- This is the first time the mobile attaches to this new SGSN.
- The new SGSN's SGSN Number has changed since the last GPRS Detach. Recall that the SGSN Number is used by non-IP protocols inside the PS CN domain to identify a SGSN (Section 2.1.1.3).

The SGSN initiates location update to the HLR by sending an Update Location message to the mobile's HLR. The HLR, upon receiving the Update Location message from the new SGSN, will instruct the old SGSN to cancel the location state maintained for this mobile. If the old SGSN has active PDP contexts for this mobile, the old SGSN will delete these PDP contexts by sending a Delete PDP Context Request to the GGSN.

The HLR will also send the mobile's GPRS service subscription information (i.e., what GPRS services the user has subscribed) to the new SGSN in an Insert Subscriber Data message. The Insert Subscriber Data message also contains information that allows the SGSN to determine whether the mobile should be authorized for accessing this particular SGSN. The new SGSN responds to the Insert Subscriber Data message by returning an Insert Subscriber Data ACK message. This will trigger the HLR to send an Update Location ACK message back to the SGSN to complete the location update procedure.

To complete a successful GPRS Attach, the SGSN sends an Attach Accept message to the mobile. The Attach Accept message may carry a new P-TMSI if a new P-TMSI is assigned to the mobile by the new SGSN. In this case, the Attach Accept message will also carry a P-TMSI Signature for the new P-TMSI. However, The Attach Accept message will not carry the mobile's IMSI. Therefore, if the Attach Accept message is intercepted by a malicious user, it will not be able to learn the mapping between the mobile's IMSI and P-TMSI.

If the new P-TMSI is different from the old P-TMSI used by the mobile, the mobile will need to acknowledge the change explicitly by returning an Attach Complete message to the new SGSN, which completes the GPRS Attach procedure.

2.1.9   PDP Context Activation and Modification

PDP Context Activation has the following main functions:

- *PDP Address allocation*: The network allocates a PDP address to the mobile if needed.
- *CN Bearer Establishment*: The network creates and activates the PDP context on GGSN and SGSN and establishes all the necessary bearers between SGSN

and GGSN for transporting user and signaling traffic for the activated PDP context.

- *RAB Assignment*: The network establishes the Radio Access Bearers to carry user traffic.

PDP context activation and modification can be initiated by the mobile or the network, which will be discussed next.

2.1.9.1 Mobile-Initiated PDP Context Activation and Modification

Figure 2.13 illustrates the procedure for mobile-initiated PDP context activation [22], assuming the radio access network is UTRAN. A mobile initiates PDP context activation by sending an Activate PDP Context Request to the SGSN. The message carries the following main information elements:

- *PDP Address*: The PDP Address field can either contain a PDP address specified by the mobile or 0.0.0.0 to indicate to the network that the mobile wishes to acquire a PDP address from an external IP network.
- *Network-layer Service Access Point Identifier (NSAPI)*: An unused NSAPI used by the mobile to send user packets originated from the PDP address and to receive user packets addressed to the PDP address.
- *PDP Type*: Indicates the type of the PDP used by the mobile, such as IP, Point-to-Point Protocol (PPP) [51], [29], or X.25.
- *Access Point Name (APN)*: Contains the APN requested by the mobile.
- *QoS Requested*: Contains the QoS profile requested by the mobile.
- *PDP Configuration Options*: Used by a mobile to communicate optional PDP parameters directly with the GGSN (i.e., these parameters will not be interpreted by the SGSN).

The mobile uses the PDP Address field in the Activate PDP Context Request message to inform the network whether it wishes to use a static PDP address or to be assigned a dynamic PDP address. To use a static PDP address, the mobile will

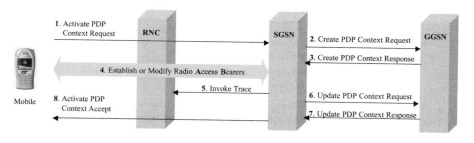

Fig. 2.13 *3GPP mobile-initiated PDP context activation*

provide the static address in the PDP Address field of the Activate PDP Context Request. To have a PDP address dynamically assigned by the visited PS domain or an external packet network, the mobile will set the PDP Address field of the Activate PDP Context Request to zero.

The mobile uses an APN to select a service (or a GGSN) in the PS domain or a contact point in an external packet network. An APN contains two main parts:

- *APN Network Identifier*: This part of an APN is mandatory. It identifies to which external packet network the GGSN is connected to or a PS domain service requested by the mobile.
- *APN Operator Identifier*: This part of an APN is optional. It identifies the PLMN in which the GGSN is located.

An APN has the same Name Syntax as the Internet Domain Names [42]. That is, an APN takes the form of *Label1.Label2.Label3*. Each label can contain only alphabetic characters (A–Z and a–z), digits (0–9), and hyphens. A label can start only with an alphabetic character or a digit. The APN Network Identifier will contain at least one label. A Domain Name System (DNS) [42] can be used to translate an APN to an IP address.

An SGSN uses the APN received from the mobile and the configuration information stored on the SGSN to select a GGSN to be the mobile's serving GGSN. Different GGSNs may have different capabilities and provide different services to the mobiles. For example, some GGSNs may be configured by the network operator to support dynamic PDP address assignment by external packet networks, whereas other GGSNs may be configured to support only local dynamic PDP address assignment.

The SGSN knows, from its configuration information, the PDP Types subscribed by each subscriber and the APN configured by the network operator to support each PDP Type. When the APN received from the mobile (in the Activate PDP Context Request) matches an APN in the mobile's subscribed APNs, the SGSN will use the APN to query the DNS to discover the IP address of the GGSN for the mobile. If no APN is provided in the Activate PDP Context Request, the SGSN may use an APN in the mobile's subscription profile that is configured by the network operator to support the PDP Type specified in the Activate PDP Context Request. If the APN provided in the Activate PDP Context Request matches no APN subscribed by the mobile, the SGSN will reject the mobile's Activate PDP Context Request.

Once the SGSN selects the mobile's serving GGSN, the SGSN sends a Create PDP Context Request to the selected GGSN to request the GGSN to establish a PDP context for the mobile and the GTP tunnel between the SGSN and the GGSN to transport user packets for this PDP context. The Create PDP Context Request message carries the following main information elements:

- *NSAPI*: Copied from the same field in the Activate PDP Context Request message received from the mobile.

- *PDP Type*: Copied from the same field in the Activate PDP Context Request message received from the mobile.
- *PDP Address*: The PDP address from the Activate PDP Context Request message received from the mobile.
- *APN*: Contains the APN Network Identifier of the APN selected by the SGSN for the mobile.
- *QoS negotiated*: QoS profile the SGSN agrees to support for the mobile.
- *Tunnel Endpoint Identifier (TEID)*: A TEID created by the SGSN based on the mobile's IMSI and on the NSAPI in the Activate PDP Context Request message received from the mobile. This TEID identifies the SGSN side of the GTP tunnel to be established for the mobile between the SGSN and the GGSN. This TEID will be used by the GGSN to tunnel PDP packets over the GTP tunnel to the SGSN.
- *Selection Mode*: Indicates whether the APN carried in the Create PDP Context Request message was an APN subscribed by the mobile, or a nonsubscribed APN selected by the SGSN.
- *Charging Characteristics*: Indicate what kind of charging the PDP context is liable for.
- *PDP Configuration Options*: Copied from the Activate PDP Context Request message received from the mobile.

Upon receiving the Create PDP Context Request, the GGSN uses the APN in the Create PDP Context Request to either activate a service provided by the GGSN (e.g., dynamic PDP address assignment by the GGSN) or find a contact point in an external packet network (e.g., a PDP address server). The APN may be mapped to a service on the GGSN in the following ways:

- When an APN corresponds to the domain name of a GGSN, the APN Network Identifier will be interpreted by the GGSN as a request for a service identified by the same APN Network Identifier.
- The first label of an APN Network Identifier can be a Reserved Service Label. Each Reserved Service Label is mapped to a service provided by a GGSN.

The GGSN will then create a new PDP context and insert it into its PDP context table. The GGSN will also establish a CN Bearer (i.e., a GTP tunnel) between the GGSN and the SGSN for the PDP context. Then, the GGSN sends a Create PDP Context Response message back to the SGSN. The Create PDP Context Response message carries the following main information elements:

- *TEID*: The TEID created by the GGSN to identify the GGSN side of the GTP tunnel set up for the mobile's PDP context. This TEID will be used by the SGSN to tunnel user packets over the GTP tunnel to the GGSN.

- *PDP Address*: The PDP Address filed can contain a PDP address assigned by the GGSN to the mobile or 0.0.0.0 if the mobile indicated its request to acquire a PDP address from an external network.
- *QoS Negotiated*: Contains the QoS profile agreed on by the GGSN.
- *PDP Configuration Options*: This information element is relayed by intermediate nodes (i.e., the SGSN or nodes in a RAN) transparently to the mobile (i.e., it will not be interpreted by intermediate network nodes).

If a mobile wishes to obtain a dynamic PDP address from an external packet network, the SGSN and the GGSN will first create for the mobile an active PDP context that does not contain a valid PDP address. This active PDP context will allow a mobile to send PDP packets through the PS domain to contact a PDP address server in the external network to acquire a PDP address. The GGSN will monitor the packets between the mobile and the external PDP address server to learn the PDP address assigned by the external network to the mobile. Once the GGSN learns the PDP address assigned to the mobile, the GGSN will inform the SGSN of the assigned PDP address in the PDP Address field of the Create PDP Context Response message sent back to the SGSN in response to the received Create PDP Context Request. The SGSN will in turn inform the mobile of the assigned PDP address in the PDP Address field of the Activate PDP Context Accept sent back to the mobile in response to the received Activate PDP Context Request. Section 2.1.12 will discuss in greater detail the procedures for supporting dynamic PDP address assignment by external networks.

Upon receiving the Create PDP Context Response, the SGSN will trigger the process to establish the Radio Access Bearers (RABs) for the mobile, using the RAB Assignment procedure in Section 2.1.10.

When the RABs have been established successfully, the SGSN may send an Invoke Trace message to the RNC to request the RAN to start collecting statistics on the network resources used for the PDP context.

In case the QoS attributes of any established RAB are worse than the QoS profile in the PDP context stored on the GGSN, the SGSN will inform the GGSN by sending an Update PDP Context Request to the affected GGSN. Upon receiving a positive Update PDP Context Response from the GGSN, the SGSN will inform the mobile of successful PDP context activation by sending an Activate PDP Context Accept message to the mobile.

2.1.9.2 Network-Requested PDP Context Activation
When a GGSN has user packets to send to an PDP address but does not have an active PDP context for the PDP address, the GGSN may use the Network-Requested PDP Context Activation procedure to establish and activate a PDP context for the PDP address in order to deliver the packets to the mobile. A network operator can determine the policy used by a GGSN to determine whether the GGSN should initiate Network-Requested PDP Context Activation for each specific PDP address based on, for example, the user's service subscriptions.

To support Network-requested PDP Context Activation for a PDP address, the GGSN must have *static* information about the PDP address. For example, the GGSN needs to know the mobile's IMSI in order to query the HLR to determine which SGSN is currently serving the destination mobile. What static information to store for a PDP address, where to store them, and how a GGSN retrieves such static information are regarded as implementation issues and therefore not standardized by 3GPP.

As illustrated in Figure 2.14, the GGSN starts Network-Requested PDP Context Activation by sending a Send-Routing-Information-for-GPRS message to the HLR to retrieve the address of the destination mobile's serving SGSN. This message carries the mobile's IMSI, which will be used by the HLR to determine whether the request can be granted and to search the HLR database for the requested information regarding this mobile. The HLR replies with Send-Routing-Information-for-GPRS-ACK message. This message will carry the address of the mobile's serving SGSN if the HLR determines to grant the Send-Routing-Information-for-GPRS received from the GGSN, or an error cause if the HLR cannot grant the Send-Routing-Information-for-GPRS message.

The GGSN will then send a PDU Notification Request message to the mobile's serving SGSN to ask the SGSN to instruct the mobile to initiate PDP context activation. The PDU Notification Request message will carry the mobile's IMSI, the PDP Type, the PDP Address for which a PDP context should be activated, and the APN that the SGSN and the mobile should use to determine which GGSN to use.

Upon receiving the PDU Notification Request, the SGSN will first inform the GGSN that it will honor the GGSN's request by returning a PDU Notification Response message to the GGSN. Then, the SGSN will send a Request PDP Context Activation message to the mobile to instruct the mobile to start the Mobile-initiated PDP Context Activation procedure described in Figure 2.13 to activate the PDP

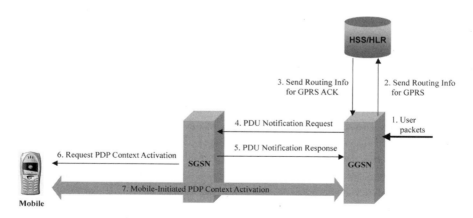

Fig. 2.14 *3GPP network-requested PDP context activation*

context for the PDP address specified in the Request PDP Context Activation message. The Request PDP Context Activation message will also carry the APN the SGSN has received from the GGSN. This APN will then be used by the mobile in the Mobile-initiated PDP Context Activation procedure.

2.1.9.3 PDP Context Modification

Information maintained in an active PDP context may be modified. The main information elements in a PDP context that often need to be modified include the PDP address and the QoS profile(s). Release 5 allows only a GGSN to initiate the process to modify the PDP address in an active PDP context. Modification to the QoS profile(s), however, may be initiated by the mobile, GGSN, SGSN, or the RAN.

Here, we describe the GGSN-initiated PDP Context Modification procedure, which is illustrated in Figure 2.15.

A GGSN initiates the PDP Context Modification procedure by sending an Update PDP Context Request message to the SGSN. This message carries the following information elements:

- *TEID*: Contains the TEID that identifies the SGSN end of the GTP tunnel associated with the PDP context to be modified.
- *NSAPI*: Contains the NSAPI that identifies the PDP context to be modified.
- *PDP Address*: Contains a new PDP address if the GGSN wishes to modify the PDP Address. This field is optional.
- *QoS Requested*: Contains the new QoS profile suggested by the GGSN.

If the GGSN requests a change of the PDP address, the SGSN will make the requested change in the corresponding PDP context it maintains. If the GGSN requests a change of the QoS profile, the SGSN will put together a QoS Negotiated profile to offer to the mobile. The QoS Negotiated profile will be constructed based on the SGSN's capabilities, current load, the QoS profile subscribed by user, and

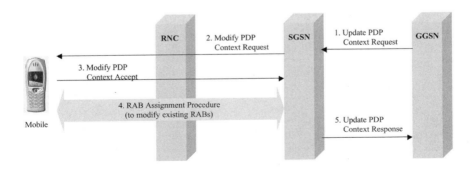

Fig. 2.15 *3GPP GGSN-initiated PDP context modification*

the QoS Requested profile received in the Update PDP Context Request message from the GGSN.

The SGSN then sends a Modify PDP Context Request message to the mobile. The Modify PDP Context Request message carries, among other information elements, the PDP Address and the QoS Negotiated profile. The PDP Address field is optional and, if present, will contain a new PDP address to be used to replace the PDP address in the existing PDP context.

Upon receiving the Modify PDP Context Request, the mobile can either accept or reject the requested changes. The mobile accepts the requested changes by returning a Modify PDP Context Accept message to the SGSN. It rejects the requested changes by deactivating the affected PDP context using the Mobile-Initiated PDP Context Deactivation procedure.

Upon receiving a Modify PDP Context Accept message indicating that the mobile has accepted the requested changes to the PDP context, the SGSN may initiate the RAB Assignment procedure to modify the RABs if they need to be modified to support the newly agreed on QoS profile. After the completion of the RAB Assignment procedure, the SGSN sends on Update PDP Context Response message back to the GGSN to inform the GGSN of the successful completion of the GGSN-initiated PDP Context Activation procedure. The Update PDP Context Response message carries the following information elements:

- *TEID*: Contains the TEID that identifies the GGSN end of the GTP tunnel that is associated with the PDP context being modified.
- *QoS Negotiated*: Contains the new QoS profile that has been agreed on by the SGSN and the mobile.

2.1.10 Radio Access Bearer Assignment

Assignment, modification, and release of Radio Access Bearers are performed using the RAB Assignment procedure. With Release 5, the RAB Assignment procedure cannot be initiated directly by a mobile; it can only be initiated by the network. In particular, RABs for packet-switched services are initiated by the SGSN upon triggered by other network entities in the CN or the RAN. For example, the SGSN may initiate the RAB Assignment procedure upon receiving a Create PDP Context Response message from a GGSN informing the SGSN that the PDP context has been successfully activated on the GGSN for the mobile during a PDP context activation process (Section 2.1.9.1).

As shown in Figure 2.16, the SGSN initiates the RAB Assignment procedure by sending a RAB Assignment Request message over the I_u interface to an RNC to request the RNC to establish, modify, or release one or more RABs. Upon receiving the RAB Assignment request, the RNC will initiate the process to set up, modify, or release the Radio Bearers for the RABs indicated in the RAB Assignment Request message. Radio Bearers are established using procedures specific to each radio system. For example, in a GERAN or a UTRAN, the Radio Resource Control (RRC) protocol will be used to establish, maintain, and release the Radio Bearers.

The RNC uses RAB Assignment Responses to inform the SGSN of the results of the RAB Assignment Request. Multiple RAB Assignment Responses may be sent back to the SGSN for each RAB Assignment Request to report the progress and status of the actions taken by the RNC to satisfy the RAB Assignment Request. For example, the request to establish a RAB may be queued by the RNC temporarily because the RNC is processing other RABs. In this case, the RNC may send a first RAB Assignment Response to inform the SGSN that the request is queued, and then a second RAB Assignment Response when the Radio Bearer is successfully established for the requested RAB.

During the RAB assignment process, the SGSN negotiates with the RAN about the QoS profile for the mobile. In particular, if the RAB Assignment Responses indicate that a requested RAB cannot be established because the radio network cannot support the requested QoS profile, the SGSN may send a new RAB Assignment Request with a different QoS profile to retry the establishment of this RAB. The number of times the SGSN attempts to establish a RAB and how the SGSN modifies the QoS profile for each attempt is implementation dependent and is often a configurable parameter that can be controlled by a network operator.

2.1.11 Packet-Switched Domain Protocol Stacks

This section describes the protocol stacks used for the main protocol interfaces in a 3GPP PS domain. These interfaces include the G_n, G_p, I_u, G_i, G_s, G_c, and G_r interfaces as illustrated in Figure 2.5 in Section 2.1.2. In addition to these individual interfaces, we will also describe the protocols used between a mobile and a GGSN.

2.1.11.1 G_n and G_p Interfaces and the GPRS Tunneling Protocol   The G_n interface is used between SGSN and GGSN as well as between SGSNs in the same PLMN. The G_p interface is used between an SGSN and a GGSN in a different PLMN.

Figure 2.17 (a) and (b) illustrate the user-plane (for transporting user-packets) and the control-plane (for signaling and control) protocol stacks of the G_n and G_p

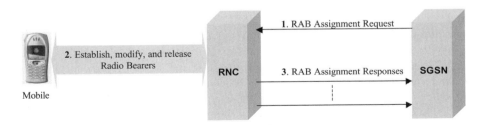

Fig. 2.16   *3GPP Radio Access Bearer Assignment*

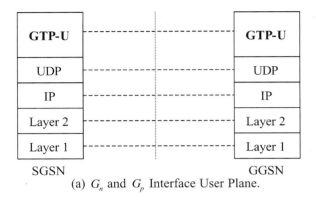

(a) G_n and G_p Interface User Plane.

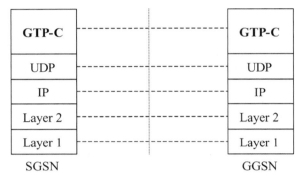

(b) G_n and G_p Interface Control Plane.

Fig. 2.17 *3GPP G_n and G_p interface protocol stacks*

interfaces, respectively. The protocol for both the user plane and the control plane over the G_n and G_p interfaces is the *GPRS Tunneling Protocol* (GTP). GTP is also used to tunnel user packets (i.e., as the user-plane protocol) between a RAN and the CN (i.e., an SGSN in the CN).

The IETF has defined several protocols for tunneling IP packets over non-IP or IP networks. These include the Generic Routing Encapsulation (GRE) protocol [33] and the IP-in-IP Tunneling protocol [52]. Therefore, a natural and important question is: *Can we replace GTP with a standard IP tunneling protocol already defined by the IETF?* The answer requires careful thought.

GTP is not just a tunneling protocol. It is also the signaling protocol used to support an essential part of GPRS: PDP context activation, deactivation, and modification. GTP is used for mobility management inside the PS CN domain (i.e., by the SGSN and the GGSN to maintain host-specific CN routes for mobiles while they are on the move). Therefore, if GTP is replaced with a standard IP tunneling protocol defined by the IETF, mobility management in the PS CN domain will need to be supported by other means. Without GTP, some potential mobility management alternatives in the PS CN domain are as follows:

- *Continue to use PDP contexts*: PDP contexts may continue to be used when a standard IP tunneling protocol is used to tunnel user packets between SGSN and GGSN and between SGSNs. The portion of the GTP protocol used for handling PDP contexts can continue to be used to handle the PDP contexts on the SGSN and the GGSN. However, this approach does not seem to bring significant benefit compared to the current way of using GTP.
- *Do not use PDP contexts*: If PDP contexts are no longer used, a protocol will be needed to support mobility management in the PS CN domain. For example, this protocol should ensure that a GGSN always knows each mobile's serving SGSN. Multiple alternative protocols exist for mobility management in a PS CN domain. For example, Mobile IP [44], Cellular IP [28], or HAWAII [46].

It is interesting to note that the 3GPP2 packet data network architecture (Section 2.2) uses GRE to tunnel user packets through the packet-switched CN domain. Mobile IP is then used for mobility management inside the packet-switched CN.

Next, let's turn our attention to the details of GTP. GTP consists of two sets of messages and procedures. One set of messages and procedures forms the control plane or GTP-C. GTP-C is used for managing (create, modify, and release) GTP-U (GTP User Plane) tunnels, for managing PDP contexts, for location management, and for mobility management. The other set of GTP messages and procedures forms the user plane or GTP-U. GTP-U is used to establish and manage GTP tunnels that will be used to tunnel user packets.

One GTP-U tunnel between SGSN and GGSN will be established for every active PDP context. However, multiple PDP contexts with the same PDP address will share a common GTP-C tunnel.

GTP's main functionality, and therefore its messages, can be grouped into the following categories:

- *Tunnel Management*: The GTP Tunnel Management messages are used to activate, modify, and remove PDP Contexts and their associated GTP tunnels between an SGSN and a GGSN.
- *Location Management*: GTP Location Management messages can be used by a GGSN to retrieve location information from the HLR. However, existing HLRs are mostly designed for circuit-switched mobile networks and, therefore, use the Mobile Application Part (MAP) protocol originally designed for circuit-switched mobile networks to communicate with other network nodes. Therefore, a protocol converter will be needed to convert between the GTP Location Management messages and the MAP messages.
- *Mobility Management*: The GTP Mobility Management messages are used between the SGSNs to transfer mobility-related information from one SGSN

to another SGSN. Transferring mobility-related information from one SGSN to another may be used, for example, when a mobile is performing GPRS Attach, performing Routing Area Update with a new SGSN, or performing handoff from one SGSN to another.

- *Path Management*: The Path Management messages are used by a node to determine if a peer node is alive and to inform the peer node of what GTP header extensions it can support.

Next, we further describe the GTP Tunnel Management messages, many of which will be used throughout the book, for example, in PDP context activation, routing area update, and handoff procedures. The GTP Tunnel Management messages include:

- *Create PDP Context Request/Response*: Create PDP Context Request is sent by an SGSN to a GGSN as part of the PDP Context Activation procedure (Section 2.1.9) to request the establishment of PDP context. Create PDP Context Response is the GGSN's response to a received Create PDP Context Request.
- *Update PDP Context Request/Response*: An Update PDP Context Request can be sent by a GGSN to an SGSN to request the SGSN to modify a PDP context. For example, the GGSN may learn the PDP address after the PDP context has been activated and then use Update PDP Context Request to inform the SGSN of the PDP context. The GGSN can also use Update PDP Context Request to ask an SGSN to modify the QoS profile of a PDP context. Update PDP Context Response is the message sent in reply to an Update PDP Context Request.
- *Delete PDP Context Request/Response*: A Delete PDP Context Request is sent from an SGSN to a GGSN as part of the GPRS Detach procedure or the GGSN-initiated PDP Context Deactivation procedure to request the deletion of a PDP context. Delete PDP Context Response is the message sent in reply to a Delete PDP Context Request.
- *PDU Notification Request/Response*: PDU Notification Request is sent by a GGSN to an SGSN in the Network-Requested PDP Context Activation procedure to request the SGSN to initiate the Network-Requested PDP Context Activation procedure. PDU Notification Response is the message sent in response to a PDU Notification Request.
- *PDU Notification Reject Request/Response*: If the SGSN receives a PDU Notification Request from the GGSN but cannot successfully perform a Network-Requested PDP Context Activation procedure, the SGSN will send a PDU Notification Reject Request back to the GGSN. PDU Notification Reject Response is the GGSN's response to the PDU Notification Reject Request.
- *Error Indication*: When a GSN receives a user packet for which no active PDP context or RAB exists, the GSN will return an Error Indication message to the network entity from which this packet is received.

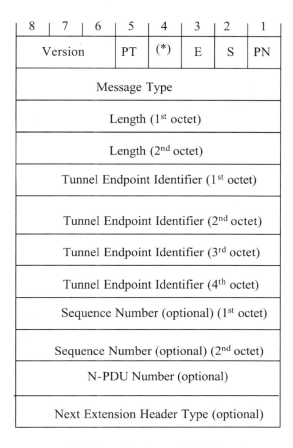

8	7	6	5	4	3	2	1
Version			PT	(*)	E	S	PN
Message Type							
Length (1st octet)							
Length (2nd octet)							
Tunnel Endpoint Identifier (1st octet)							
Tunnel Endpoint Identifier (2nd octet)							
Tunnel Endpoint Identifier (3rd octet)							
Tunnel Endpoint Identifier (4th octet)							
Sequence Number (optional) (1st octet)							
Sequence Number (optional) (2nd octet)							
N-PDU Number (optional)							
Next Extension Header Type (optional)							

Fig. 2.18 *GPRS Tunneling Protocol (GTP) header format*

GTP-C and GTP-U use the same message header format shown in Figure 2.18. This message header format contains the following flags and fields:

- *Version*: Contains the version of the GTP and should be set to 1 for the current version.
- *PT (Protocol Type)*: Indicates whether the GTP protocol is the one defined for 3GPP CN or the one for GPRS/GSM.
- *E (Extension Header Flag)*: Indicates whether the Next Extension Header is present. The Extension Header allows the message format to be extended to carry information not defined in the basic format shown in Figure 2.18.
- *S (Sequence Number Flag)*: Indicates if the Sequence Number field is present.

- *PN (N-PDU Number Flag)*: Indicates whether the N-PDU Number field is present. The N-PDU Number field is used in the inter-SGSN Routing Area Update procedure and some intersystem handoff procedures for coordinating data transmission between a mobile terminal and an SGSN.
- *Message Type*: Indicates the type of the GTP message.
- *Length*: Indicates the length in octets of the payload. The payload is considered to start immediately after the first eight octets of the GTP header. The first eight octets of the GTP header are a mandatory part of the GTP header. The part of the header after the first eight octets are optional header fields and are considered part of the payload.
- *Tunnel Endpoint Identifier (TEID)*: A TEID uniquely identifies a tunnel endpoint on the receiving end of the tunnel. The receiving end of a GTP tunnel assigns a local TEID value. This TEID will then be communicated to the sending end of the GTP tunnel via GTP-C (or RANAP over the I_u-PS interface) to be used by the sending end of the tunnel to send messages through the tunnel.
- *Sequence Number*: Used by the GTP-C to match a response message to a request message and used by the GTP-U to ensure packet transmission order.

2.1.11.2 The I_u-PS Interface
The I_u-PS interface connects a RAN to the PS CN domain. It provides the following main capabilities [19], [20]:

- *Tunnel Management*: The I_u-PS interface provides procedures for establishing, maintaining, and releasing the GTP tunnels between an RNC and an SGSN. Recall that GTP tunnels are used to transport user packets between an RNC and an SGSN.
- *Radio Access Bearer Management*: The I_u-PS interface provides procedures for establishing, maintaining, and releasing RABs. Recall that a RAB is a connection, with its assigned radio resources, between a mobile and the CN that can be used for the mobile to exchange user or signaling data with the CN. It is the CN (more precisely, the SGSN) that controls toward the radio access network the establishment, modification, and release of RABs.
- *Radio Resource Management*: When an RNC receives, over the I_u-PS interface, a request from the CN to establish or modify a RAB, the RNC will analyze the radio resources currently available in the radio access network to determine whether to accept or reject the request. This process is called *Radio Resource Admission Control* in 3GPP.
- *Mobility Management*: The I_u-PS interface provides procedures for supporting handoffs between RNCs. For example, the I_u-PS interface provides the procedures for *Serving RNS Relocation*. Serving RNS Relocation is to move the RNS side of an RANAP connection from one RNS to another. This procedure is needed for supporting a handoff between RNSs. The I_u-PS interface also provides paging functions. Furthermore, the I_u-PS interface

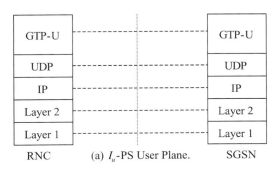

(a) I_u-PS User Plane.

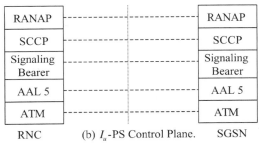

(b) I_u-PS Control Plane.

Fig. 2.19   *3GPP I_u-PS interface protocol stacks*

provides functions for reporting mobiles' geographical locations to the CN to support positioning services.

The Radio Access Network Application Part (RANAP) protocol [20] is used as the signaling protocol to support the above capabilities over the I_u-PS interface. In addition to the capabilities described above, the RANAP protocol can also encapsulate higher layer signaling messages so that these messages can be carried over the I_u-PS interface transparently.

Figure 2.19 (a) and (b) illustrate the user-plane and the control-plane protocol stacks, respectively, for the I_u-PS interface. The RANAP runs over the SS7 Signaling Connection Control Part (SCCP) protocol. SCCP is a transport-layer protocol that provides capabilities similar to the capabilities provided by UDP and TCP over an IP network. But, SCCP runs over ATM (Asynchronous Transfer Mode) using the ATM Adaptation Layer 5 (AAL5). A separate SCCP connection is used for each individual mobile to transport the RANAP messages related to this mobile.

On the user plane, GTP-U tunnels are used over the I_u-PS interface to transport user packets. The GTP-U tunnels are implemented over UDP/IP. The UDP/IP protocol stack for transporting GTP-U packets can be implemented over any lower

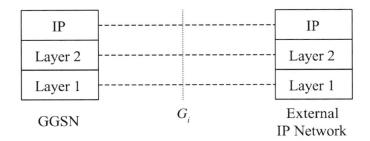

Fig. 2.20   *3GPP G$_i$ interface protocol stack*

layer technologies. It is important to note that even though GTP-C already provides the signaling capabilities for establishing and managing GTP-U tunnels, Release 5 uses the RANAP protocol instead of GTP-C to control the GTP-U tunnels over the I_u-PS interface.

2.1.11.3   G$_i$, G$_r$, G$_c$, and G$_s$ Interfaces   The G_i interface is used by the GGSN to connect to any external IP network either in the same or a different administrative domain. Standard IP routing protocol is used over the G_i interface as illustrated in Figure 2.20. The G_i interface makes the GGSN look like a standard IP router to the external IP network. As standard IP routing protocol is used over the G_i interface, the data and control planes are identical.

The G_r interface between SGSN and HLR and the G_c interface between GGSN and HLR use an identical control-plane protocol stack, which is shown in Figure 2.21. The MAP protocol is used for signaling over the G_r and G_c interfaces. MAP is implemented over the SS7 Transaction Capabilities Application Part (TCAP) protocol. TCAP messages are transported over the SS7 SCCP, which is in turn implemented over ATM connections.

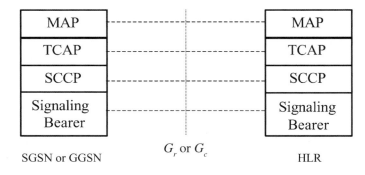

Fig. 2.21   *3GPP control-plane protocol stack between SGSN (or GGSN) and HLR*

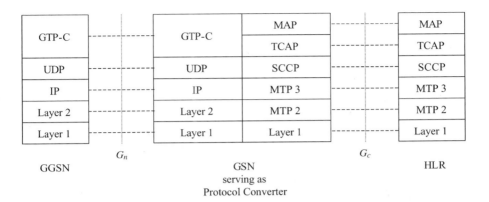

Fig. 2.22 *3GPP control-plane protocol stack between GGSN and HLR based on GTP*

A GGSN may also use GTP-C to interact with the HLR. In this case, protocol conversion between GTP-C and MAP will be needed for most existing HLRs that do not understand GTP-C. Figure 2.22 illustrates the protocol stack for supporting this scenario.

The G_s interface between SGSN and MSC/VLR also uses SS7 signaling protocols. As illustrated in Figure 2.23, the SS7 Base Station System Application Part+ (BSSAP+) protocol is used for signaling between SGSN and MSC. The BSSAP+ messages are transported over the SS7 SCCP protocol.

2.1.11.4 *Mobile-to-GGSN Protocol Stacks*

Figure 2.24 shows the user-plane protocol stacks between the mobile and the GGSN.

Over the air interface (i.e., the U_u interface), the Packet Data Convergence Protocol (PDCP) [17] is used to transport higher layer packets. The current version

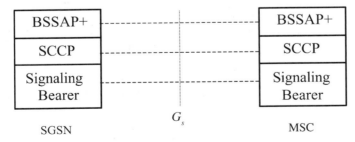

Fig. 2.23 *3GPP control-plane protocol stack between SGSN and MSC/VLR*

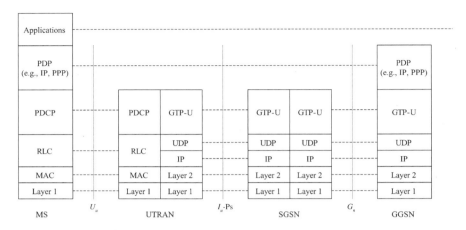

Fig. 2.24 *3GPP user-plane protocol stack between mobile and GGSN*

of PDCP supports PPP [51], IPv4, and IPv6 as higher-layer protocols. The PDCP performs three main functions:

- Header compression for higher layer data streams.
- Mapping higher layer data into the underlying radio interface protocols.
- Maintaining data transmission orders for upper layer protocols that have such a requirement.

Among the three functions described above, header compression is the primary function of PDCP because IP protocols could introduce large header overheads, which could cause a significant waste of the scarce radio bandwidth. For example, speech data for IP telephony will most likely be carried by the Real-time Transport Protocol (RTP) [26], which runs over UDP/IP. Therefore, a packet will have an IP header (20 octets for IPv4 and 40 octets for IPv6), a UDP header (8 octets), and an RTP header (12 octets). This results in a total of 40 and 60 octets in header overheads for IPv4 and IPv6 networks, respectively. The payload of a voice packet is typically 15 to 20 octets in length, depending on the speech coding and lower-layer frame sizes used. PDCP relies on mechanisms and protocols defined by the IETF for header compression. For example, IP Header Compression (IPHC) [30] can be used to compress IP, UDP, and TCP headers. The more recent Robust Header Compression (ROHC) [27] can be used to compress RTP/UDP/IP, UDP/IP, and ESP/IP headers. ESP (Encapsulating Security Payload) [36] defines a header that is used to carry a mixture of security-related information. The ESP header is inserted after the IP header and before the upper layer protocol header or before an encapsulated IP header.

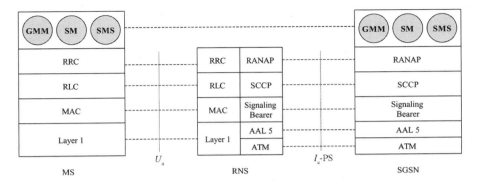

Fig. 2.25 *3GPP control-plane protocol stack between mobile and SGSN*

The Radio Link Control (RLC) protocol provides logical link control over the radio interfaces. A mobile can have multiple RLC connections. The Medium Access Control (MAC) protocol controls the access for the radio channels.

Figure 2.25 shows the control-plane protocol stacks between the mobile and the SGSN. The Radio Resource Control (RRC) protocol is the signaling protocol for controlling the allocation of radio resource over the air interface [16], [18]. The RRC protocol supports a range of critical control functions:

- Broadcast information related to the RAN and the CN to the mobiles.
- Establish, maintain, and release RRC connections between mobiles and the RAN. Recall that an RRC connection is the set of dedicated signaling and traffic channels between a mobile and the RAN (e.g., the RNC in UTRAN).
- Establish, maintain, and release Radio Bearers for transporting user packets.
- Paging.
- Radio power control.
- Control of radio measurement and reporting to be performed by the mobiles.
- Control of the on and off of ciphering between the mobile and the RAN.

Over the RRC layer, GPRS-specific applications are used to support mobility management, session management, and the Short Message Services (SMS):

- *GPRS Mobility Management (GMM)*: GMM supports mobility management functions, including GPRS Attach and Detach operations, security, and routing area update procedure.
- *Session Management (SM)*: SM supports PDP context activation, modification, and deactivation.
- *SMS (Short Message Service) [24]*: SMS supports short messages to and from mobile terminals.

These applications communicate directly with their peer applications on the SGSN.

2.1.12 Accessing IP Networks through PS Domain

A mobile may use a 3GPP PS domain as an access network to access an external IP network. Here, an external IP network refers to any IP network that is not part of the 3GPP PS domain. For example, it may be an IP intranet that belongs to the same network provider that operates the 3GPP PS domain, the Internet, an Internet Service Provider's network, or a mobile's enterprise IP network.

In order for a mobile to access an external IP network through a 3GPP PS domain, some or all of the following interactions between the mobile and the IP network need to take place:

- User registration (e.g., authentication and authorization) with the external IP network.
- Dynamic assignment of IP addresses to the mobile by the external IP network.
- Encryption of user data transported between the mobile and the external IP network.

As discussed earlier, the GGSN in the PS domain is responsible for interconnecting the PS domain with other IP networks via the G_i interface, as illustrated in Figure 2.26. Depending on whether the GGSN in the PS domain participates in the procedures described above, 3GPP defines two basic ways to access an external IP network through a PS domain [14]:

- *Transparent Access*: The GGSN does not participate in any interaction between the mobile and the external IP network except transporting user packets.
- *Non-transparent Access*: The GGSN participates in at least one of the interactions between the mobile and the external IP network described above.

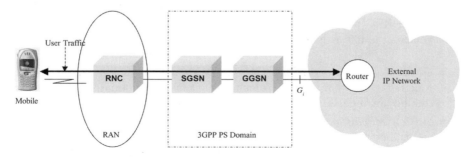

Fig. 2.26 *Access another IP network through 3GPP PS domain*

Next, we describe the following:

- Transparent access.
- Non-transparent access using Mobile IP.
- Acquiring IP address dynamically from external IP networks.
- Dialup through PS domain to external IP network (non-transparent inter-working).

2.1.12.1   *Transparent Access*   With transparent access, a mobile will first gain access to a GGSN in the local (visited) PS CN so that the mobile can send and receive user IP packets over the PS CN to and from the external IP network. This can be achieved using the GPRS Attach and then the PDP Context Activation procedures described in previous sections.

The mobile also needs to acquire an IP address from the local PS domain to use as its PDP address to send and receive IP packets over the local PS CN domain. This local IP address can be statically configured, for example, at the time of service subscription with the local 3GPP network. Alternatively, the mobile's local IP address can be dynamically assigned by the local PS CN domain during the PDP Context Activation process.

Once the mobile can send and receive IP packets through the local PS CN domain, it can proceed to register with the external IP network in order to gain access to the external IP network. Registration with an external IP network may use any IP-based protocols. The local PS CN domain will not participate in any interaction between the mobile and the external IP networks, except relaying user IP packets between the mobile and the external IP network.

It is important to note that, with transparent access, the local PS domain needs to determine, without contacting the external IP network, whether to grant network access to the mobile and to assign an IP address to the mobile during the PDP

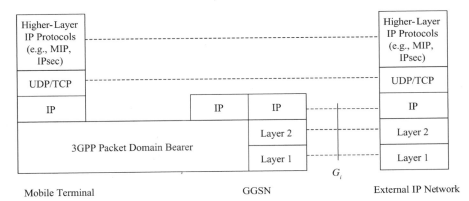

Fig. 2.27   *Protocol stacks for transparent to IP networks through 3GPP PS CN*

Context Activation procedure. This requires the local PS domain to maintain sufficient user subscription information needed to authenticate and authorize the mobile.

Figure 2.27 illustrates the protocol stacks for transparent access. IP is used as the PDP over the PS domain between a mobile and the GGSN. IP is also the network-layer routing protocol between the GGSN and the external IP network. Any IP-based protocol may be used for the mobile to register with the external IP network, and to secure user IP traffic. For example, IPsec [37] may be used for authentication, and for securing the user traffic between the mobile and the external IP network.

2.1.12.2 Non-Transparent Access Using Mobile IP

With transparent access discussed earlier, each mobile will need a unique IP address in the local PS domain. Therefore, a large number of IP addresses will be required as the number of 3GPP network users grow. However, if IPv4 is used as the PDP, the IPv4 address space may not be large enough to support the expected number of mobile users in the future.

3GPP specifies a non-transparent access method using MIPv4 [14]. With this approach, a mobile uses its MIP home address as its PDP address to send and receive user IP packets over the local PS domain. Each GGSN also serves as a MIPv4 Foreign Agent (FA). Every mobile served by the GGSN uses the IP address of the GGSN as its FA Care of Address (CoA) inside the local PS domain. Therefore, the local PS domain does not have to allocate any IP address to the visiting mobiles, except offering its own IP address(es) for the mobiles to use as their MIPv4 CoA. A mobile uses regular MIPv4 messages and procedures to register this FA CoA with its MIPv4 home agent inside the mobile's home network, which may be an external IP network.

IP packets addressed to the mobile's MIPv4 home address will be routed to the mobile's Home Agent (HA) as in regular MIPv4 operation. The mobile's HA will then tunnel these packets to the mobile's current CoA. In this case, these packets will be tunneled to the MIPv4 FA on the GGSN. The GGSN extracts the payload IP

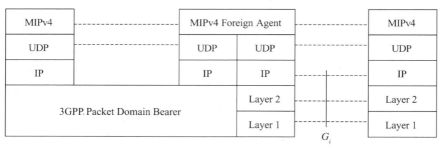

Fig. 2.28 *Protocol stacks for non-transparent access to IP networks through the PS CN domain*

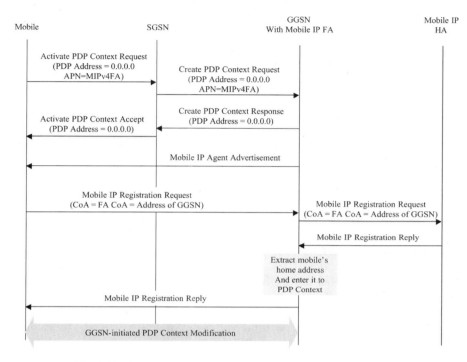

Fig. 2.29 *Non-transparent access to IP networks using Mobile IPv4*

packets out of the MIPv4 tunnel and forwards the payload packets to the mobile along the path identified by the mobile's PDP context.

Figure 2.28 illustrates the protocol stacks for non-transparent access using MIPv4. Figure 2.29 [14] illustrates the signaling message flows for non-transparent access using MIPv4. The SGSN and GGSN in the local PS domain do not rely on the mobile to provide its home address during the PDP Context Activation. This is because a mobile may not always have a valid home address during the PDP Activation process, which is the case, for example, when a mobile uses a dynamic home address assigned by its home network. The GGSN will, therefore, learn a mobile's home address by intercepting and inspecting the MIP signaling messages sent back to the mobile from the mobile's MIP home agent.

To intercept MIPv4 signaling messages sent by a mobile's HA back to the mobile, the GGSN needs to contain a MIPv4 Foreign Agent. As every mobile uses only FA CoA, the MIP signaling messages and all user packets addressed to the mobile's home address will be routed first to the FA on the GGSN.

The mobile will first need to gain access to the GGSN using the mobile-initiated PDP Context Activation procedure. When initiating the PDP Context Activation process, the mobile sets the PDP Address filed in its Activate PDP Context Request to zero, even if the mobile already has a permanent MIP home address. The SGSN will use the APN carried in the Activate PDP Context Request to select a GGSN that is configured by the network operator to have a MIPv4 FA and to support dynamic IP address assignment by the external IP network using MIPv4.

The SGSN then sends a Create PDP Context Request to the selected GGSN to request the GGSN to set up a PDP context for the mobile. The PDP Address field of the Create PDP Context Request will also be zero, indicating to the GGSN that the mobile's home address should be reset by the HA after the PDP context is activated. The GGSN will set up a PDP context for the mobile without assigning any PDP address to the mobile. The GGSN then returns a Create PDP Context Response message to the SGSN to acknowledge the Create PDP Context Request from the SGSN. The SGSN will in turn acknowledge the mobile's Activate PDP Context Request by returning a Activate PDP Context Accept message to the mobile.

After the PDP context is activated for the mobile, the FA on the GGSN will start to send MIPv4 Agent Advertisement messages to the mobile. These messages will be delivered in *limited IP broadcast* packets (i.e., IP packets that use a limited broadcast destination address 255.255.255.255.). However, the messages will not be actually broadcast over the air, but instead, they will be sent over the point-to-point tunnel from the GGSN to the mobile.

Upon receiving the MIPv4 Agent Advertisement messages, the mobile uses the FA's IP address as its CoA and performs MIPv4 registration with its HA by sending a MIPv4 Registration Request to the HA. If the mobile uses a permanent home address, the MIP Registration message will carry either this permanent home address or the mobile's NAI as the mobile's identifier. If the mobile wishes to obtain a dynamic home address from its MIPv4 HA, the MIPv4 Registration Request will carry a null home address and carries the mobile's NAI as the mobile's identifier for the MIPv4 registration.

Upon receiving the MIPv4 Registration Request, the GGSN will record the mobile's home address or NAI (or both) and map them to the TEID of the GTP-U tunnel used for delivering user packets to the mobile. This is needed by the GGSN to map the messages coming back from the HA to the corresponding mobile. The FA on the GGSN will then forward the MIPv4 Registration Request to the mobile's HA.

Upon receiving the MIPv4 Registration Request, the HA will send a MIPv4 Registration Reply back to the mobile's CoA, i.e., the address of the GGSN. The FA on the GGSN will intercept this MIPv4 Registration Reply and extract the mobile's home address from the Registration Reply before forwarding the message to the mobile. Then, the GGSN will insert the mobile's home address to the PDP context and trigger the GGSN-Initiated PDP Context Modification procedure (Section 2.1.9.3) to update the PDP address on the SGSN.

2.1.12.3 Acquiring IP Address Dynamically Using DHCP from an External Network

To use the Dynamic Host Configuration Protocol (DHCP) [31] to acquire an IP address from an external IP network, the mobile needs to communicate with a DHCP server in the external IP network. This means that the mobile needs to send user data over the local PS CN before its PDP context in the PS CN has a valid PDP address. Therefore, a PDP context for the mobile needs to be activated without an IP address initially.

After the PDP context is activated, the mobile can communicate with the external DHCP server to acquire an IP address. This requires the following issues to be resolved:

- Before an IP address is assigned to the mobile by the external IP network, the PS CN domain should be able to relay DHCP messages between the mobile and external DHCP server.
- When an IP address is assigned to the mobile by the external IP network, the mobile's PDP contexts on the SGSN and the GGSN need to be updated to include the mobile's IP address.

To allow DHCP messages to be transported over the PS CN domain between a mobile and a DHCP server in the external IP network before the mobile's PDP context contains a valid PDP address, the GGSN acts as a DHCP Relay Agent [31]. A DHCP Relay Agent can relay DHCP messages between the DHCP client on the mobile and a DHCP server that is on a different IP subnet from the subnet of the DHCP client. The GGSN can learn the IP addresses of the external DHCP servers based on the APN received from the SGSN during PDP context activation. The DHCP Relay Agent determines to which mobile DHCP messages coming back from a DHCP server should be forwarded based on the mobile's hardware address (i.e., layer-2 address) carried in these DHCP messages.

Recall that with the current 3GPP specifications, only a GGSN can initiate the process to update the PDP addresses in active PDP contexts on the GGSN and SGSN (Section 2.1.9.3). Therefore, to support dynamic IP address assignment by external IP network via DHCP, the GGSN will have to learn the IP address assigned by the external network to the mobile and then initiate the procedure to update the PDP address in the mobile's PDP context on the GGSN and the SGSN. To learn the IP address assigned to the mobile by a DHCP server in an external IP network, the DHCP Relay Agent on the GGSN interprets and examines the DHCP messages from the DHCP server to the mobile.

Figure 2.30 [14] illustrates the protocol stacks for supporting dynamic IP address allocation by an external network using DHCP.

Figure 2.31 illustrates a sample signaling flow for supporting IP address assignment by an external network using DHCP. The mobile will first need to gain access to the GGSN using the mobile-initiated PDP Context Activation procedure. When initiating the PDP Context Activation process, the mobile sets the PDP Address of its Activate PDP Context Request message to zero. The SGSN selects, based on the APN in the received Activate PDP Context Request message and configuration information on the GGSN, a GGSN that is configured by the network operator to support external address assignment using DHCP. Then the SGSN sends a Create PDP Context Request message to the selected GGSN.

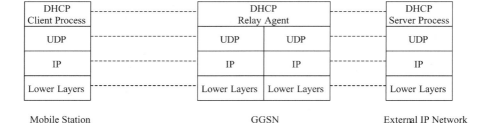

Fig. 2.30 *3GPP protocol stacks for supporting IP address assignment by external network using DHCP*

Upon receiving the Create PDP Context Request message from the SGSN, the GGSN creates and activates a PDP context for the mobile with a null PDP address. The GGSN then sends a Create PDP Context Response message without any PDP address back to the SGSN, indicating the completion of the PDP context activation.

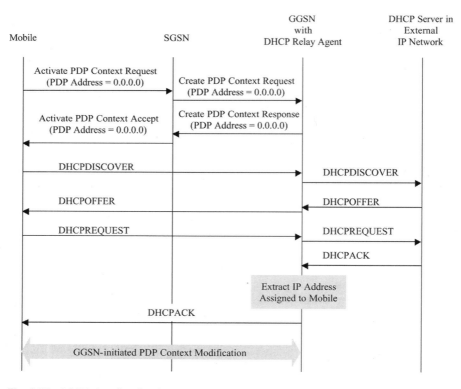

Fig. 2.31 *3GPP signaling flow for supporting IP address assignment by external network using DHCP*

The SGSN in turn notifies the mobile of the completion of the PDP Activation procedure by sending an Activate PDP Context Accept to the mobile.

At this point, the mobile's active PDP context enables the mobile to send user data over the PS CN domain, but only to the DHCP server identified by the APN in the mobile's PDP context on the GGSN. This allows the mobile to communicate with the external DHCP server(s) to acquire an IP address. The mobile does so by sending a DHCPDISCOVER message to the external DHCP servers. The mobile will set the destination IP address of the DHCPDISCOVER message to the limited broadcast address (all 1 s) as in the regular operation of DHCP. The DHCP Relay Agent on the GGSN will pass the DHCPDISCOVER message to the external DHCP server(s) identified by the APN in the mobile's PDP context.

Upon receiving the DHCPDISCOVER message, the DHCP server will reply with a DHCPOFFER message that will carry the IP address assigned to the mobile. This message will be forwarded by the DHCP Relay Agent on the GGSN to the mobile. The mobile may receive multiple DHCPOFFER messages from multiple DHCP servers. It selects one offered IP address and sends a DHCPREQUEST to the DHCP servers to confirm its acceptance of the selected IP address. This message will again be forwarded by the DHCP Relay Agent on the GGSN to the corresponding DHCP servers. The DHCP server receiving the DHCPREQUEST message indicating that the mobile has accepted the IP address it offered will reply with a DHCPACK message that will contain all the configuration parameters assigned to the mobile.

The DHCP Relay Agent extracts the IP address assigned to the mobile from the DHCPACK message before forwarding the message to the mobile. The DHCP Relay Agent then passes the IP address to the GGSN. The GGSN inserts the IP address into the mobile's PDP context and initiates the GGSN-initiated PDP Context Modification procedure (Section 2.1.9.3) to update the IP address in the PDP context on the SGSN.

2.1.12.4   Dial-up Access Using PPP   The PS domain can support dialup connections to an external IP network. Dialup refers to the process of establishing a link-layer connection to an IP network. Once the dialup connection is established, IP can run over the connection as if the mobile is connected to the external network via a local area IP network.

An end-to-end link-layer dialup connection may be implemented using different protocols in different intermediate networks traversed by the connection. PPP is a most widely used protocol for setting up a link-layer connection along a point-to-point link over a non-IP network [51]. The Layer-2 Tunneling Protocol (L2TP) [54] defined by the IETF can be used to extend a PPP connection over an IP network to a network access server inside the IP network.

As discussed in Section 2.1.6, the PS domain does not use standard IP protocols for routing of user IP packets between a mobile and a GGSN, but instead, it uses point-to-point host-specific routes established and maintained by the GPRS protocols. Therefore, a PPP connection is a natural choice for implementing the portion of a dialup connection over the PS domain (i.e., between the mobile and its

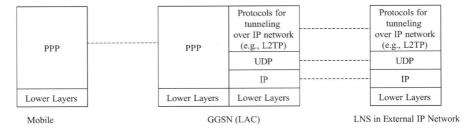

Fig. 2.32 *Protocol stacks for dialup through 3GPP packet domain to an IP network*

serving GGSN). Then, L2TP may be used to extend the PPP connection from the GGSN to the mobile's contact point inside the external IP network.

The L2TP endpoint on the IP network side is referred to as the L2TP Network Server (LNS). The L2TP endpoint on the user side is referred to as the L2TP Access Concentrator (LAC). Therefore, to use L2TP to extend a PPP connection from the GGSN to the mobile's contact point in an external IP network, the GGSN will need to contain a LAC and the mobile's contact point in the external IP network will need to be an LNS.

Figures 2.32 and 2.33 show a sample protocol stack and a sample signaling flow, respectively, for supporting dialup through the PS domain to an external IP network [14]. For illustration purpose, Figure 2.33 assumes that L2TP is used to implement the layer-2 connection between the GGSN and the external IP network.

The mobile will first need to gain access to the GGSN using the mobile-initiated PDP Context Activation procedure. When a mobile uses a layer-2 tunnel over the PS domain to connect to an external IP network, the mobile will have to rely on the

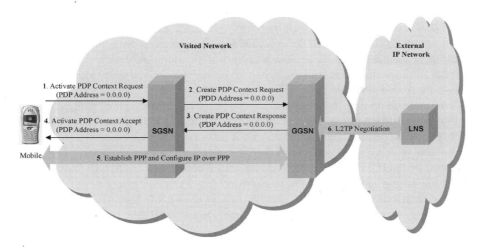

Fig. 2.33 *Signaling flows for dialup through 3GPP packet domain to an IP network*

external IP network for IP address assignment. The mobile may use a static address or a dynamic address. If a static address is used, the mobile will provide the address in the PDP Address field of its Activate PDP Context Request message sent to the SGSN. Otherwise, the PDP Address field will be set to zero. The SGSN will then send a Create PDP Context Request message to the GGSN.

The GGSN, upon receiving the Create PDP Context Request message, will derive from the APN in the Activate PDP Context Request the address of the LNS in the external IP network. The GGSN will activate the PDP context and return a Create PDP Context Response message to the SGSN. The SGSN in turn sends an Activate PDP Context Accept message to the mobile to complete the PDP context activation process.

After the PDP Context Activation procedure, the mobile can initiate the process to establish a PPP connection between the mobile and the GGSN and configure IP as the network layer protocol over the PPP connection. After the PPP connection is configured, the GGSN will set up the L2TP tunnel to the LNS in the external IP network to complete the dialup connection to the external IP network.

2.2 3GPP2 PACKET DATA NETWORKS

This section discusses the third-generation wireless network architecture defined by 3GPP2. In particular, we will discuss the following:

- *3GPP2 network architecture* (Section 2.2.1)
- *3GPP2 packet data network architecture* (Section 2.2.2)
- *Protocol reference model for 3GPP2 packet data networks* (Section 2.2.3)
- *Access to 3GPP2 packet data network* (Section 2.2.4)
- *User packet routing and transport* (Section 2.2.5)
- *Protocol stacks for packet data services* (Section 2.2.6)

2.2.1 3GPP2 Network Architecture

The 3GPP2 conceptual network architecture is illustrated in Figure 2.34 [4]. It shares significant similarity with the 3GPP network architecture at a high level view. In particular, the 3GPP2 network consists of Radio Networks (RNs) interconnected by a core network. The core network is composed of a circuit-switched domain and a packet-switched domain. Circuit-switched and packet-switched traffic from mobile users are split at the Base Station (BS) inside the RN and then routed to the corresponding domains in the core network.

The Radio Network is based on circuit-switched technologies and is used for both circuit-switched and packet-switched services. Its main components are Base Stations (BSs). A BS provides the means for a mobile to access the network using radio. It consists of a Base Station Controller (BSC) and a Base Transceiver System (BTS). The BTS provides radio transmission capabilities. The BSC provides

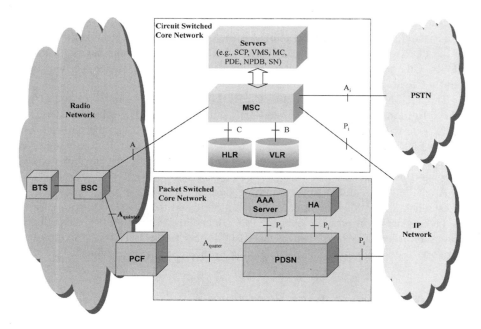

Fig. 2.34   *3GPP2 conceptual network architecture*

control and management functions for one or multiple BTSs and connects the BTSs to the circuit-switched core network.

The 3GPP2 RN uses the cdma2000 radio technologies. The cdma2000 base stations are organized into *Systems* and *Networks*. A network is a subset of a system. Each system is identified by a System ID (SID). Each network within the system is identified uniquely within the system by a Network ID (NID). The pair (SID, NID) uniquely identifies a network.

2.2.1.1   Circuit-Switched Core Network   The circuit-switched core network has a similar architecture as the core network of a second-generation wireless network (e.g., GSM). It supports circuit-switched services and a range of critical functions needed by the packet-switched portion of the network. For example, mobility within the 3GPP2 packet-switched core network relies heavily on the existing circuit-switched network entities (e.g., the MSC) and protocols. Mobility management will be discussed in detail in Chapter 4, "Mobility Management."

The circuit-switched core network has the following main components:

- *Switching and call control components*: The Mobile Switching Center (MSC) provides switching functions for circuit-switched calls and supports basic call control functions and mobility management functions. The MSC also interacts with service control platforms in the circuit-switched core network for providing advanced call control functions.

- *Information Servers*: The main information servers include the following:
  - Home Location Registrar (HLR) stores service subscription and location information of the home subscribers.
  - Visitor Location Registrar (VLR) stores a visiting mobile's location and subscription information temporarily (i.e., while the visiting mobile is within the service area of the VLR).
  - Equipment Identity Registrar (EIR) stores user equipment identities.
- *Service control servers*: These servers provide support for supplementary services. The main network entities include
  - *Service Control Point (SCP)* is a real-time transaction processing system and database that provides service control.
  - *Voice Message System (VMS)* stores received voice messages, data messages (e.g., e-mail), and provides methods for users to retrieve these stored messages.
  - *Message Center (MC)* stores and forwards short messages to support SMS.
  - *Position Determining Entity (PDE)* helps determine a mobile's geographical position.
  - *Number Portability Database (NPDB)* stores information for portable directory numbers to support telephone number portability. Telephone number portability is the ability to move a user's telephone number from one location to another or from one network provider to another.
  - *Service Node (SN)* provides service control, service data, and specialized resources to support circuit-switched bearer-related services.
  - *Intelligent Peripherals* plays announcements, performs speech-to-text or text-to-speech conversion, recording and storing voice messages, and facsimile services.

2.2.2 3GPP2 Packet Data Network Architecture

2.2.2.1 Functional Architecture Figure 2.35 [2] illustrates the 3GPP2 packet data network reference model. The Packet Data Serving Node (PDSN) is the main network node for supporting packet data services. The PDSN provides the following main functions:

- Route IP packets between the 3GPP2 network and any external IP networks.
- Route IP packets between mobile terminals inside the same operator's 3GPP2 network.
- Act as an IP address server to assign IP address to mobiles.

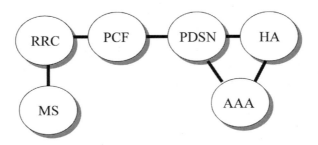

Fig. 2.35   *3GPP2 packet data network functional architecture*

- Act a PPP server for mobiles (i.e., establish, maintain, and terminate PPP [51] session to a mobile terminal).
- Provide mobility management functions. The PDSN may support MIPv4 Foreign Agent functionality to allow mobiles to roam into the 3GPP2 network.
- Perform AAA (Authentication, Authorization, and Accounting) functions for mobile terminals. To authenticate or authorize a mobile user, the PDSN may need to communicate with an AAA server.

The Packet Control Function (PCF) supports interworking between the radio network and the PDSN. The PCF performs the following specific functions:

- Establish, maintain, and terminate layer-2 connections to the PDSN.
- Maintain reachability information for mobile terminals.
- Relay IP packets between an RN and the PDSN.
- Tracks status of radio resources.
- Communicate with the RRC function on the BSC to manage radio resources.

The RRC function is part of the Radio Network and typically resides on the BSC. It has the following main functions:

- Establish, maintain, and terminate radio connections to mobiles and management radio resources allocated to these connections.
- Broadcast system information to mobiles.
- Maintain status of mobile terminals (e.g., active, dormant).

The Mobile IP HA in the visited network is not required. But, it can be used by a network provider to allow a mobile subscriber to remain addressable by the same IP home address regardless of the mobile's current location.

A mobile station (MS) comprises User Identity Module (UIM) and Mobile Equipment (ME) as shown in Figure 2.36 [10]. The UIM, which may be removable

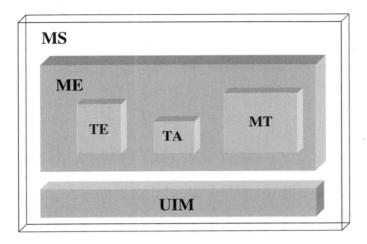

Fig. 2.36 *Functional architecture of a mobile station (MS)*

or integrated into ME, contains subscription information. Similiar to 3GPP, ME comprises Terminal Equipment (TE), Mobile Terminal (MT), and Terminal Adapter (TA).

2.2.2.2 *Reference Network Architecture* A mobile has two methods to access a 3GPP2 packet data network:

- *Simple IP Access*: A mobile is assigned an IP address dynamically by the visited access network provider (more precisely, by a PDSN in the visited network). The mobile can use this IP address to send and receive IP packets through the packet-switched core network while moving within the geographical area served by the PDSN that assigned the address. However, when the mobile moves to a new PDSN, it will have to obtain a new IP address. This is because different PDSNs use disjoint IP address spaces.
- *Mobile IP Access*: Mobile IP (v4 or v6) may be used to support mobility between PDSNs. The mobile may use a static or dynamically assigned Mobile IP home address. Mobile IP enables a mobile to continue to receive packets addressed to its home address regardless of its current attachment point to the network.

Figure 2.37 [2], [1] illustrates the 3GPP2 reference packet data network architecture for supporting Mobile IP access. The reference network architecture for simple IP access will be the same except that no Mobile IP Home Agent will be needed.

A Home Access Provider Network is the home cellular network for the mobile. A Home IP Network is the home network that provides IP services to the mobile. A

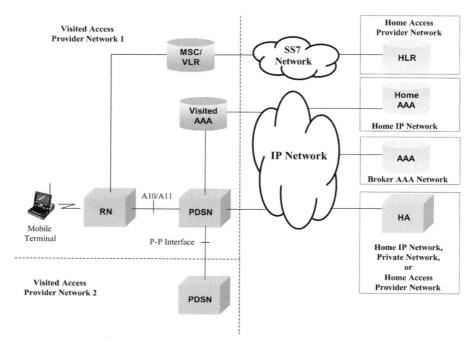

Fig. 2.37 *3GPP2 packet data network reference physical architecture*

Broker AAA Network is a network that contains *broker AAA servers* that have security relationships with visited AAA servers and home AAA servers. A Broker AAA Network is used to securely transfer AAA signaling messages between visited and home AAA servers. A Private Network is a Home IP Network that resides behind an IP firewall and that may use private IP addresses. The IP Network shown in the figure can be the public Internet or a private IP network used to interconnect the Visited Access Provider Network and other networks.

The 3GPP2 packet data network supports both IPv4 and IPv6. For IPv6, the PDSN will act as an IPv6 access router. After the PPP connection is established between a mobile and a PDSN, IPv6 can be configured as a network-layer protocol using IPv6 over PPP [29]. Then, the PDSN will send Router Advertisement messages to the mobile. The Router Advertisement messages contain IPv6 network prefixes that can be used by a mobile to construct a local IPv6 address. The mobile can use IPv6 stateless autoconfiguration [50] to construct and configure a local IPv6 address based on the IPv6 network prefixes received from the Router Advertisement messages.

For IPv4 or IPv6, all mobiles currently served by a PDSN are on the same IP subnet. Mobiles served by different PDSNs will be on different IP subnets. Therefore, when Mobile IPv4 is used to support mobility between PDSNs, a Mobile IPv4 Foreign Agent needs to be colocated on each PDSN.

For Mobile IPv4 or Mobile IPv6, a mobile's home agent may be in the mobile's Home Access Provider Network, Home IP Network, or a Private Network. Any of these networks may be inside the same administrative domain as the mobile's current Visited Access Provider Network.

Many critical capabilities needed in the 3GPP2 packet data network rely on the support of the circuit-switched network. The capabilities include handoff, paging, and connection setup for packet data services. However, as illustrated in Figure 2.37, the 3GPP2 packet data core network does not directly interface with the circuit-switched core network. Instead, circuit-switched procedures are initiated by the BSC inside the radio network upon receiving data or requests from the PCF.

This differs significantly from the 3GPP Packet-Switched (PS) core network architecture. As discussed in Section 2.1, the SGSN in the 3GPP packet-switched core network interfaces directly with the MSC/VLR in the circuit-switched core network in order to perform location management and network access control. The GGSN in the PS domain may also need to interface directly with the HLR in the CS domain to support network-requested procedures such as network-requested PDP context activation.

2.2.3   Protocol Reference Model

Figure 2.38 shows the protocol reference model for a 3GPP2 packet data network [9]. The main protocol reference points are as follows:

- *A Reference Point*: The A reference point consists of the interfaces between a BSC and an MSC. It includes three individual interfaces: A1, A2, and A5.

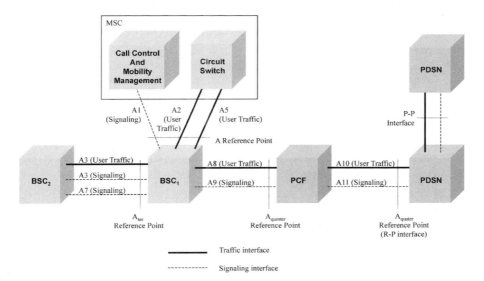

Fig. 2.38   *3GPP2 protocol reference model*

- Interface A1 carries signaling traffic between the Call Control and Mobility Management functions of the MSC and the Call Control function of the BSC.
- Interface A2 and A5 carry different types of user traffic between the switch component of the MSC and Selection and Distribution Unit (SDU) on the BSC. The SDU is used to support soft handoff, a process whereby a mobile receives copies of the same data from multiple BTSs. In the forward direction (from network to mobile), the SDU is responsible for distributing copies of the user data to the BSs involved in a soft handoff process. It is also responsible for collecting copies of the same user data sent by the mobile toward the core network via multiple BTSs and selecting one copy for forwarding to the core network.

- A_{ter} *Reference Point*: The A_{ter} reference point consists of interfaces between the BSCs. It includes two separate interfaces: A3 and A7. The A3 and the A7 interfaces are used primarily for supporting inter-BSC soft handoffs.

  - Interface A3 carries signaling and user traffic between the SDU on a source BSC and a target BTS for supporting soft handoff. The A3 interface consists of two parts: signaling and user traffic. Signaling information and user traffic are carried over separate logical channels. A3 signaling controls the allocation and use of A3 user traffic channels.
  - Interface A7 carries other signaling information not carried by the A3 interface between a source and a target BS.

- $A_{quinter}$ *Reference Point*: This reference point consists of the signaling and traffic interfaces between a BSC and a PCF. It is composed of the following two individual interfaces: the A8 interface for transporting user data traffic and the A9 interface for signaling between a BSC and a PCF to support packet data services. The A8 and A9 interfaces are also used to support mobility between BSCs under the same PCF.

- $A_{quarter}$ *Reference Point (R-P Interface)*: This reference point consists of the interfaces between a PCF and a PDSN. It includes two individual interfaces: the A10 interface used to provide a path for user traffic and the A11 interface used for signaling between the PCF and the PDSN to support packet data services. The A10 and A11 interfaces are also used to support mobility between PCFs under the same PDSN. The A10 and A11 interfaces are often referred to as the R-P interface, indicating that A10 and A11 are interfaces between a radio network (i.e., a PCF) and the packet core network (more precisely, a PDSN).

- *P-P Interface (optional)*: The PDSN-to-PDSN or P-P Interface is used to support a fast handoff between PDSNs.

The A reference point uses existing protocols developed for circuit-switched networks and, hence, will not be discussed further. More details on the A interface can be found in [5]. The $A_{quinter}$, $A_{quarter}$, and the P-P reference points are unique for supporting packet data services and will be discussed in more detail in Section 2.2.6.

2.2.4 Access to 3GPP2 Packet Data Network

Figure 2.39 illustrates the procedure for a mobile to use MIPv4 to access a 3GPP2 network [3]. This procedure contains the following main steps:

- *Step 1: Gain access to PDSN.* This step consists of the following three substeps.
  - *Step 1-A: Gain access to the Radio Network.* The mobile initiates the service activation process by sending an Origination Message to the BSC, indicating that it is requesting for packet data services and that it has data packets ready to send. The BSC acknowledges the Origination Message by sending back a Base Station Acknowledgement Order to the mobile. The BSC then sends a Connection Management (CM) Service Request message to the MSC to ask for permission to establish traffic radio channels for the mobile. The MSC, upon verifying that the user is authorized to access the network, will send an Assignment Request to

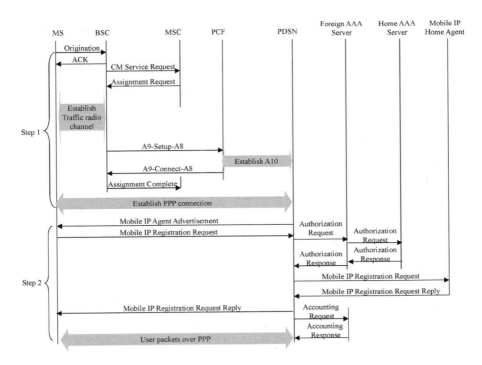

Fig. 2.39 *3GPP2 Packet Service Activation (using Mobile IP)*

the BSC to instruct the BSC to set up the required radio channels for the mobile. Upon receiving the Assignment Request, the BSC will initiate the process to establish traffic radio channels for the mobile.

- *Step 1-B: Setting up resources between the BSC and the PDSN.* After the radio channels are set up in Step 1, the BSC will initiate the process to set up the user traffic connection between the BSC and the PCF (i.e., an A8 connection) by sending an A9-Setup-A8 message to the PCF. Upon receiving the A9-Setup-A8 request, the PCF will first initiate the procedure to set up a user traffic connection between the PCF and the PDSN (i.e., an A10 connection). The PCF will establish an A8 connection for the mobile only after an A10 connection is successfully set up.

  Sometimes, for mobility management purposes, the BSC may not want to set up any A8 connection, but instead it may want to trigger the PCF to establish an A10 connection for the mobile (see Chapter 4, "Mobility Management"). For example, when a mobile moves to a new PCF but does not have data to send at the moment, a new A10 connection still needs to be set up between the new PCF and the PDSN so that the PDSN always knows which PCF is the current serving PCF for the mobile. The same A9-Setup-A8 message is also used for this purpose. The BSC indicates to the PCF that it is asking the PCF to set up an A10 connection, not an A8 connection, by setting the Data-Ready-to-Send field in the A9-Setup-A8 message to zero, indicating that the mobile has no data to send at this moment.

- *Step 1-C: Establish PPP connection between mobile and PDSN.* PPP is used as the data link layer protocol between each mobile and a PDSN for transporting user IP packets. After the A8 and A10 connections are set up, the mobile can establish a PPP connection to the PDSN along with the following resources allocated to it by the network: the traffic radio channel to the BSC, the A8 connection between the BSC and the PCF, and the A10 connection between the PCF and the PDSN. With the PPP connection, the user can begin to send user packets to other IP hosts via the PDSN.

- *Step 2: MIPv4 registration.* After the PPP connection is set up, the mobile will have to use MIPv4 to acquire a temporary care-of address from the visited network and register this new care-of address with its Home Agent (in the mobile's Home IP Network) so that the mobile's MIPv4 Home Agent will start to tunnel packets destined to the mobile's home address to the mobile's current location.

  To support MIPv4, the PDSN will act as a MIPv4 Foreign Agent (FA). Immediately following the establishment of PPP between the mobile and the PDSN, the PDSN will begin transmission of a predetermined number of MIPv4 Agent Advertisement messages. These messages will let the mobile know the IP address of the MIPv4 FA and allow the mobile to configure a local care-of address. The number of Agent Advertisement messages to be sent is determined by the network operator. To minimize the number of Agent Advertisement messages that will have to be sent over the air, the PDSN will stop sending any more Agent Advertisement messages once the pre-determined number is reached. Furthermore, the PDSN will not periodically send Agent Advertisement messages to mobiles to their Agent Advertisement lifetime timers.

When a mobile needs Agent Advertisement messages from the network, it can also send Agent Solicitation messages to the PDSN to trigger the Mobile FA on the PDSN to send an Agent Advertisement message.

2.2.5   User Packet Routing and Transport

Similar to 3GPP, 3GPP2 also uses host-specific routes to forward packets between a mobile and the packet-switched CN (more precisely, a PDSN). In particular, each mobile maintains a PPP connection to a PDSN that acts as the mobile's *serving PDSN*. All user packets from the mobile will be carried over this PPP connection to the mobile's PDSN first, regardless of the destinations of the packets. The mobile's serving PDSN is responsible for forwarding the user IP packets toward their final destinations. All packets destined to a mobile will also have to be routed to the mobile's serving PDSN first, which will in turn transport the packets over the PPP connection to the mobile.

Figure 2.40 illustrates the user traffic flow between two mobiles. Figure 2.40 assumes that the source mobile's serving PDSN is PDSN 1 and the destination mobile's serving PDSN is PDSN 2. PDSN 1 and PDSN 2 may be the same PDSN.

Consider user IP packets sent by the source mobile to the destination mobile. The source mobile sends these IP packets over the PPP connection to its serving PDSN, i.e., PDSN 1. The resulting PPP frames are tunneled over an A8 connection between BSC 1 and PCF 1 and then over an A10 connection between PCF 1 and PDSN 1 to PDSN 1. The A8 and A10 connections are implemented as IP tunnels using the Generic Routing Encapsulation (GRE) protocol [33] defined by the IETF.

The PDSNs are interconnected via a standard IP network. Therefore, upon receiving the user IP packets destined to the destination mobile, PDSN 1 routes these IP packets via regular IP routing, with the assistance of an IP-layer mobility management protocol (e.g., Mobile IP), to the destination mobile's serving PDSN (i.e., PDSN 2). Mobility in 3GPP2 networks will be discussed in detail in Chapter 4, "Mobility Management".

PDSN 2 will then forward these IP packets over its PPP connection to the destination mobile. The resulting PPP frames will be tunneled over an A10 connection between PDSN 2 and PCF 2, then over an A8 connection between PCF 2 and BSC 2, and then over the radio interface to the destination mobile.

A single PPP connection is used between a mobile and its serving PDSN. If MIPv4 is used by the mobile to access IP services, multiple IP addresses can be used for the same mobile over a single PPP connection. On the other hand, if

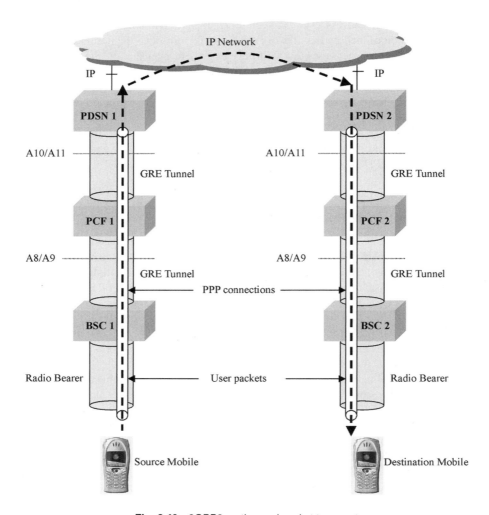

Fig. 2.40 *3GPP2 routing and packet transport*

Simple IP access is used, only one IP address can be used for each mobile over one PPP connection.

2.2.6 Protocol Stacks for Packet Data Services

As discussed in Section 2.2.3, the main protocol reference points in a 3GPP2 packet data network are the *Aquinter* reference point between BSC and PCF, *Aquater* reference point (i.e., the R-P interface) between PCF and PDSN, and the P-P interface between two PDSNs. This section will discuss the protocol stacks used over these protocol reference points.

Recall that the *Aquinter* reference point consists of a traffic interface A8 and a signaling interface A9 and that the *Aquater* reference point consists of a traffic interface A10 and a signaling interface A11. Section 2.2.6.1 describes the protocols used over the signaling interfaces A9 and A11. Section 2.2.6.2 describes the protocols used over the traffic interfaces A8 and A10. Section 2.2.6.3 describes the protocols over the P-P interface. Section 2.2.6.4 describes the protocol stacks for signaling and user data transport between a mobile and its serving PDSN.

2.2.6.1 Protocol Stacks over A9 and A11 Interfaces Figure 2.41
illustrates the protocol stacks over the A9 and the A11 interfaces [6], [7], [8].

The A9 interface is used by a BSC and a PCF to set up and tear down A8 connections between the BSC and the PCF. The A9 interface is also used by a BSC to trigger a PCF to set up A10 connections between PCF and PDSN. The A9 signaling protocol has the following main messages [7]:

- *A9-Setup-A8* and *A9-Connect-A8*: The A9-Setup-A8 message is sent by a BSC to a PCF to set up an A8 connection or trigger the PCF to establish an A10 connection. The A9-Connect-A8 message is sent by the PCF to the BSC in response to an A9-Setup-A8 message from the BSC.
- *A9-Release-A8* and *A9-Release-A8 Complete*: The A9-Release-A8 message is sent by a BSC to a PCF to request the release of an A8 connection. It is also sent by a PCF to a BSC to reject the BSC's A9-Setup-A8 request. The A9-Release-A8 Complete message is sent by the PCF to the BSC in response to an A9-Release-A8 message.
- *A9-Disconnect-A8*: This message is sent by a PCF to a BSC to request the release of an A8 connection.
- *A9-Update-A8* and *A9-Update-A8 Ack*: The A9-Update-A8 message is sent by a BSC to a PCF to send accounting and status information to the PCF. The A9-Update-A8 Ack is the PCF's reply to the A9-Update-A8 message.

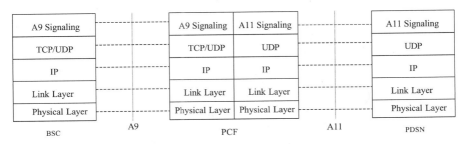

Fig. 2.41 *3GPP2 protocol stacks for the A9 and A11 interfaces*

- *A9-Air Link (AL) Connected* and *A9-Air Link (AL) Connected Ack*: The A9-Air Link (AL) Connected message is sent by a BSC to a PCF to notify the PCF that an air link is successfully established between the BSC and a mobile and that the PCF can start to send user traffic over the A8 connection established for the user. The A9-Air Link (AL) Connected Ack message is sent by a PCF to a BSC to indicate the results of the PCF's processing of the A9-Air Link (AL) Connected message.

- *A9-Air Link (AL) Disconnected* and *A9-Air Link (AL) Disconnected Ack*: The A9-Air Link (AL) Disconnected message is sent by a BSC to a PCF to notify the PCF that the air link for a mobile has been disconnected temporarily. The PCF returns an A9-Air Link (AL) Disconnected Ack message to the BSC in response to an A9-Air Link (AL) Disconnected message.

The A11 interface is used by the PCF and the PDSN to set up and tear down A10 connections. An A10 connection is an IP tunnel, implemented using a standard IETF protocol GRE, between the PDSN and the PCF for carrying user IP packets. A11 signaling protocol is modeled after the Mobile IPv4 protocol. In particular, the PDSN acts as if it was a MIPv4 Home Agent, and the PCF acts as if it was a MIPv4 Foreign Agent. The A11 signaling protocol uses the following messages [8]:

- *A11 Registration Request*: The A11 Registration Request message is sent by the PCF to the PDSN to request the establishment of an A10 connection.
- *A11 Registration Reply*: The A11 Registration Reply message is the PDSN's response to an A11 Registration Request received from a PCF.
- *A11 Registration Update*: The A11 Registration Update message is sent by a PDSN to a PCF to tear down an A10 connection.
- *A11 Registration Acknowledge*: The Registration Update Acknowledge message is the PCF's response to an A11 Registration Update message received from a PDSN.

When a mobile connects to a PCF, the PCF sends an A11 Registration Request message to the selected PDSN to request the PDSN to establish an A10 connection to the PCF. Upon receiving an A11 Registration Request, the PDSN will establish a GRE tunnel to the PCF and replies to the A11 Registration Request with an A11 Registration Reply message. This works in the same manner as a mobile using MIPv4 Registration Request to inform its Home Agent of its current Care-of Address and to trigger the Home Agent to set up a Mobile IP tunnel to the mobile's current Care-of Address.

Therefore, the A11 Registration Request and A11 Registration Reply messages use the same formats as the MIPv4 Registration Request and MIPv4 Registration Reply messages, respectively. Formats of the MIPv4 Registration Request and Registrations Reply are described in detail in Chapter 4, "Mobility Management," where MIPv4 is discussed. Several key fields in an A11 Registration Request message need to be set in a particular way. These fields include the Home Agent field, Home Address field, and the Care-of Address field. The PCF will set the Care-of Address field to the PCF's own IP address. The Home Address field and the Home Agent field need to be set based on whether the mobile is in the process of a fast inter-PDSN handoff. Handoffs and how A11 signaling works during handoffs will be discussed in detail in Chapter 4, "Mobility Management."

The UDP protocol will be used as the transport protocol for A11 signaling messages because MIPv4 signaling messages are transported over UDP.

The PDSN uses "soft states" to maintain A11 connections. A soft state will time out after a predetermined time period. Therefore, the PCF needs to refresh each A10 connection by periodically sending A11 Registration Request messages to the PDSN.

Any physical layer can be used over the A9 and A11 interfaces. The link layers shown in Figure 2.41 can use any protocol that is appropriate for the physical layer in use and that can transport IP packets.

2.2.6.2   Protocol Stacks over A8 and A10 Interfaces
Figure 2.42 illustrates the protocol stacks for carrying user traffic over the A8 and the A10 interfaces [6], [7], [8].

As discussed earlier, user IP packets are encapsulated and tunneled over an A10 connection between PDSN and PCF and over an A8 connection between PCF and BSC. Encapsulation is done using the GRE protocol defined by the IETF. To help understand how user packets are transported over the A8 and A10 interfaces and how traffic connections are implemented over these interfaces, we need to first understand how GRE works.

GRE encapsulates a user packet by adding a GRE header to the user packet. Figure 2.43(a) shows the GRE header as defined in [33], and Figure 2.43(b) shows how the header is used for encapsulation over the A8 and A10 interfaces. The GRE header has the following flags and fields:

- *C*: indicates whether the Checksum field is present.
- *R*: indicates whether the Routing field is present.
- *K*: indicates whether the Key field is present.
- *S*: indicates whether the Sequence Number field is present.

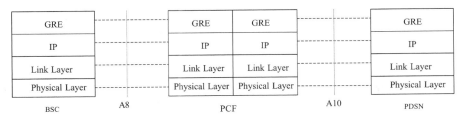

Fig. 2.42   *3GPP2 protocol stacks for the A8 and A10 interfaces*

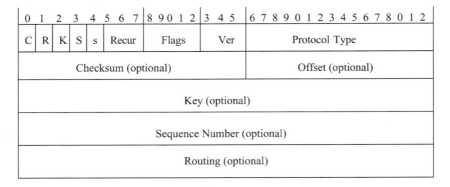

(a) GRE header format.

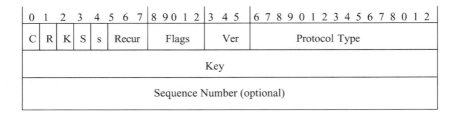

(b) Format of GRE header used for tunneling between PCF and PDSN or between BSC and PCF.

Fig. 2.43 *Generic Routing Encapsulation (GRE) protocol header*

- *s*: indicates whether Strict Source Route is present (Strict Source Route is a special type of source route).
- *Recur*: designed for future use and is currently required to be set to zero.
- *Flags*: used for flags to be introduced in the future.
- *Ver*: contains the version number of the GRE protocol that is required to be set to zero for the current version of GRE.
- *Protocol Type*: contains the protocol type of the payload packet.
- *Checksum*: contains the check sum of the GRE header and payload packet.
- *Offset*: contains the octet offset from the start of the Routing field to the first octet of the active Source Route Entry.
- *Key*: contains a four-octet number that was inserted by the encapsulator.

- *Sequence Number*: contains an unsigned 32-bit integer that is inserted by the encapsulator. It may be used to establish the order in which the packets have been transmitted from the encapsulator.
- *Routing*: contains a list of Source Route Entries; each identifies a source route to a destination.

GRE encapsulation over the A8 and A10 interfaces uses the Key field and the Sequence Number field in the GRE protocol header. The Sequence Number field is used to help ensure packet delivery order when the protocol being encapsulated requires packets to be delivered in sequence. The Key field is used to identify the IP packets to and from each mobile terminal:

- For packets going from the PDSN to the PCF over the A10 interface, the Key should be set to the PCF Session Identifier (PCF SID) used by the PDSN to uniquely identify a session (call) from the PDSN to a mobile terminal.
- For packets going from the PCF to the PDSN, the Key field should contain the PDSN Session Identifier (PDSN SID) that is used by the PCF to uniquely identify a session from a mobile to the PDSN. The PDSN SID and the PCF SID for the same session should be identical.

2.2.6.3 *Protocol Stacks over P-P Interface* The P-P interface is an optional interface used to support fast inter-PDSN handoff. Consider an inter-PDSN handoff from PDSN S to PDSN T. Suppose that PDSN S was the mobile's serving PDSN before the handoff. PDSN T will be called a target PDSN. Without fast inter-PDSN handoff, the mobile's serving PDSN will need to be changed to PDSN T during the handoff. This requires the mobile to establish a new PPP connection to PDSN T during the handoff process. Establishing a new PPP connection and configuring the network protocol (i.e., IPv4 or IPv6) over the PPP connection could take a long time, therefore, introducing long handoff delays. Furthermore, changing serving PDSN often means that the mobile also needs a new IP address to receive IP packets from the new serving PDSN. When Mobile IP is used for mobility management, this means that the mobile will have to register its new IP address with its Mobile IP home agent, further increasing handoff delay.

With fast inter-PDSN handoff, the mobile's serving PDSN remains unchanged during and even after the handoff. The mobile continues to use its existing PPP connection, that terminates on its serving PDSN S, to send and receive user IP packets over the core network while the mobile is being handed off to PDSN T and even after the mobile is handed off to the PDSN T. As long as a mobile uses the same PPP connection, it does not have to change its IP address. As a result, the mobile does not have to perform Mobile IP registration with its home agent.

IP packets destined to the mobile will continue to be routed first to the mobile's serving PDSN. The serving PDSN forwards these IP packets over its PPP connection to the mobile. The resulting PPP frames will be tunneled by the mobile's serving PDSN S to PDSN T first. PDSN T de-tunnels the PPP frames received from the mobile's serving PDSN and then tunnels them over an A10 connection to the target PCF, that is connected to PDSN *T* and is currently serving the mobile. This target PCF in turn tunnels the packets over an A8 connection to the target BSC that is connected to the PCF and is currently serving the mobile. The BSC de-tunnels the PPP frame and forwards them to the mobile over the Radio Bearer. Tunneling PPP frames between a serving PDSN and a mobile via a target PDSN allows the mobile to maintain the PPP connection to its serving PDSN even when the mobile moves to a target PDSN that uses a disjoint IP address space.

The P-P interface provides the protocols and procedures for signaling and for tunneling user traffic between PDSNs to support fast inter-PDSN handoff. The P-P interface consists of two individual interfaces:

- *P-P Bearer Interface*: The bearer interface implements P-P traffic connections (bearers) to tunnel user packets between the PDSNs. A P-P traffic connection will be implemented as a GRE tunnel.

- *P-P Signaling Interface*: The signaling interface provides the signaling messages and procedures for managing the P-P traffic connections.

The P-P signaling protocol is similar to the A11 signaling protocol (between PCF and PDSN) and is also modeled after Mobile IPv4 signaling. The serving PDSN acts as if it was a Mobile IPv4 home agent, and the target PDSN acts as if it was a proxy for a mobile. The P-P signaling protocol uses the following four messages that have the same message formats as the respective A11 signaling messages:

- *P-P Registration Request*: The P-P Registration Request is sent by a target PDSN to a serving PDSN to request the serving PDSN to establish a P-P traffic connection to the target PDSN.

- *P-P Registration Reply*: The P-P Registration Reply message is the serving PDSN's response to a P-P Registration Request.

- *P-P Registration Update*: The P-P Registration Update message is sent by a serving PDSN to a target PDSN to tear down a P-P connection.

- *P-P Registration Acknowledge*: The P-P Registration Acknowledge message is a target PDSN's response to a P-P Registration Update message.

As signaling over the A11 interface, the P-P Registration Request and P-P Registration Reply message also use the same formats of the Mobile IPv4 Registration Request and Mobile IP Registration Reply messages, respectively.

Figure 2.44 illustrates the control-plane and the user-plane protocol stacks for the P-P interface. IPsec [37] can be used to establish a secure connection between a target PDSN and a serving PDSN for transporting signaling and user packets. P-P signaling messages are transported over UDP just as MIPv4 signaling messages are transported over UDP.

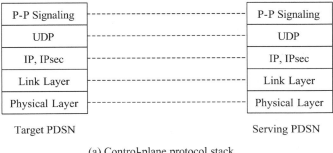

(a) Control-plane protocol stack

(b) User-plane protocol stack

Fig. 2.44   *Protocol stacks for the P-P interface*

2.2.6.4 *Protocol Stacks Between Mobile and PDSN*   Now let's put the protocol stacks for the individual interfaces together to picture the protocol stacks for end-to-end transport of user data between a mobile and a correspondent host (CH). A correspondent host is any terminal that is communicating with the mobile. A correspondent host can be located in any IP network that is reachable via IP from the 3GPP2 packet network. We also illustrate the protocol stacks for signaling between a mobile and its serving PDSN.

First, let's consider the scenario in which the mobile is not in the process of a fast inter-PDSN handoff. The protocol stacks for user data transport between a mobile and a correspondent terminal are illustrated in Figure 2.45 [1], [3]. In the figure, the Link Access Control (LAC) manages the establishment, use, modification, and removal of radio links.

Now, assume that the P-P interface is used by the mobile's serving PDSN to tunnel user traffic via a target PDSN to the mobile. Figure 2.46 shows the protocol stacks for end-to-end user traffic transport between a mobile and a correspondent host (CH).

Next, we consider the protocol stacks for signaling between a mobile and its serving PDSN. A mobile needs to exchange signaling messages with its serving

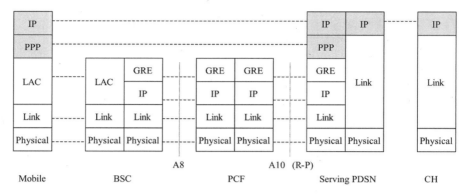

Fig. 2.45 *3GPP2 protocol stacks for user data between mobile terminal and PDSN (without P-P interface)*

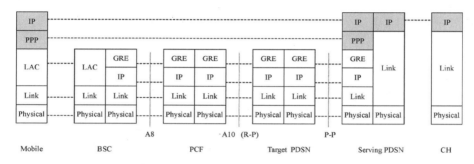

Fig. 2.46 *Protocol stacks for end-to-end user traffic transport when P-P interface is used*

PDSN to set up the PPP connection and to perform MIPv4 registrations. Signaling messages between a mobile and its serving PDSN are transported over the traffic bearers (connections) throughout the RAN and the packet data core network, as illustrated in Figure 2.47 [1], [2], [7], [8].

2.3 MWIF ALL-IP MOBILE NETWORKS

MWIF seeks to develop an end-to-end all-IP wireless network [39] that will use IETF protocols to support all networking functions at the network layer and higher layers, including naming and addressing, signaling, service control, routing, transport, mobility management, quality of service mechanisms, security, accounting, and network management. Unlike the 3GPP and 3GPP2 networks, the MWIF architecture will no longer rely on protocols or network entities in circuit-switched core networks.

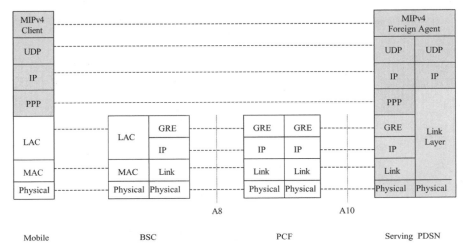

Fig. 2.47   *3GPP2 protocol stacks for signaling between mobile terminal and PDSN*

This section discusses the following:

- MWIF network architecture (Section 2.3.1)
- Access to MWIF networks (Section 2.3.2)
- MWIF session management architecture (Section 2.3.3)

2.3.1   Network Architectures

Similar to the 3GPP and 3GPP2 networks, the MWIF architecture is also composed of Access Networks interconnected by a Core Network. The Core Network is designed to be independent of access-specific technologies used in different Access Networks so that the same Core Network can be used to support different Access Networks. The Core Network is also designed to be all-IP network using standard IETF protocols. This is different from the 3GPP and the 3GPP2 core network designs that use non-IP protocols specific to the 3GPP and 3GPP2 networks.

MWIF uses a layered functional architecture with four functional layers, as illustrated in Figure 2.48 [38]:

- *Transport Layer (in both Access Network and Core Network)*: The transport layer provides the basic IP routing and IP packet transport functions. It also contains gateways needed for (1) the access network to connect to the core network and (2) the core network to interact with external networks.
- *Control Layer*: The control layer provides functionality needed for the control of IP transport. The control layer supports, for example, mobility

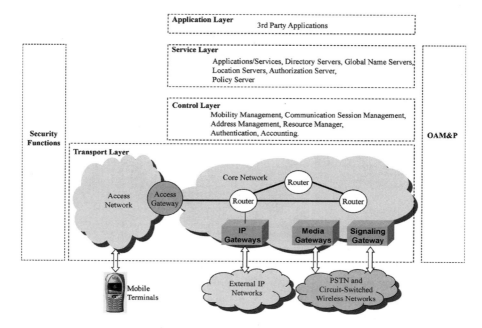

Fig. 2.48   *MWIF-layered functional architecture*

management, communication session management, resource management, address management, authentication, and accounting functions.

- *Service Layer*: The service layer provides service control functions for (1) services provided to the end users and (2) services needed by the network to make effective use of the lower functional layers. The service layer contains, for example, control logics for services provided to the users, applications provided to the users, directory and name servers, location servers, policy servers, and authentication and authorization servers.

- *Application Layer*: The application layer is composed of third-party applications and services provided to the users. The application layer may interact with any lower functional layer directly without having to go through the intermediate functional layers. A third-party refers to a provider that is not the owner of the other three functional layers.

The security and the OAM&P (Operation, Administration, Maintenance and Provisioning) functions may span across multiple functional layers. The OAM&P functions are responsible for ensuring proper operation and maintenance of network transport, control, and service infrastructure. Main OAM&P functionality includes network provisioning, network configuration management, fault management, performance management, and billing functions.

The layered functional architecture reflects the design principle of separation of concerns. In particular, control and user traffic transport are logically separated. Within the control layer, session management and mobility are also logically separated. Service control is also logically separated from session management and data transport.

Figure 2.49 illustrates the network reference architecture that shows the main functional entities and their interactions [41].

An access network provides radio connectivity to enable a mobile to gain radio access to the core network. It also provides mobility management capabilities to enable a mobile to move inside the access network. Each access network is connected to a core network via an Access Gateway. An Access Gateway contains the following functional entities [41]:

- *Access Transport Gateway*: The Access Transport Gateway relays packets between the access network and the core network. It performs access control functions (e.g., admission control) and enforces QoS agreements. It also collects usage information needed for accounting and billing.
- *IP Address Manager*: The IP Address Management is responsible for assigning IP addresses to mobiles dynamically.
- *Mobile Attendant*: A Mobile Attendant provides IP-layer mobility management functions inside the access network. For example, it may act as a Mobile IP Foreign Agent (if Mobile IP v4 is used). It may request mobiles to supply information needed for authentication and authorization. It acts as a proxy to relay mobility management messages between a mobile and their Home Mobility Manager (e.g., a Mobile IP Home Agent).

As in 3GPP and 3GPP2, the MWIF architecture also implements most of its Control Layer and Service Layer capabilities inside the core network. These capabilities are supported by the following servers or functional entities in the core network [41]:

- *Mobility Management Servers*: Main servers in the core network for mobility management include the following:
  - *Home Mobility Manager (HMM)*: Supports the movement of a mobile from one Access Gateway to another in the same or a different administrative domain. Enables packets sent to the mobile's home network to be forwarded to the mobile's current location.
  - *Home IP Address Manager*: Assigns home address dynamically to mobiles.
  - *IP Address Manager*: Assigns local IP addresses dynamically to mobiles.
- *Communication Session Management Servers*: Communication Session Management is responsible for managing real-time voice or multimedia sessions and will be discussed in Section 2.3.3.

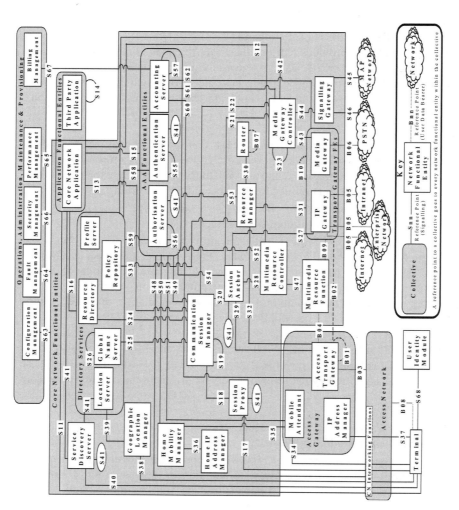

Fig. 2.49 MWIF network reference architecture

- *Resource Management Servers*: Media and resource control entities include the following:
  - *Resource Manager*: Manages the overall quality of services provided by the core network. For example, it monitors the qualities of services supported by the core network and performs QoS admission control.
  - *Multimedia Resource Function (MRF)*: Provide resources to support a multimedia session. For example, an MRF performs the following functions: transport bearer traffic, plays announcements, generate dial tones and other tones, provide conference bridging services such as transcoding.
  - *Multimedia Resource Controller (MRC)*: Manages an MRF and controls the resources available in an MRF in response to requests from a Session Anchor and network applications.
- *Registration and AAA Servers*: These include Authentication Server, Authorization Server, and Accounting Server.
- *Gateways*: MWIF defines several gateways and gateway controllers:
  - *IP Gateway*: Provides controlled access between the core network and other IP networks (e.g., the Internet, intranets, or enterprise networks).
  - *Media Gateway (MG)*: An MG interconnects the core network to the PSTN or other circuit-switched networks. For example, it enforces policies on volume, delay, and delay variance; provides firewall functionality; and translate between media formats.
  - *Media Gateway Controller (MGC)*: An MGC controls the bearer paths through an MG. For example, it interacts with an MG to activate and deactivate the firewalls enforced by the MG; performs application-layer signaling interworking (e.g., converts between SIP signaling messages and SS7 ISUP messages).
- *Information Servers:* Information servers maintain and provide information and translate between different information items for supporting the operation of the network. The MWIF network uses the following main information servers: Service Discovery Server, Location Server, Geographical Location Manager (GLM), Global Name Server (GNS), Profile Server, Policy Repository, and Resource Directory.

2.3.2 Access to MWIF Networks

MWIF uses a multi-tiered service activation process that consists of the following main phases:

- *Access Network Registration*: This will allow a mobile to gain access to the radio access network. The MWIF architecture assumes that access network registration is performed by methods specific to each access network. The

MWIF architecture, therefore, does not have recommendations on how such a registration should be carried out.

- *Basic Registration for Core Network*: The basic registration will enable a mobile to gain access to the core network and to send and receive IP packets over the core network.

- *SIP Registration*: SIP is the protocol of choice for session and service management over the MWIF network architecture. SIP registration will enable a user to use SIP to initiate and receive multimedia communications. SIP registration can be performed at the time the user wishes to set up a voice or multimedia application session. In other words, SIP registration will be an integral part of session and service management and will be discussed in Section 2.3.3.

Figure 2.50 illustrates the Basic Registration procedure. It contains three main steps, where the first two steps are mandatory and the third step is optional:

- *Step 1*: Obtain local IP address.
- *Step 2*: Ensure authentication and authorization by both visited and home network.
- *Step 3*: QoS procedure: e.g., resource reservation.

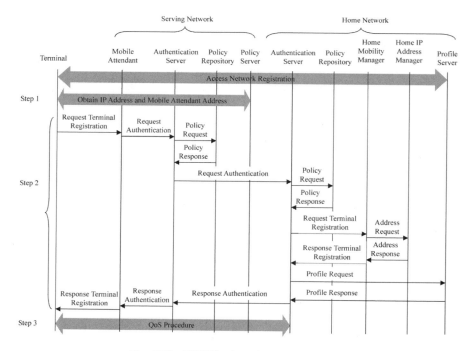

Fig. 2.50   *MWIF basic registration procedure*

In Step 1, the mobile acquires a local IP address from the network. The mobile will also obtain, from the network, other addresses the mobile needs in order to contact the network servers in the visited network, for example, the Mobile Attendant, Session Proxy, and Service Discovery Server.

In Step 2, the mobile performs Mobile IP registration with its Home Agent in its home network. It initiates Mobile IP registration by sending a Request Terminal Registration message, which will be a Mobile IP Registration Request if Mobile IPv4 is used, to the Mobile Attendant. Upon receiving the Request Terminal Registration message, the Mobile Attendant initiates AAA procedure with the local authentication server. The local authentication server may retrieve policy information from the local Policy Repository to help determine whether the mobile should be authorized. The local authentication server may then forward the Mobile IP and AAA requests to the mobile's home network, for example, to an authentication server in the mobile's home network. The home authentication server, after positive verification of the policy applicable to the mobile, will forward the Mobile IP registration request to the Home Mobility Manager (which is typically a Mobile IP Home Agent). The Home Mobility Manager may assign a new home address to the mobile. After recording the information received in the Mobile IP registration request, the Home Mobility Manager will send a Response Terminal Registration message (e.g., a Mobile IP Registration Reply message if Mobile IPv4 is used) to the home authentication server. The home authentication server in turn sends a response back to the authentication server in the visited network. The authentication server in the visited network then forwards the response to the Mobile Attendant, which will then generate a Response Terminal Registration message and send it to the mobile, completing the registration process.

In the optional Step 3, the mobile and the network may negotiate QoS terms and set up the resources needed to support the agreed on QoS levels.

2.3.3 Session Management

We will first describe the functional entities, protocol reference points, and protocol stacks for supporting session management. We will then illustrate the signaling flows for setting up a mobile-originated session.

2.3.3.1 *Functional Entities, Protocol Reference Points, and Stacks* The MWIF use three main functional entities for session management:

- *Communication Session Manager (CSM)*: The CSM controls the communication sessions for each user. A communication session is an association of (or logical connection between) two network entities. The CSM is responsible for establishing, monitoring, maintaining, modifying, adding and removing endpoints, tearing down of a communication session, as well as gathering statistics regarding the communication sessions.

- *Session Anchor*: A Session Anchor is an agent that allocates media resources to a given session.
- *Session Proxy*: A Session Proxy serves as the proxy for all session management requests between a mobile terminal and the CSM in the mobile's home network. The Session Proxy is the first contact point in the core network for session control requests from a mobile.

Other supporting functional entities include the following:

- *Global Name Server*: A Global Name Server provides name and address mapping (translation) services. For example, it can map between the following identifiers for a specific subscriber: Subscriber URL, E.164 telephone number, IP address, and Subscriber Identity (e.g., E.212 IMSI numbers). It can map IP domain names to IP addresses, map E.164 telephone numbers to IP addresses, and map URLs to applications.
- *Resource Directory*: A Resource Directory maintains information on the available network resources such as Media Gateways.
- *Service Directory Server*: A Service Directory Server enables discovery of network services. Network servers and services can register with a Service Directory Server the information that can be used by a network entity or user terminal to discover the servers and services. Such information includes, for example, address of a server or service and the attributes of a server or service. The Service Directory Server can then supply such information to a network entity or user terminal upon request.

The MWIF architecture uses SIP for session and service management. Therefore, SIP will be the signaling protocol between a mobile and a Session Proxy, between a Session Proxy and CSM, and between the CSMs.

Figure 2.51 illustrates the functional entities and protocol reference points for session management. The three main session control entities and the AAA functional entities are highlighted. The protocol reference points between the main session management functional entities are marked by the protocol reference number defined by the MWIF. The MWIF recommended protocols for the reference points shown on Figure 2.51 are [40] as follows:

- *Reference points S17, S18, S19, S20, and S22*: These protocol reference points are used to exchange session management signaling messages. The MWIF recommends SIP and the Session Description Protocol (SDP) as the signaling and control protocols over these interfaces. SIP is used for controlling the sessions. SDP [43] is used to describe multimedia sessions to support SIP session initiation. SDP uses a text format description that provides many details of a multimedia session, including the originator of the session, a URL related to the session, the connection address for the session media(s), and optional attributes for the session media(s). SIP and SDP may be transported

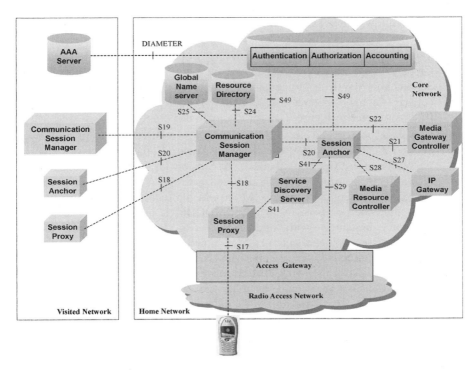

Fig. 2.51 *MWIF functional entities for session and service management*

over TCP/IP, UDP/IP, or SCTP/IP. SCTP (Stream Control Transmission Protocol) [45] is a transport-layer protocol designed to transport PSTN signaling messages over IP networks. SIP and SDP are described in Chapter 3.

- *Reference point S21*: This protocol reference point is used for a Session Anchor to send requests to a Media Gateway Controller. The MWIF recommends to use the same protocol stacks used for S17 for session management over S21 and to use the Telephony Routing over IP Protocol (TRIP) [48] for notification of media gateway reachability. TRIP is an interadministrative domain protocol for distributing the reachability of telephony destinations and for advertising attributes of the routes to those destinations. TRIP is modeled after the Border Gateway Protocol (BGP-4) [47] that is used to distribute IP routing information between administrative domains.

- *Reference point S24*: This protocol reference point is used by a Session Manager to access the Resource Directory to locate resources required to support a session. The MWIF recommends the Lightweight Directory Access Protocol (LDAP) [34] as the directory access protocol over this interface.

LDAP allows a network entity to read from and write to a directory on an IP network. LDAP is transported over TCP/IP.

- *Reference points S25 and S26*: This protocol reference point is used by a Global Name Server to request a name translation from another Global Name Server. The MWIF recommends LDAP or the DNS protocol as the protocol between Global Name Servers. As LDAP, DNS is also transported over TCP/IP. Please refer to Figure 2.49 for the reference point S26.

- *Reference points S27 and S29*: These two reference points are used by a Session Anchor to control the IP-based gateways: the IP Gateway and the Access Transport Gateway. It allows the Session Anchor to send, for example, firewall control messages to an IP-based gateway. The MWIF recommends the Middlebox Communications (MIDCOM) Protocol to be used for signaling over reference points S27 and S29. MIDCOM is a protocol for a network entity to communicate with "middleboxes" (e.g., Firewalls, Network Address Translator servers, and gateways) in a network. MIDCOM messages are transported directly over IP. The MIDCOM protocol requirements can be found in [53]. But the MIDCOM protocol is still under development on the IETF and has not yet become an IETF RFC.

- *Reference point S41*: This protocol reference point allows a core network entity to register itself with the Service Discovery Server and to request service addresses from a Service Discovery Server. For example, the following core network entities are expected to register themselves with the Service Discovery Server: Session Proxy, Session Anchor, Authentication Server, Authorization Server, Location Server, and Core Network Applications. The MWIF recommends the Service Location Protocol (SLP) [32], DHCP, or DNS to be used as the protocol over this reference point. SLP allows a terminal or network entity to discover network services by discovering information about the existence, location, and configuration of these networked services.

2.3.3.2 *Mobile-Initiated Call Setup*   Figure 2.52 illustrates the procedure for setting up a mobile-originated SIP call [41]. The mobile requests for a SIP session setup by sending a SIP INVITE to the Session Proxy in the serving (visited) network. The Session Proxy forwards the SIP INVITE to the CSM in the originating user's home network. The Home CSM initiates user authentication by contacting the home Authentication Server. Upon positive authentication of the user, the home CSM contacts the home Authorization Server to authorize the SIP session requested by the user. The home Authorization Server utilizes the user profile information maintained by the Policy Repository to determine whether the user session should be authorized and return the results to the home CSM.

Upon positive authorization, the user's home CSM will forward the SIP INVITE to the destination CSM, which will in turn forward the SIP INVITE to the destination user. The destination user responds to the SIP INVITE message by sending back a SIP 183 Call Processing message to the originating user's home

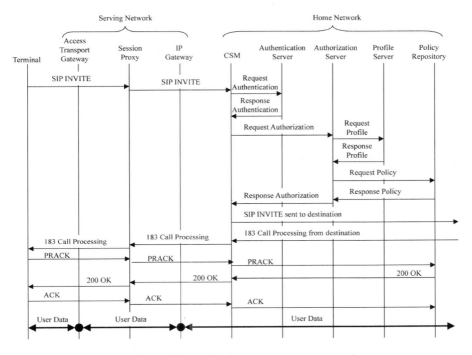

Fig. 2.52 *MWIF mobile-originated call setup procedure*

CSM. This home CSM forwards the SIP 183 Call Processing message to the Session Proxy in the serving network, which will in turn forward the message to the originating user. Upon receiving the SIP 183 Call Processing message, the originating user will send back a PRACK message to the Session Proxy in the serving network. The Session Proxy will then forward the PRACK message to the originating user's home CSM, which will in turn forward the message to the destination.

After sending the PRACK message, the originating mobile may initiate the QoS procedures to reserve the resources needed in the serving network to support the SIP session.

Upon receiving the PRACK message, the destination will respond by sending back a SIP 200 OK message to the originating user's home CSM. The originating user's home CSM will then forward the SIP 200 OK message to the Session Proxy in the serving network, which will forward the message to the originating mobile. The originating mobile sends a SIP ACK message to the Session Proxy to confirm the setup of the SIP session. The Session Proxy will forward the SIP ACK message to the originating user's home CSM, which will then forward the message to the destination to complete the call setup.

REFERENCES

1. 3rd Generation Partnership Project 2 (3GPP2). Wireless IP network standard. 3GPP2 P.S0001, Version 1.0, December 1999.

2. 3rd Generation Partnership Project 2 (3GPP2). Wireless IP architecture based on IETF protocols. 3GPP2 P.R0001, Version 1.0.0, July 2000.

3. 3rd Generation Partnership Project 2 (3GPP2). 3GPP2 access network interfaces interoperability specification, revision A (3G-IOS v4.1.1). 3GPP2 A.S0001-A, Version 2.0, June 2001.

4. 3rd Generation Partnership Project 2 (3GPP2). Network reference model for cdma2000 spread spectrum systems. 3GPP2 S.R0005-B, Version 1.0, Revision B, April 2001.

5. 3rd Generation Partnership Project 2 (3GPP2). 3GPP2 interoperability specification (IOS) for cdma2000 access network interfaces—part 4 (A1, A2, and A5 interfaces). 3GPP2 A.S0014-0, Version 2.0, May 2002.

6. 3rd Generation Partnership Project 2 (3GPP2). 3GPP2 interoperability specifications (IOS) for cdma2000 access network interfaces—part 2 Transport. 3GPP2 A.S0012-0, Version 2.0, May 2002.

7. 3rd Generation Partnership Project 2 (3GPP2). 3GPP2 interoperability specifications (IOS) for cdma2000 access network interfaces—part 6 (A8 and A9 interfaces). 3GPP2 A.S0016-0, Version 2.0, May 2002.

8. 3rd Generation Partnership Project 2 (3GPP2). 3GPP2 interoperability specifications (IOS) for cdma2000 access network interfaces—part 7 (A10 and A11 interfaces). 3GPP2 A.S0017-0, Version 2.0, May 2002.

9. 3rd Generation Partnership Project 2 (3GPP2). Interoperability specification (IOS) for cdma2000 access network interfaces— part 1 Overview. 3GPP2 A.S0011-0, Version 2.0, May 2002.

10. 3rd Generation Partnership Project 2 (3GPP2). IP network architecture model for cdma2000 spread spectrum systems. 3GPP2 S.R0037-0, Version 2.0, May 2002.

11. 3rd Generation Partnership Project (3GPP), Technical Specification Group Core Network. Customized applications for mobile network enhanced logic (CAMEL) phase 4—stage 2 (release 5). 3GPP TS 23.078, Version 5.0.0, June 2002.

12. 3rd Generation Partnership Project (3GPP), Technical Specification Group Core Network. Mobile application part specification (release 5). 3GPP TS 29.002, Version 5.2.0, June 2002.

13. 3rd Generation Partnership Project (3GPP), Technical Specification Group Core Network. GSM—UMTS public land mobile natwork (PLMN) access reference configuration, release 5. 3GPP TS 24.002, Version 5.1.0, March 2003.

14. 3rd Generation Partnership Project (3GPP), Technical Specification Group Core Network, Packet Domain. Interworking between the public land mobile network (PLMN) supporting packet based services and packet data networks (PDN), release 5. 3GPP TS 29.061 Version 5.2.1, July 2002.

15. 3rd Generation Partnership Project (3GPP), Technical Specification Group, Core Networks. Numbering, addressing and identification (release 5). 3GPP TS 23.003, Version 5.3.0, June 2002.

16. 3rd Generation Partnership Project (3GPP), Technical Specification Group GSM/EDGE. Radio access network, overall description—stage 2, release 5. 3GPP TS 43.051, Version 5.6.0, April 2002.

17. 3rd Generation Partnership Project (3GPP), Technical Specification Group, Radio Access Network. Packet data convergenceprotocol (PDCP) specification, release 5. 3GPP TS 25.323, Version 5.1.0, June 2002.

18. 3rd Generation Partnership Project (3GPP), Technical Specification Group Radio Access Network. Radio resource control (RRC); protocol specification, release 5. 3GPP TS 25.331, Version 5.1.0, June 2002.

19. 3rd Generation Partnership Project (3GPP), Technical Specification Group, Radio Access Network. UTRAN Iu interface: general aspects and principles, release 5. 3GPP TS 25.410, Version 5.1.0, June 2002.

20. 3rd Generation Partnership Project (3GPP), Technical Specification Group, Radio Access Network. UTRAN Iu interface RANAP signaling, release 5. 3GPP TS 25.413, Version 5.1.0, June 2002.

21. 3rd Generation Partnership Project (3GPP), Technical Specification Group, Services and System Aspects. Architecture requirements, release 5. 3GPP TS 23.221, Version 5.5.0, June 2002.

22. 3rd Generation Partnership Project (3GPP), Technical Specification Group, Services and System Aspects. General packet radio service (GPRS) service description, stage 2, release 5. 3GPP TS 23.060, Version 5.2.0, June 2002.

23. 3rd Generation Partnership Project (3GPP), Technical Specification Group, Services and System Aspects. Network architecture, release 5. 3GPP TS 23.002, Version 5.7.0, June 2002.

24. 3rd Generation Partnership Project (3GPP), Technical Specification Group Terminals. Technical realization of the short message service (SMS), release 5. 3GPP TS 23.040, Version 5.4.0, June 2002.

25. IEEE 802.11 wireless local area networks. http://www.grouper.org/groups/802/11/index.html.

26. H. Schulzrinne, S. Casner, R. Frederick, and V. Jacobson. RTP: a transport protocol for real-time applications. IETF RFC 1889, January 1996.

27. C. Bormann, C. Burmeister, M. Degermark, H. Fukushima, H. Hannu, L-E. Jonsson, R. Hakenberg, T. Koren, K. Le, Z. Liu, A. Martensson, A. Miyazaki, K. Svanbro, T. Wiebke, T. Yoshimura, and H. Zheng. Robust header compression (ROHC): framework and four profiles: RTP, UDP, ESP, and uncompressed. IETF RFC 3095, July 2001.

28. A.T. Campbell, J. Gomez, S. Kim, A.G. Valko, C-Y. Wan, and Z. Turanyi. Design, implementation and evaluation of cellular IP. *IEEE Personal Communications*, August 2000.

29. D. Haskin and E. Allen. IP version 6 over PPP. IETF RFC 2472, December 1998.

30. M. Degermark, B. Nordgren, and S. Pink. IP header compression. IETF RFC 2507, February 1999.

31. R. Droms. Dynamic host configuration protocol. IETF RFC 2131, March 1997.

32. E. Guttman, C. Perkins, J. Veizades, and M. Day. Service location protocol, version 2. IETF RFC 2608, June 1999.

33. S. Hanks, T. Li, D. Farinacci, and P. Traina. Generic routing encapsulation (GRE). IETF RFC 1701, October 1994.

34. J. Hodges and R. Morgan. Lightweight directory access protocol (v3): technical specification. IETF RFC 3377, September 2002.

35. IEEE Std 802.11: Wireless LAN medium access control (MAC) and physical layer (PHY) specifications, November 1997.

36. S. Kent and R. Atkinson. IP encapsulating security payload (ESP). IETF RFC 2406, November 1998.

37. S. Kent and R. Atkinson. Security architecture for the Internet protocol. IETF RFC 2401, November 1998.

38. Mobile Wireless Internet Forum. Layered functional architecture. Draft Technical Report MTR-003 Release 1.0, August 2000.

39. Mobile Wireless Internet Forum. Architecture principles. Technical Report MTR-001 Release 1.7, February 2001.

40. Mobile Wireless Internet Forum. Architecture requirements. Technical Report MTR-002 Release 1.7, February 2001.

41. Mobile Wireless Internet Forum. Network reference architecture. Technical Report MTR-004 Release 2.0, May 2001.

42. P. Mockapetris. Domain names—implementation and specification. IETF RFC 1035, November 1987.

43. S. Olson, G. Camarillo, and A.B. Roach. Support for IPv6 in session description protocol (SDP). IETF RFC 3266, June 2002.

44. C. Perkins. IP mobility support for IPv4. IETF RFC 3344, August 2002.

45. R. Stewart, Q. Xie, K. Morneault, C. Sharp, H. Schwarzbauer, T. Taylor, I. Rytina, M. Kalla, L. Zhang, and V. Paxson. Stream control transmission protocol. IETF RFC 2960, October 2000.

46. R. Ramjee, T.L. Porta, L. Salgarelli, S. Thuel, K. Varadhan, and L. Li. IP-based access network infrastructure for next generation wireless data networks. *IEEE Personal Communications*, August 2000.

47. Y. Rekhter and T. Li. A border gateway protocol 4 (BGP-4). IETF RFC 1771, March 1995.

48. J. Rosenberg, H. Salama, and M. Squire. Telephony routing over IP (TRIP). IETF RFC 3219, January 2002.

49. J. Rosenberg, H. Schulzrinne, G. Camarillo, A. Johnston, J. Peterson, R. Sparks, M. Handley, and E. Schooler. SIP: session initiation protocol. IETF RFC 3261, June 2002.

50. S. Thomson and T. Narten. IPv6 stateless address autoconfiguration. IETF RFC 2462, December 1998.

51. W. Simpson. The point-to-point protocol (PPP). IETF RFC 1661, July 1994.

52. W. Simpson. IP in IP tunneling. IETF RFC 1853, October 1995.

53. R.P. Swale, P.A. Mart, P. Sijben, S. Brim, and M. Shore. Middlebox communications (midcom) protocol requirements. IETF RFC 3304, August 2002.

54. W. Townsley, A. Valencia, A. Rubens, G. Paul, G. Zorn, and B. Palter. Layer two tunneling protocol (L2TP). IETF RFC 2661, August 1999.

3

IP Multimedia Subsystems and Application-Level Signaling

This chapter examines the signaling and session management architectures and protocols for supporting multimedia applications in IP networks and in the 3G IP Multimedia Subsystems (IMS) defined by 3GPP and 3GPP2. Multimedia applications may be any real-time applications, such as speech, audio, video, and text applications. An IMS is a subsystem in a 3G packet-switching domain that supports IP multimedia services. Signaling and session management in the IMSs defined by both 3GPP and 3GPP2 are based on today's signaling protocol of choice for the Internet—the IETF Session Initiation Protocol (SIP) [27]. Therefore, this chapter first examines the signaling protocols developed by the IETF for IP networks, focusing on SIP, and then discusses the signaling architectures and protocols specified in 3GPP and 3GPP2 IMSs.

Why choose SIP

3.1 SIGNALING IN IP NETWORKS

Two major standards have emerged for the signaling and session control of packet telephony. The first one is H.323 [21], supported by the ITU-T. The second one is Session Initiation Protocol (SIP) [27] developed by the IETF. These two protocols provide similar signaling functionality. However, SIP is a lightweight protocol that provides a simpler implementation and a greater degree of efficiency than H.323 [30]. SIP has been chosen by both 3GPP and 3GPP2 as the signaling protocol in packet-switching domain.

IP-Based Next-Generation Wireless Networks: Systems, Architectures, and Protocols,
By Jyh-Cheng Chen and Tao Zhang. ISBN 0-471-23526-1 © 2004 John Wiley & Sons, Inc.

3.1.1 Session Initiation Protocol (SIP)

SIP is an application-layer protocol that can establish, modify, and terminate multimedia sessions (conferences) over the Internet. A multimedia session is a set of senders and receivers and the data streams flowing from the senders to the receivers. For example, a session may be a telephony call between two parties or a conference call among more than two parties. SIP can also be used to invite a participant to an on-going session such as a conference. SIP messages could contain session descriptions such that participants can negotiate with media types and other parameters of the session. SIP provides its own mechanisms for reliable transmission and can run over several different transport protocols such as TCP, UDP, and SCTP (Stream Control Transmission Protocol) [22]. SIP is also compatible with both IPv4 and IPv6.

SIP provides the following key capabilities for managing multimedia communications:

- *Determine destination user's current location.*
- *Determine whether a user is willing to participate in a session.*
- *Determine the capabilities of a user's terminal.*
- *Set up a session.*
- *Manage a session.* This includes modifying the parameters of a session, invoking service functions to provide services to a session, and terminating a session.

Like HTTP, SIP is a client-server protocol that uses a request and response transaction model. A SIP client (or client, for simplicity) is any network element that generates SIP requests and receives SIP responses. A SIP server (or server, for simplicity) is a network element that receives SIP requests in order to service them and sends back responses to these requests. There are four major components in the SIP architecture:

- **SIP user agent:** A user agent (UA) is an Internet endpoint, such as IP phone, PC, or conference bridge, that is used to establish, modify, and terminate sessions. A UA could act as both a user agent client (UAC) and user agent server (UAS). A UAC is a logical entity that initiates a request. A UAS, on the other hand, generates a response to a SIP request.
- **SIP redirect server:** A redirect server is a UAS that generates a response to redirect a request to other location.
- **SIP proxy server:** A proxy server assumes the roles of both UAC and UAS. It acts as an intermediary entity between other user agents to route SIP messages to the destination user.
- **SIP registrar:** A registrar is a UAS that processes SIP *REGISTER* requests. It maintains mappings from SIP user names to addresses and is the front end of

the *location service*, which will be discussed in Section 3.1.1.3. Typically, it is consulted by a SIP server to route messages.

Next, we discuss the following key aspects of SIP:

- *Naming and addressing*
- *Messages*
- *Location registration*
- *Session establishment and termination*

3.1.1.1 *Naming and Addressing*

Each SIP user is uniquely identified by a SIP Uniform Resource Identifier (URI). A SIP URI is similar to an email address. It typically has a user name and a host name. For example, a SIP user Tao may have a SIP URI of

`sip : tao@research.telcordia.com`

where `research.telcordia.com` is the Internet domain name of a SIP server in Tao's SIP service provider's network.

A secure SIP URI, called SIPS URI, specifies a secure and encrypted way to transport SIP messages. Usually, TLS (Transport Layer Security) [17] is employed to secure the SIP messages. Specific security mechanism, however, depends on the policy of each domain. The format of SIPS URI is the same as SIP URI, except that the prefix `sip` is replaced by `sips`. For instance, `sips:tao@research.telcordia.com` is the SIPS URI for the example shown above. Although this section discusses only SIP URI, the same rules apply to SIPS URI as well.

The SIP URI follows the guidelines defined in [14]. In general, a SIP URI has a format of

`sip : user : password@host : port; uri-parameters?headers`

The first part could be either `sip:` or `sips:`, specifying whether the URI is a SIP URI or SIPS URI. Next comes an optional *userinfo*, which comprises username and password in the format of `user:password@`. If the userinfo is absent, the destination host is the resource that is being identified. Therefore, there is no userinfo for this URI. Otherwise, the `user` field identifies a particular user at the host, and the `password` field specifies the password associated with the user. The `user` field is not necessary to be a human user. It essentially is a *resource* that is being addressed at the host. For instance, `user` could be a telephone number as specified in [32]. Although SIP URI allows a password to be present, the use of it is not recommended because the URI is transmitted in cleartext. Although the *userinfo* is optional, the field of `user` cannot be empty if the sign @ is present in the SIP URI. The userinfo is case sensitive. Other fields of the SIP URI are not case sensitive unless explicitly defined.

The `host` identifies the host that provides the resource identified by the SIP URI. A host can be identified by either an IPv4/IPv6 address or a FQDN (Fully Qualified Domain Name), although FQDN is preferable. The port number where the request should be sent is defined in the `port` field. The `uri-parameters` specifies the URI parameters, which include `transport, maddr, ttl, user, method,` and `lr` parameters. The parameters are defined by `parameter-name=parameter-value` and are separated by semicolons. Details are itemized as follows:

- *transport:* This parameter specifies the transport protocol used to transport SIP messages. The transport protocol might be UDP, TCP, SCTP, TLS, or other protocols. If UDP is adopted, the URI parameter takes the form of `transport=udp`. Because it is not case sensitive, `transport=udp` is equivalent to `Transport=UDP`.

- *maddr:* The maddr parameter provides a simple form of loose source routing. This parameter can be used to identify a proxy that the SIP message must traverse on its way to the destination. It overrides the address specified in the `host` field and directs the request to the server specified by the maddr parameter. For example, a URI with `maddr=140.114.79.60` indicates that the request must be delivered to the server with IP address of 140.114.79.60 first before it is sent to the address specified in the `host` field.

- *ttl:* The ttl parameter contains the time-to-live (TTL) value of a multicast packet and therefore is used only when the maddr is a multicast address and the transport protocol is UDP. For example, `ttl=15` defines that the maximum number of hops a UDP multicast packet can travel is 15.

- *user:* As mentioned earlier, the `user` field in *userinfo* of a SIP URI could be a telephone number. However, it is also possible that a user has a name look like a telephone number. The parameter of `user` is utilized to distinguish a real telephone number from a user name that resembles a telephone number. For instance, the URI of

  `sip:+886-3-574-2961:1234@cs.nthu.edu.tw;user=phone`

  specifies that the user is a real telephone-subscriber with the number of *+886-3-574-2961*. It also shows a password of 1234, which could be a PIN number.

- *method:* The method parameter specifies the *method* of the SIP request. It indicates the primary function of the message. For example, `method=REGISTER` indicates that it is a SIP request for registration. The default method is *INVITE*. Details of methods are discussed in Section 3.1.1.2.

- *lr:* The lr parameter is used when a specific SIP routing mechanism is implemented. We will not discuss the SIP routing mechanisms further. Interested readers please refer to [27] for details.

Following the URI parameters is the `headers` field specified with "?". It could be utilized to indicate the *Header Fields* that are requested to be included in the URI.

Header fields are attributes that provide additional information about a SIP message. For example, the following URI

sip:jcchen@cs.nthu.edu.tw?subject = Wiley%20Book&priority = urgent

includes two header fields of *subject* and *priority*, which are concatenated by a & sign. The header fields indicate a subject of *Wiley Book* and an *urgent* priority. The %20 represents a *space* character that must be escaped in the SIP URI. SIP follows the rules defined in [14] for escaped characters. More header fields will be discussed in Section 3.1.1.2.

A final note of SIP URI is that only the host field is mandatory in a SIP URI. Therefore, a simplest form of a SIP URI could be

```
sip:wire.cs.nthu.edu.tw
```

3.1.1.2 Messages   Each SIP message is either a request message or a response message. A request message is sent from a client (UAC) to a server (UAS) to invoke a particular operation or function. The function invoked by a request is referred to as a *method*. A response message is sent from a UAS to a UAC to indicate the status of a request. The following methods have been defined in SIP:

- **INVITE:** Used by a user to invite another user to establish a SIP session [27].
- **ACK:** Used to confirm final response [27].
- **BYE:** Used to terminate a session [27].
- **CANCEL:** Used to cancel a SIP request [27].
- **OPTIONS:** Used to query servers about their capabilities [27].
- **REGISTER:** Used by a user to register information (e.g., the user's current location) with a server [27].
- **INFO:** Used to carry session-related control information such as ISUP and ISDN signaling messages [29].
- **SUBSCRIBE:** Used to request current state and state updates from a remote node [23].
- **NOTIFY:** Used to notify a SIP node that an event which has been requested by an earlier SUBSCRIBE method has occurred [23].
- **PRACK:** Used to provide a reliable Provisional Response ACKnowledgement (PRACK) [26].
- **UPDATE:** Used to update parameters of a session such as the set of media streams and their codecs [24].
- **MESSAGE:** Used to transfer Instant Messages (IM) [15].
- **REFER:** Used to direct a recipient to other resources by using the contact information provided in the REFER request. It could be used for call transfer [31].

SIP is a text-based protocol. In other words, its messages are written in text formats. As shown in Table 3.1, every SIP message consists of the following:

- *A start-line*
- *One or more header fields*
- *An empty line indicating the end of the header fields*
- *An optional message body*

The start-line is used to distinguish a request message from a response message. In particular, the start-line is interpreted as a *Request-Line* in a request message but is read as a *Status-line* in a response message.

A Request-Line contains a method name, a Request-URI, and the SIP protocol version. A Request-URI is a SIP URI that indicates a user or a service that this request is addressed to. For example, the Request-Line for a SIP INVITE request used to invite user tao@research.telcordiac.com to a SIP session may look like: INVITE sip:tao@research.telcordia.com SIP/2.0.

A Status-Line consists of the SIP protocol version followed by a numerical Status-Code and the textual explanation of the Status-Code. A Status-Code is a 3-digit integer used to indicate the results of a request. The first digit of the Status-Code defines the class of the response. SIP defines the following six classes of Status-Code:

- *1xx*: Provisional—indicate a request is received and is being processed.
- *2xx*: Success—indicate the method invoked by a request is successfully accepted.

TABLE 3.1 Structure of a SIP message

Start-line	INVITE sip:tao@research.telcordia.com SIP/2.0
Header Field(s)	Via: SIP/2.0/UDP fly.cs.nthu.edu.tw:5060;branch=z9hG4bK776asdhds Max-Forwards: 70 To: Tao ⟨sip:tao@research.telcordia.com⟩ From: Jyh-Cheng ⟨sip:jcchen@cs.nthu.edu.tw⟩;tag=1928301774 Call-ID: a84b4c76e66710@fly.cs.nthu.edu.tw CSeq: 123456 INVITE Contact: ⟨sip:jcchen@fly.cs.nthu.edu.tw⟩ Content-Type: application/sdp Content-Length: 132
Empty Line	
Message Body (Optional)	v=0 t=2873397496 2873404696 m=audio 49170 RTP/AVP 0 ......

- *3xx*: Redirection—further action needs to be taken by the sender of the corresponding sender in order to complete the request.
- *4xx*: Client error—the request contains syntax error or cannot be fulfilled at this server.
- *5xx*: Server error—the server failed to fulfill an apparently valid request.
- *6xx*: Global failure—the request cannot be fulfilled at any server.

For example, when user `tao@research.telcordia.com` receives a SIP INVITE request and decides to answer the call, his application will send a SIP response message to the sender of the SIP INVITE message. The Status-Line of the response message may look like `SIP/2.0 200 OK`.

The header fields are used to carry information needed to route a request or to manage a SIP session. A header field consists of a field name followed by a colon (":") and a field value. The following are some of the important header fields:

- *To*: The *To* header field specifies the desired recipient of the request. For example, *To: Tao ⟨sip:tao@research.telcordia.com⟩* is a valid *To* header field destining to Tao.
- *From*: The *From* header field identifies the initiator of the request, for instance, *From: Jyh-Cheng ⟨sip:jcchen@cs.nthu.edu.tw⟩*.
- *Subject*: The *Subject* header field specifies the subject of the session.
- *Call-ID*: The Call-ID header field contains a unique identifier of the call (session). It is used to identify messages that belong to the same call (session).
- *Via*: The *Via* header field indicates the transport-layer protocol (e.g., UDP) used for the transaction and identifies the location where the response message for this request should be sent to.
- *Contact*: The *Contact* header field contains a SIP URI that could be used for subsequent requests. Comparing with the *Via* header field, which indicates where to send the response, the *Contact* header field informs other user agents where to send future requests. It is also used in a response message from a SIP redirect server to provide a location for the originator of the request to contact next in order to connect to the destination.
- *Max-Forwards*: The *Max-Forwards* contains an integer that indicates the maximum number of hops a request can traverse on its way to the destination.
- *CSeq*: Also called *Command Sequence*, the *CSeq* contains an integer and a method name. It essentially is a sequence number and is incremented for each new request.

Please be advised that some header fields apply only to request messages, and some header fields could only be used for response messages.

Following the header fields, both request and response messages may include a message body. There are various types of message bodies. They are specified by different header fields. For example, a message body of Internet media could be specified by the header field of *Content-Type*. The *Content-Type: application/sdp* indicates that the message body is a description of the session in the format of Session Description Protocol (SDP) [19]. An SDP may include type of media, codec, and sampling rate. Section 3.1.2 provides more information about SDP. If the body is encoded or compressed, it is specified by the *Content-Encoding* header field. The *Content-Encoding: gzip*, for instance, indicates that the message body is compressed by gzip. Although SIP message is text-based, the message body could be based on either text or binary.

3.1.1.3 Location Registration

SIP provides an inherent way for location service. A special logical SIP UAS referred to as a *SIP Registrar* residing in a user's SIP service provider's network is responsible for maintaining information on the user's current location. Whenever a user changes its location, it will register its new location with the SIP Registrar.

Figure 3.1 illustrates the signaling message flow for SIP registration. The UA of user *Tao* sends a REGISTER message to a registrar. The address of the registrar could be preconfigured. If there is no registrar preconfigured, the host part of the *address-of-record* should be used. For example, the user *sip:tao@research.telcordia.com* will send his REGISTER to *sip:research.telcordia.com*. In addition to preconfiguration and address-of-record, the UA can also multicast the REGISTER to the well-known multicast address of *all SIP servers*. In IPv4, the address of 224.0.1.75 has been allocated to `sip.mcast.net` as the multicast address of all SIP servers. SIP UAs may listen to that address to learn the address of local servers. No IPv6 multicast address has been allocated for this purpose yet.

A sample REGISTER message from Tao to his registrar is as follows:

REGISTER sip:registrar.research.telcordia.com SIP/2.0
Via: SIP/2.0/UDP tao-pc.research.telcordia.com:5060;branch=z9hG4bKnashds7
Max-Forwards: 70
To: Tao ⟨sip:tao@research.telcordia.com⟩
From: Tao ⟨sip:tao@research.telcordia.com⟩
Call-ID: 843817638423076@989sddhas09
CSeq: 2660 REGISTER
Contact: ⟨sip:tao@128.96.60.187⟩
Expires: 3600
Content–Length: 0

The header field *Expires* specifies that this registration is valid only for 3600 *seconds*, i.e., one hour. The maximum number of hops this message can travel is 70 as defined by *Max-Forwards*. There is no message body because the *Content-Length* is 0. Other header fields will be discussed in Section 3.1.1.4.

After receiving the REGISTER message, the registrar processes the request. The registrar may authenticate the user and then decide whether the user is authorized to

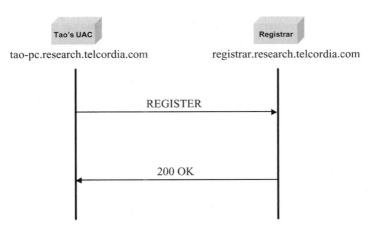

Fig. 3.1 *SIP registration*

modify the registration record. A 403 *Forbidden* will be returned if the user fails the authentication and authorization process. Otherwise, the registrar updates the user's location binding information with the information carried in the header fields. Once the user's binding is updated successfully, a *200 OK* message is returned to the user. A sample 200 OK message is shown as follows:

SIP/2.0 200 OK
Via: SIP/2.0/UDP tao-pc.research.telcordia.com:5060
   ;branch=z9hG4bKnashds7;received=128.96.60.187
To: Tao ⟨sip:tao@research.telcordia.com⟩
From: Tao ⟨sip:tao@research.telcordia.com⟩
Call-ID: 843817638423076@989sddhas09
CSeq: 2660 REGISTER
Contact: ⟨sip:tao@128.96.60.187⟩
Expires: 3600
Content-Length: 0

3.1.1.4 *Session Establishment and Termination* A SIP session can be established using a peer-to-peer mode or a server mode. In a peer-to-peer mode, a caller establishes a call to a callee directly without going through any SIP server. This requires that the caller knows the callee's current location. In server mode, on the other hand, a caller directs its SIP request messages to a SIP server in the caller's or callee's SIP service provider's network. The SIP server then assists the caller in the establishment of the SIP session. A SIP server may forward the received SIP request toward its final destination on behalf of the caller. In this case, the SIP server is referred to as a SIP *Proxy Server*. A proxy server may rewrite specific parts of the message before forwarding it. Alternatively, a SIP server may not relay the SIP request toward its final destination. Instead, it will respond to a request with the callee's contact information to indicate where the caller should contact next. In this case, the SIP server is called a SIP *Redirect Server*.

Consider a scenario where a caller *Jyh-Cheng Chen* uses a SIP UA to establish a SIP session with a callee *Tao Zhang*. Assuming that Jyh-Cheng's SIP URI is *sip:jcchen@cs.nthu.edu.tw* and Tao's SIP URI is *sip:tao@research.telcordia.com*. Figure 3.2 illustrates a sample message flow when the SIP server in Tao's SIP service provider's network is a proxy server. The caller first sends a SIP INVITE message to the callee `tao@research.telcordia.com` to request the establishment of a SIP session. The INVITE message may look like:

INVITE sip:tao@research.telcordia.com SIP/2.0
Via: SIP/2.0/UDP fly.cs.nthu.edu.tw:5060;branch=z9hG4bK776asdhds
Max-Forwards: 70
To: Tao ⟨sip:tao@research.telcordia.com⟩
From: Jyh-Cheng ⟨sip:jcchen@cs.nthu.edu.tw⟩;tag=1928301774
Call-ID: a84b4c76e66710@fly.cs.nthu.edu.tw
CSeq: 123456 INVITE
Contact: ⟨sip:jcchen@fly.cs.nthu.edu.tw⟩
Content-Type: application/sdp
Content-Length: 132

The first line is a Request-Line that indicates that the method is INVITE, and the Request-URI is `sip:tao@research.telcordia.com`. It also shows that the

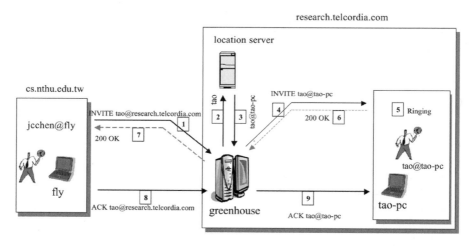

Fig. 3.2   *SIP in proxy mode*

version of SIP protocol is 2.0. The second line specifies that the transport layer protocol is UDP. It also identifies the address (fly.cs.nthu.edu.tw) and port number (5060) that Jyh-Cheng is expecting to receive the response. The branch parameter is used to identify the transaction created by this request. As discussed earlier, *Max-Forwards* indicates the maximum number of hops the INVITE message can traverse on its way to the destination. The *To* and *From* fields specify the source and destination URIs, respectively. A *tag* is a random string that is used for identification purposes. The *Call-ID* is a globally unique identifier for this call. The *CSeq* contains a sequence number that is incremented for each new request. The *Contact* field indicates the URI that Jyh-Cheng can be contacted directly. As mentioned earlier, the *Contact* field shows the URI for future requests, whereas the *Via* field specifies the location to send back the response for this request. The *Content-Type* and *Content-Length* indicate that the message body is a SDP message with a length of 132 bytes. In the example above, the message body is not shown. In addition to the Request-Line, *To, From, CSeq, Call-ID, Max-Forwards*, and *Via* are mandatory in all SIP requests formulated by a SIP UAC. Messages in Figure 3.2 are depicted with few header fields only.

The INVITE message is routed to a SIP server in the callee's SIP service provider's network. This SIP server is identified by the domain name research.telcordia.com in the callee's SIP URI tao@research. telcordia.com. The SIP server in the callee's SIP service provider's network determines the callee's current location, which may be done by querying a *location server*. Figure 3.2 shows that the location server returns Tao's current location of tao@tao-pc.research.telcordia.com to the SIP server of *greenhouse*. Once the callee is located, the SIP server of *greenhouse* relays the INVITE message to the callee. The callee's phone rings. Once Tao answers the call, his SIP UA replies a SIP response message with a Status-Code 200 OK (commonly referred to as a 200 OK messages or an OK message, for simplicity) to indicate the acceptance of the call. The 200 OK message may look like the following:

SIP/2.0 200 OK
Via: SIP/2.0/UDP greenhouse.research.telcordia.com:5060
    ;branch=z9hG4bKnashds8;received=207.3.230.150
Via: SIP/2.0/UDP fly.cs.nthu.edu.tw:5060
    ;branch=z9hG4bK776asdhds;received=140.114.79.59
To: Tao ⟨sip:tao@research.telcordia.com⟩;tag=a6c85cf
From: Jyh-Cheng ⟨sip:jcchen@cs.nthu.edu.tw⟩;tag=1928301774
Call-ID: a84b4c76e66710@fly.cs.nthu.edu.tw
CSeq: 123456 INVITE
Contact: ⟨sip:tao@tao-pc.research.telcordia.com⟩
Content-Type: application/sdp
Content-Length: 121

The first line of the response indicates the response code (200) and the reason phrase (OK). There are two *Via* headers. The first one is added by the proxy server of *greenhouse* in `research.telcordia.com`. The second line is copied from the INVITE request with an additional parameter of *received* to indicate where the request is originated. The 200 OK message above shows that the INVITE message is received from 207.3.230.150 (`greenhouse.research.telcordia.com`), which received it from 140.114.79.59 (`fly.cs.nthu.edu.tw`). This information can be utilized to send back the response. After *Via* headers, the *To*, *From*, *Call-ID*, and *CSeq* header fields are copied from the INVITE request. A *tag* is added by Tao in the end of the *To* header. A peer-to-peer SIP relationship between Jyh-Cheng and Tao thus can be identified by *Call-ID*, *To tag*, and *From tag*. This peer-to-peer SIP relationship is referred to as a *dialog*, which will be included in all future requests and responses in this call. The remaining lines including *Contact*, *Content-Type*, and *Content-Length* are the same as those in INVITE. Similarly, the message body of SDP is not shown.

Based on the *Via* headers, the 200 OK message is first routed to the SIP server of `greenhouse.research.telcordia.com`, which then relays the message to the caller. As illustrated in Figure 3.2, an ACK is then sent from caller to callee through the proxy. Thus, the three-way handshake including INVITE, 200 OK, and ACK for session setup is completed. The ACK message sending from Jyh-Cheng to Tao may look like the following:

ACK sip:tao@research.telcordia.com SIP/2.0

Via: SIP/2.0/UDP fly.cs.nthu.edu.tw:5060;branch=z9hG4bK776asdhds

Max-Forwards: 70

To: Tao ⟨sip:tao@research.telcordia.com⟩;tag=a6c85cf

From: Jyh-Cheng ⟨sip:jcchen@cs.nthu.edu.tw⟩;tag=1928301774

Call-ID: a84b4c76e66710@fly.cs.nthu.edu.tw

CSeq: 123456 ACK

Content-Length: 0

After the caller sends back the ACK message to the callee, the media stream between them may or may not go through any proxy. In general, the media stream takes a different path from the SIP signaling messages. To terminate the session a BYE message is originated. A 200 OK message is then returned to respond to the BYE message. The format of the BYE message is similar to that of the ACK message. The following is an example of BYE message assuming that user Tao originates the BYE request to Jyh-Cheng:

BYE sip:jcchen@cs.nthu.edu.tw SIP/2.0

Via: SIP/2.0/UDP tao-pc.research.telcordia.com;branch=z9hG4bKnashds10

Max-Forwards: 70

From: Tao ⟨sip:tao@research.telcordia.com⟩;tag=a6c85cf

To: Jyh-Cheng ⟨sip:jcchen@cs.nthu.edu.tw⟩;tag=1928301774

Call-ID: a84b4c76e66710@fly.cs.nthu.edu.tw

CSeq: 231 BYE

Content-Length: 0

In the above discussion, we assume that all SIP *signaling* messages are relayed by the SIP proxy server. However, the end hosts may acquire each other's address from the *Contact* header field based on the initial INVITE and 200 OK exchange. Other SIP signaling messages, therefore, may be sent to each other directly.

Figure 3.3 illustrates a sample SIP signaling flow when the SIP server in Tao's SIP service provider's network is a redirect server. In this case, upon receiving the SIP INVITE message, the redirect server finds the callee's contact information (i.e., the callee's location or another SIP server that can forward the request further to the callee) and returns it to the caller. For instance, the SIP redirect server returns a *302 Moved Temporarily* message in response to the *INVITE sip:tao@research.telcordia.com*, assuming that Tao is now with `wire.cs.nthu.edu.tw`. As depicted in Figure 3.3, the *Contact* header field of the *302 Moved Temporarily* message specifies the new URI of `sip:tao@wire.cs.nthu.edu.tw`. Upon receiving the callee's contact information from the redirect server, the caller returns an ACK message to the redirect server and then sends a SIP INVITE message to the location

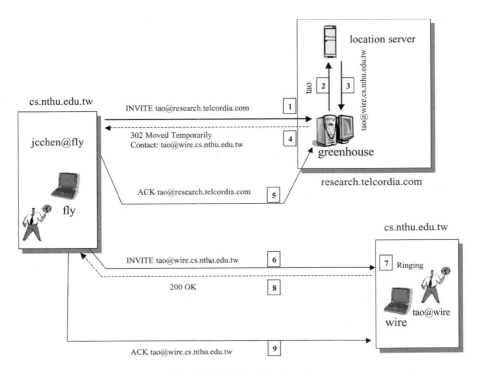

Fig. 3.3   *SIP in redirect mode*

indicated by the contact information directly. The 200 OK and ACK messages then follow to complete three-way handshake to establish the session.

3.1.2 Session Description Protocol (SDP)

The Session Description Protocol (SDP) was designed by the IETF to describe multimedia sessions. SDP conveys information of media streams so prospective participants of multimedia sessions could learn the relevant setup information. SDP is not intended for negotiation of media encodings. It simply describes the session. It does not incorporate any transport protocol. Therefore, a common usage of SDP is to embed SDP in the payload (message body) of other protocols. As discussed in Section 3.1.1.2, for instance, the payload of a SIP INVITE message may include SDP by indicating it in the *Content-Type: application/sdp.* As specified in the IETF RFC 2327 [19], SDP includes the following:

- Name and purpose of the session
- Activation time of the session
- Media comprising the session
- Information, such as address, port number, and format, to receive the media

Other information may also be included to ensure SDP conveys sufficient information. An SDP session description comprises several text lines with the format

⟨type⟩ = ⟨value⟩

The ⟨type⟩ field must be one character and is case sensitive. The ⟨value⟩ field is a structured text string. It is also case sensitive unless specified elsewhere. There should be no space in either side of the = sign. There are various types of descriptions in SDP. Instead of listing all of them in a formal format, we use the following example to illustrate the most common descriptions:

v=0
o=jcchen 2890844526 2890842807 IN IP4 140.114.79.59
s=Wiley Book
i=Discussion on book writing
c=IN IP4 224.2.17.12/127
t=2873397496 2873404696
m=audio 49170 RTP/AVP 0
m=video 51372 RTP/AVP 31
m=application 32416 udp wb

The first line starting with the letter *v* indicates that the version of SDP is zero. The second line mainly specifies the owner and the session identifier in the format of o=⟨*username*⟩ ⟨*session id*⟩ ⟨*version*⟩ ⟨*network type*⟩ ⟨*address type*⟩ ⟨*address*⟩. The example shows that the user name is *jcchen*. The usage of *session id* and *version* depends on how this line is created. The example shows that the session ID is 2890844526, which could be utilized to uniquely identify the session. The *version* is incremented when the session data is modified. If SDP is enclosed for session announcement, for example, the version could be used to detect the most recent announcement among several announcements in the same session. The *IN* means the *network type* is Internet. The rest of the line specifies that the *address type* is IPv4 with the *address* of 140.114.79.59, which is the address of the host generating the session. The address could be presented by Fully Qualified Domain Name (FQDN) as well. The next two lines starting with the letters *s* and *i* show the session name and session information, respectively. The address of 224.2.17.12 in the line starting with the letter *c* is the *connection address*. The usage of the connection address depends on applications. Typically, it is an IP multicast address. The TTL (time to live) of 127 is appended to the address by using a slash. The *t* character in the next line indicates the start time and stop time of the session in the format of NTP (Network Time Protocol) [16]. The zero stop time represents that the session is not bounded. If the start time is also zero, the session is permanent.

The next few lines starting with the letter *m* specify the *media types* and transport address. The line of *m=audio 49170 RTP/AVP 0* shows that the media type is audio, which is transported over RTP [13] through port 49170. The *AVP* indicates that the RTP is operated by using Audio/Video Profile [12]. As defined in the RTP Audio/Video Profile, payload type 0 means that the audio is a single-channel audio sampled at 8 KHz with μ-law PCM (PCMU) codec. The next line with payload type 31 stands for video in H.261 format with clock rate of 9 KHz. The line of *m=application 32416 udp wb* specifies a whiteboard (wb) application transported by UDP in port 32416.

SDP was primarily considered for multicast originally. Unlike multicast, unicast usually needs to reach an agreement for the session between two parties. For instance, the participants would need to find common codecs both of them can support. For this purpose, RFC 3264 [25] specifies an *Offer/Answer Model* with SDP. Two participants, therefore, can arrive at a common view for the multimedia session between them. In this model, one party utilizes SDP to generate an *offer* message, which comprises information such as codecs, IP address, and port number the offerer would like to use. The other party then *answers* with another SDP message, which consists of matching streams for each stream in the offer message to indicate whether the stream is accepted. Either one of the participants may generate a new offer message to update the session. If the offer is rejected, the session returns to the prior status. The Offer/Answer Model of SDP is mandatory for SIP.

For the example of session establishment discussed in Section 3.1.1.4, the message body of the SIP INVITE from Jyh-Cheng may look like the following:

v=0
o=Jyh-Cheng 2890844526 2890844526 IN IP4 fly.cs.nthu.edu.tw
s=

c=IN IP4 fly.cs.nthu.edu.tw
t=0 0
m=audio 49170 RTP/AVP 0
m=video 51372 RTP/AVP 31
m=application 32416 udp wb

Jyh-Cheng specifies an audio stream, a video stream, and a whiteboard application for the multimedia session. Assuming that Tao, the recipient, does not want to send and receive the video of type 31 (H.261 video with clock rate of 9 KHz), the payload of the 200 OK message may look like this:

v=0
o=Tao 2890844730 2890844730 IN IP4 tao-pc.research.telcordia.com
s=
c=IN IP4 tao-pc.research.telcordia.com
t=0 0
m=audio 49920 RTP/AVP 0
m=video **0** RTP/AVP 31
m=application 32416 udp wb

The rejection is indicated by a zero port number in the line of video media. Later on if Tao decides to change the media stream, he could send an SDP offer to Jyh-Cheng to request the modification.

3.2 3GPP IP MULTIMEDIA SUBSYSTEM (IMS)

This section describes the IP Multimedia Subsystem (IMS) architecture defined by 3GPP, services provided by the IMS, and how users access these services.

3.2.1 IMS Architecture

The service requirements of the 3GPP IMS are specified in 3GPP TS 22.228 [8]. The technical details are described in 3GPP TS 23.228 [10]. The IMS provides all the network entities and procedures to support real-time voice and multimedia IP applications. It uses SIP to support signaling and session control for real-time services. Figure 3.4 illustrates the IMS functional architecture [11]. The main functional entity in an IMS is the Call State Control Function (CSCF). A CSCF is a SIP server. Depending on the specific tasks performed by a CSCF, CSCFs can be divided into three different types:

- *Serving CSCF (S-CSCF),*
- *Proxy CSCF (P-CSCF),*

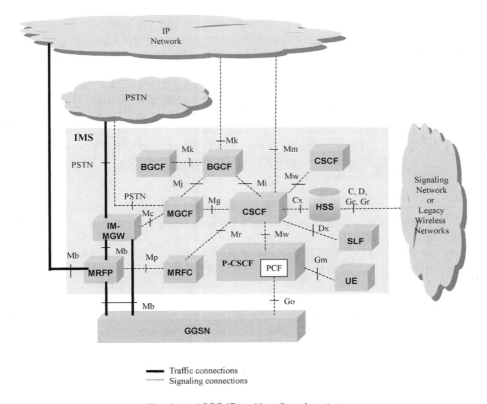

Fig. 3.4   *3GPP IP multimedia subsystem*

- *Interrogating CSCF (I-CSCF).*

An S-CSCF provides session control services for a user. It maintains session states for a registered user's on-going sessions and performs the following main tasks [10]:

- *Registration*: An S-CSCF can act as a SIP Registrar [28] to accept users' SIP registration requests and make users' registration and location information available to location servers such as the HSS (Home Subscriber Server).
- *Session Control*: An S-CSCF can perform SIP session control functions for a registered user. Relay SIP requests and responses between calling and called parties.
- *Proxy Server*: An S-CSCF may act as a SIP Proxy Server that relays SIP messages between users and other CSCFs or SIP servers.
- *Interactions with Application Servers*: An S-CSCF acts as the interface to application servers and other IP or legacy service platforms.

- _Other functions_: An S-CSCF performs a range of other functions not mentioned above. For example, it provides service-related event notifications to users and generates Call Detail Records (CDRs) needed for accounting and billing.

A P-CSCF is a mobile's first contact point inside a local (or visited) IMS. It acts as a _SIP Proxy Server_ [27]. In other words, the P-CSCF accepts SIP requests from the mobiles and then either serves these requests internally or forwards them to other servers. The P-CSCF includes a Policy Control Function (PCF) that controls the policy regarding how bearers in the GGSN should be used. The P-CSCF performs the following specific functions:

- Forward SIP REGISTER request from a mobile to the mobile's home network. If an I-CSCF is used in the mobile's home network, the P-CSCF will forward the SIP REGISTER request to the I-CSCF. Otherwise, the P-CSCF will forward the SIP REGISTER request to an S-CSCF in the mobile's home network. The P-CSCF determines where a SIP REGISTER request should be forwarded based on the home domain name in the SIP REGISTER Request received from the mobile.
- Forward other SIP messages from a mobile to a SIP server (e.g., the mobile's S-CSCF in the mobile's home network). The P-CSCF determines to which SIP server the messages should be forwarded based on the result of the SIP registration process.
- Forward SIP messages from the network to a mobile.
- Perform necessary modifications to the SIP requests before forwarding them to other network entities.
- Maintain a security association with the mobile.
- Detect emergency session.
- Create CDRs.

An I-CSCF is an optional function that can be used to hide an operator network's internal structure from an external network when an I-CSCF is used. It serves as a central contact point within an operator's network for _all_ sessions destined to a subscriber of that network or a roaming user currently visiting that network. Its main function is to select an S-CSCF for a user's session, route SIP requests to the selected S-CSCF, and generate CDRs. The I-CSCF selects an S-CSCF based primarily on the following information: (1) capabilities required by the user, (2) capabilities and availability of the S-CSCFs, and (3) topological information, such as the location of an S-CSCF and the location of the users' P-CSCFs if they are in the same operator's network as the S-CSCFs.

The Media Gateway Control Function (MGCF) and the IM Media Gateway (IM-MGW) are responsible for signaling and media interworking, respectively, between the PS domain and circuit-switched networks (e.g., PSTN). The Multimedia Resource Function Processor (MRFP) controls the bearer on the M_b interface, including processing the media streams (e.g., audio transcoding). The Multimedia Resource Function Controller (MRFC) interprets signaling information from an S-CSCF or a SIP-based Application Server and controls the media streams resources in the MRFP accordingly. The MRFC is also responsible for generating the CDRs. The Breakout Gateway Control Function (BGCF) selects to which PSTN network a session should be forwarded. It will then be responsible for forwarding the session signaling to the appropriate MGCF and BGCF in the destination PSTN network.

3.2.2   Mobile Station Addressing for Accessing the IMS

In order for a mobile user to use the services provided by a visited IMS, the mobile needs to have an IP address (i.e., the mobile's PDP address) that is logically part of the IP addressing domain of the visited IMS, as illustrated in Figure 3.5. A PDP context will be activated for this address so that the packets addressed to this IP address can be forwarded by the 3GPP packet domain to the mobile.

The IP addressing spaces used by the IMS and in a RAN may be disjoint. This, for example, is likely the case when the RAN is a wireless LAN. In such a scenario, IP packets addressed to an IP address in the IMS IP addressing space may not be routed to the mobile via regular IP routing and forwarding mechanisms.

3.2.3   Reference Interfaces

The main interfaces in the IMS (as shown in Figure 3.4) can be grouped into the following categories:

- *Interfaces for SIP-based signaling and service control*: These include interfaces M_g, M_i, M_j, M_k, M_r, and M_w, which all use SIP as the signaling protocol.
  - Interface M_g allows CSCF to interact with MGCF.
  - Interface M_i allows a CSCF to forward session signaling to a BGCF so that the session can be forwarded to PSTN networks.
  - Interface M_j allows a BGCF to forward a session signaling to a selected MGCF that will carry the session to the PSTN.
  - Interface M_k allows a BGCF to forward session signaling to another BGCF.
  - Interface M_r allows an S-CSCF to interact with an MRFC.
  - Interface M_w allows an I-CSCF to direct mobile-terminated sessions to an S-CSCF.
- *Interfaces for controlling media gateways*: These include interfaces M_c and M_p.
  - Interface M_c allows a signaling gateway to control a media gateway. For example, it is used between an MGCF and an IM-MGW, between an MSC Server and a CS-MGW, or between a GMSC Server and a CS-MGW.

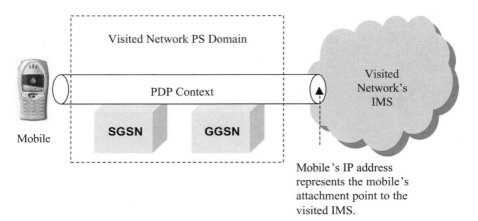

Fig. 3.5 *Mobile station addressing for accessing IMS services*

– Interface M_p allows an MRFC to control media stream resources provided by an MRFP. Signaling over interfaces M_c and M_p uses the H.248 Gateway Control Protocol [20].

- *Interfaces with the Information Servers*: Interface C_x between the CSCF and the HSS allows the CSCF to retrieve from the HSS mobility and routing information regarding a mobile user so that the CSCF can determine how to process a user's sessions.
- *Interfaces with external networks*: These include interfaces M_b, M_m, and G_o.
  – *Interface M_b* is the standard IP routing and transport interface with external IP networks. The interface M_b may be identical to the G_i interface.
  – *Interface M_m* is a standard IP-based signaling interface that handles signaling interworking between the IMS and external IP networks.
  – *Interface G_o* allows a PCF to apply policy control over the bearer usage in the GGSN.

3.2.4 Service Architecture

3GPP specifications assume that each mobile has a home network and may subscribe to a set of services with its home network (*Home Subscribed Services*). The current 3GPP specifications require that a mobile's home network provides service control for the mobile's Home Subscribed Services even when the mobile is currently in a visited network. Specifically, a mobile's S-CSCF will always be an S-CSCF in the mobile's home network. This requirement leads to two basic service IMS architectures to provide real-time services as illustrated in Figure 3.6. The two approaches differ in whether the service platform belongs to the mobile's home network or a service provider external to the mobile's home network. A service

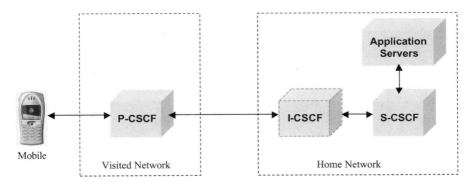

(a) Service platform in mobile's home network

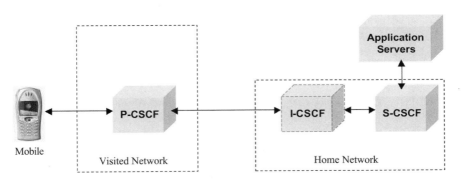

(b) External service platform

Fig. 3.6 *3GPP service architectures for real-time services*

platform provides service control for real-time services. A service platform may be a set of SIP-based application servers or a legacy service platform, e.g., the CAMEL (Customized Applications for Mobile Enhanced Logic) server. The I-CSCF in Figure 3.6 is shown as a box with dotted lines to indicate that the I-CSCF is optional.

With both service architectures, the initial SIP request from a mobile travels from the originating mobile to the visited P-CSCF first, which then forwards the request to the I-CSCF (if used) in the originating mobile's home network. This I-CSCF selects an S-CSCF in the home network for this user session and forwards the SIP request to the selected S-CSCF. From this point on, future SIP messages for this mobile's session will travel directly between the visited P-CSCF and the S-CSCF in the mobile's home network.

The S-CSCF is responsible for interfacing with internal and external service platforms as illustrated in Figure 3.7. There are three types of standardized platforms: (1) SIP application server, (2) Open Service Access (OSA) Service Capability Server (SCS), and (3) IP Multimedia Service Switching Function

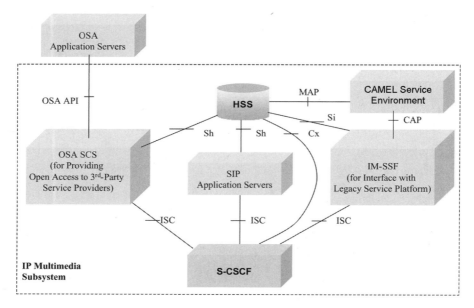

Fig. 3.7 *Interactions between S-CSCF and service platforms*

(IM-SSF). The services offered by them are value-added services (VAS) or operator-specific services. The S-CSCF uses the same interface, *IMS Service Control (ISC) interface*, to interface with all service platforms. The signaling protocol over the ISC interface is SIP. The OSA SCS and IM-SSF by themselves are not application servers. Instead, they are gateways to other service environments. As depicted in Figure 3.7, the OSA SCS and IM-SSF interface to the *OSA application server* and *CAMEL Service Environment (CSE)*, respectively. From the perspective of the S-CSCF, however, they all exhibit the same ISC interface behavior. The services are briefly described:

- *SIP Application Server:* In addition to session control, a SIP server can also provide various value-added services. A lightweight SIP-based server enables the CSCF to utilize the SIP-based services and interact with the SIP application servers without additional components.

- *CAMEL Service Environment (CSE):* The CSE provides legacy Intelligent Network (IN) services. It allows operators leverage existing infrastructure for IMS services. As specified earlier, the CSCF interacts with CSE through IM-SSF. The IM-SSF hosts the CAMEL features and interfaces with CSE by CAP (CAMEL Application Part) [6].

- *OSA Application Server:* Applications may be developed by a third party that is not the owner of the network infrastructure. The OSA application server framework provides a standardized way for a third party to secure access to the IMS. The OSA reference architecture defines an *OSA Application Server* as the service execution environment for third-party applications. The OSA application server then interfaces with the CSCF through the OSA SCS by OSA API (Application Programming Interface) [5].

3.2.5 Registration with the IMS

To access the IMS in a visited network, a mobile needs to go through the following steps:

- *Local P-CSCF Discovery*: The mobile needs to discover the IP address of a local P-CSCF in the visited IMS because the local P-CSCF will be the mobile's first contact point in the visited IMS and the mobile's proxy for registering with the IMS.
- *Registration with IMS*: The mobile needs to perform SIP registration with the visited IMS and the mobile's home IMS (which can be different from the visited IMS).

Local (or visited) P-CSCF can be discovered in one of the following two ways:

- Mobile obtains the IP address of a local P-CSCF from the visited GGSN as part of the PDP Context Activation process.
- Mobile uses DHCP [18] to discover the IP address of a local P-CSCF *after* the PDP context is activated.

If the mobile wishes to acquire the IP address of a local P-CSCF as part of the PDP Context Activation process, the mobile will indicate such a request in its Activate PDP Context Request message (Section 2.1.4 in Chapter 2). This indication will be forwarded transparently by the SGSN to the GGSN. The GGSN will acquire the IP addresses of the available P-CSCFs and include them in the Create PDP Context Response message sent back to the SGSN, which will in turn forward these addresses transparently to the mobile.

If the mobile wishes to use DHCP to discover the local P-CSCF, it first activates a PDP context with a null PDP address. As illustrated in Figure 3.8, the mobile will then query a DHCP server on the visited network to obtain the domain names of local P-CSCFs and the IP address of an IP Domain Name System (DNS) server that is capable of resolving a P-CSCF's domain name to its IP address.

The mobile does not need to know the IP address of the DHCP server before sending queries to the DHCP server. Instead, the mobile will address its initial DHCP queries to the local broadcast IP address. Once the mobile knows the domain name of a P-CSCF, it can query the DNS to retrieve the IP address of the P-CSCF.

After a mobile has discovered the local P-CSCF in the visited IMS, it can initiate the SIP registration process. The mobile's home network determines whether the mobile is authorized to use a visited IMS based on the mobile's service subscription

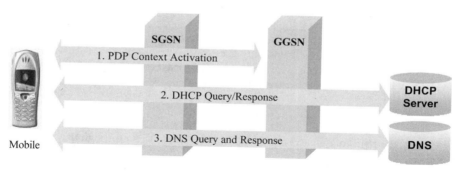

Fig. 3.8   *3GPP Local P-CSCF discovery*

and the agreement between the home network and the visited network. Therefore, a mobile's access to a visited IMS can only be authorized after the mobile successfully registers with its home network.

Figure 3.9 shows the signaling message flow for a mobile to perform IMS registration while the mobile is inside a visited network. The mobile first sends a SIP REGISTER request to the visited P-CSCF as specified by *Flow 1*. The SIP REGISTER request will carry two important pieces of information: the mobile's identity and the mobile's home network domain name.

The visited P-CSCF examines the home network domain name contained in the SIP REGISTER request received from the mobile to determine the entry point to the user's home network, e.g., an I-CSCF or an S-CSCF. As indicated by *Flow 2*, the visited P-CSCF will then forward the SIP REGISTER request to the entry point

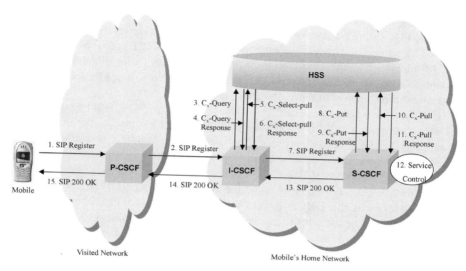

Fig. 3.9   *3GPP IP Multimedia Subsystem registration procedure*

(assuming I-CSCF) in the user's home network with added information regarding the name and address of the P-CSCF plus the identifier of the visited network.

As shown by *Flow 3*, upon receiving the SIP REGISTER request from the visited network, the I-CSCF in the mobile's home network will send the mobile's IMSI and the identifier of the visited network to the HSS in the home network in a Cx-Query message (over the Cx interface) to ask the HSS to check whether the user is authorized to use the visited IMS. The HSS returns the results to the I-CSCF in a Cx-Query Response message (*Flow 4*).

If the user is authorized to use the visited IMS and the HSS knows the name of the S-CSCF in the mobile's home network that can be used to serve the user (e.g., when the user has already registered with the home network before), the HSS will return to the I-CSCF the domain name of the S-CSCF for the user. The I-CSCF resolves the domain name of the S-CSCF to the IP address of the S-CSCF through using the DNS. The I-CSCF will then forward the user's SIP REGISTER request to the selected S-CSCF (*Flow 7*).

If the user is authorized to use the visited IMS and the HSS does not know any S-CSCF assigned to the mobile (e.g., when the user is performing registration for the first time), the I-CSCF will further query the HSS to obtain information regarding the S-CSCF capabilities required to serve the mobile as indicated by *Flow 5*. A response is returned from the HSS to the I-CSCF as shown in *Flow 6*. Then, the I-CSCF will use such information to select an S-CSCF and forward the SIP REGISTER request to the selected S-CSCF (*Flow 7*).

Upon receiving the SIP REGISTER request, the S-CSCF will send its own name together with the mobile's identities to the HSS in a Cx-Put message (over the Cx interface) as illustrated in *Flow 8*. Upon receiving a positive Cx-Put Response (*Flow 9*) from the HSS, the S-CSCF will send a Cx-Pull message (*Flow 10*) to the HSS to retrieve the mobile's service subscription information. The HSS will respond with a Cx-Pull Response message (*Flow 11*) containing the requested service subscription information. The S-CSCF caches the service subscription information and uses it to control the user's application sessions (*Flow 12*). After executing the required service control logic, the S-CSCF will respond to the SIP REGISTER request from the I-CSCF with a SIP 200 OK message (*Flow 13*). As specified by *Flow 14*, the 200 OK message is further forwarded to the P-CSCF in the visited network to indicate successful SIP registration.

Upon receiving the SIP 200 OK message from the mobile's home network, the visited P-CSCF will record the home network contact information (e.g., the address of the S-CSCF in the mobile's home network) contained in the message. The P-CSCF will then forward the SIP 200 OK message to the mobile as indicated by *Flow 15*, completing the SIP registration process.

After a successful SIP registration, the S-CSCF in the mobile's home network may simply forward subsequent SIP messages destined to the mobile directly to the P-CSCF in the visited network, i.e., without going through the I-CSCF in the mobile's home IMS. The P-CSCF may also forward subsequent SIP messages, which are not SIP REGISTER requests, to the S-CSCF in the mobile's home network directly.

The registration process discussed above is *application-level* registration for the IMS. It must be performed after the *access-level* registration is performed and the IP connectivity has been established.

3.2.6 Deregistration with the IMS

The deregistration from IMS may be *mobile initiated* or *network initiated.* The network-initiated deregistration may also be initiated by *registration timeout* or by a *network administrative function* such as HSS or S-CSCF.

Figure 3.10 depicts the signaling flows of a mobile-initiated deregistration. The process is similar to the registration procedure explained in Section 3.2.5, except that the expiration time is zero in the SIP REGISTER in *Flows 1, 2,* and *5.* When a mobile user wishes to deregister, it simply sends a new SIP REGISTER request. Because the user has registered already, the HSS indicates which S-CSCF should be contacted by *Flow 4* in response to the query from the I-CSCF (*Flow 3*). Because the expiration time is zero, the S-CSCF will perform necessary service control functions to deregister and remove all subscription information regarding this user as indicated in *Flow 6.* The S-CSCF then sends its own name together with the mobile's identities to the HSS in a Cx-Put message as illustrated in *Flow 7.* Based on the operator's choice, the Cx-Put may either ask the HSS to clear or keep the S-CSCF name for the specific user. In either case, the HSS indicates that the user is unregistered. Once the S-CSCF receives the Cx-Put Response, it clears all registration information of the user and triggers a 200 OK message to the I-CSCF. Similarly, the P-CSCF releases all registration information regarding the user upon receiving the 200 OK from the I-CSCF.

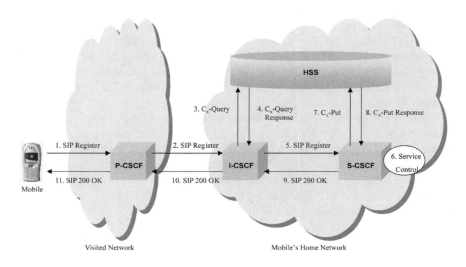

Fig. 3.10 *Mobile-initiated deregistration*

The network-initiated deregistration due to registration timeout is illustrated in Figure 3.11. Once the initial registration is completed, both P-CSCF and S-CSCF keep a same timer for the session. It assumes that the timers in P-CSCF and S-CSCF are close enough so they do not need to synchronize with each other. The timer is refreshed each time the user re-registers. When the timer expires, both P-CSCF and S-CSCF release all registration information of the user. *Flows 2*, *3*, and *4* in Figure 3.11 perform the same functionalities as *Flows 6*, *7*, and *8* in Figure 3.10.

In addition to registration timeout, the network could also deregister an application session initiated by an administrative function. This may happen, for example, when the network detects that a mobile is malfunctioning. Figure 3.12 depicts the deregistration flows initiated by HSS. From registrations, the HSS has learned the name of the S-CSCF that served the user. The HSS sends the Cx-Deregister message, which may include a reason, to the S-CSCF to initiate the deregistration. The S-CSCF then performs necessary service control functions to deregister and remove all subscription information of the user. After that, the S-CSCF removes the user and sends a Deregister message to the P-CSCF. The reason for deregistration received from the HSS, if any, is included in *Flow 3*. The P-CSCF clears the user information and informs the mobile user by *Flow 4* for the deregistration with the reason received. In parallel, the P-CSCF sends a 200 OK message back to the S-CSCF (*Flow 5*). Because the mobile may be out of radio coverage or may not function properly, the mobile may not receive the Inform message of *Flow 4* and may not issue *Flow 6* of 200 OK. Figure 3.12 indicates that the P-CSCF performs deregistration and issues 200 OK (*Flow 5*) to the S-CSCF

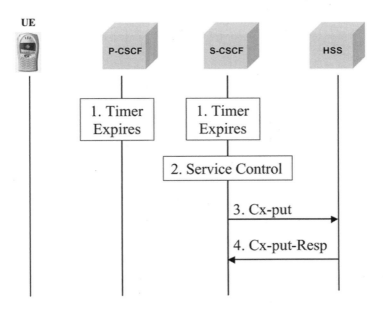

Fig. 3.11 *Network-initiated deregistration by registration timeout*

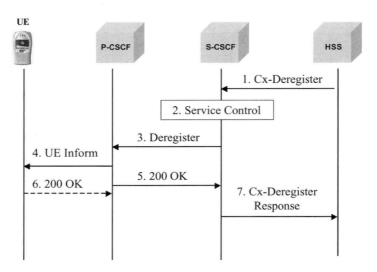

Fig. 3.12 *Network-initiated deregistration by HSS*

regardless of whether an acknowledge of *Flow 6* is received from the mobile. Upon receiving the 200 OK message, the S-CSCF sends a Cx-Deregister Response message (*Flow 7*) back to the HSS. Figure 3.13 shows the deregistration initiated by S-CSCF. Compared with Figure 3.12, here, it is the S-CSCF that informs the HSS with *Flow 6* regarding the deregistration.

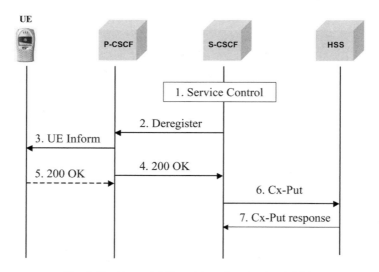

Fig. 3.13 *Network-initiated deregistration by S-CSCF*

3.2.7   End-to-End Signaling Flows for Session Control

The end-to-end signaling session control between two mobile users can be divided into the following three subflows as illustrated in Figure 3.14:

- Mobile origination flow
- Mobile termination flow
- S-CSCF to S-CSCF signaling flow

The mobile origination flow specifies the signaling path between the mobile originating the session and the S-CSCF chosen to serve the mobile. The mobile station could initiate the session in home network or visited network. Similarly, the mobile termination flow defines the procedures between the destination mobile and the S-CSCF assigned for the mobile. The destination mobile may be in its home network or in a visited network. The S-CSCF to S-CSCF signaling flow specifies the signaling between the S-CSCF in the originating home network and the S-CSCF in the destination home network. The session origination and termination may be performed by the same network operator or different network operators.

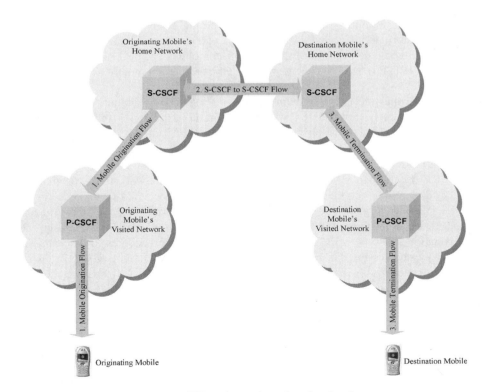

Fig. 3.14   *3GPP end-to-end session signaling flow*

Assuming that neither the originating nor the destination mobile user is in its own home network, in Figure 3.14, the mobile origination flow specified as *Flow 1* comprises the signaling flow between the originating mobile user and the P-CSCF in the visited network, and the signaling flow between the P-CSCF and the S-CSCF in the originating mobile's home network. The S-CSCF to S-CSCF signaling flow is the signaling flow between the originating mobile user's S-CSCF and the destination mobile user's S-CSCF identified as *Flow 2*. The mobile termination signaling flow comprises the signaling flow between the destination mobile user and the P-CSCF in the visited network, and the signaling flow between the P-CSCF and the S-CSCF in the destination mobile's home network. It is specified as *Flow 3* in Figure 3.14. Details of each flow are illustrated in Figures 3.15–3.17.

Figure 3.15 illustrates the mobile origination flow. The originating mobile sends a SIP INVITE request, which contains an initial SDP, to the P-CSCF in the visited network (*Flow 1*). The P-CSCF then forwards the INVITE request to the originating mobile's home S-CSCF (*Flow 2a*). Alternatively, the P-CSCF could forward the INVITE request to the originating mobile's home I-CSCF, which then further forwards the request to the S-CSCF if the home network would like to hide its network configuration (*Flows 2b1 & 2b2*). Upon receiving the SIP INVITE, the originating mobile's S-CSCF will perform any service control needed for the mobile based on the mobile's service subscription profile (*Flow 3*). The S-CSCF will use the information carried in the address part of the SIP INVITE request to determine the home network of the destination mobile. The S-CSCF then forwards the SIP INVITE request to the entry point in the destination mobile's home network (*Flow 4*).

Assume that the entry point in the destination mobile's home network is an I-CSCF, which is shown as I-CSCF #2 in Figure 3.16. *Flow 4* in Figure 3.15 is represented by *Flow 3*, which comprises *Flows 3a, 3b1*, and *3b2*, in Figure 3.16. The INVITE request could be sent to the I-CSCF #2 directly from the S-CSCF (referred to as S-CSCF #1) in the originating network. Similarly, it could be delivered to the I-CSCF (referred to as I-CSCF #1) in the originating network if the originating network would like to hide the internal configuration. Once the I-CSCF #2 receives the INVITE request, it may query the HSS for the location of the S-CSCF (*Flows 4 and 5*). Upon receiving the SIP INVITE request of *Flow 6*, the S-CSCF #2 in the destination home network will perform authentication procedures to determine if the requested session can be connected to the destination mobile (*Flow 7*). If the authentication result is positive, in *Flow 8*, the S-CSCF #2 will forward the SIP INVITE request to the P-CSCF serving the destination mobile either directly or by going through I-CSCF #2. The S-CSCF #2 learned the address of the P-CSCF in the destination mobile's visited network during the IMS registration performed by the destination mobile.

Flow 8 in Figure 3.16 is represented by *Flow 3*, which comprises *Flows 3a, 3b1*, and *3b2*, in Figure 3.17. As shown in Figure 3.17, the INVITE request is either sent from the S-CSCF #2 directly or relayed by the I-CSCF #2 to the P-CSCF in the destination mobile's visited network. The P-CSCF then forwards the SIP INVITE

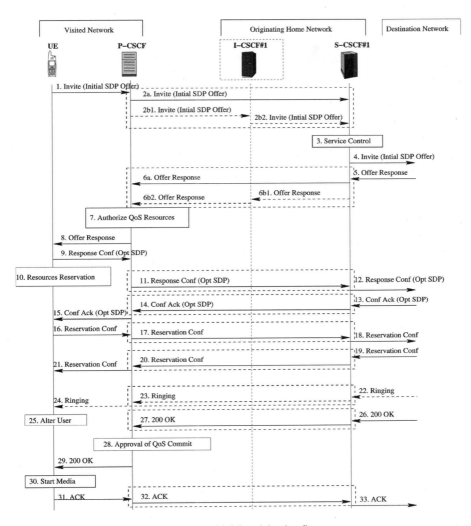

Fig. 3.15 *Mobile origination flow*

request to the destination mobile (*Flow 4*). This completes the mobile termination signaling flow for the SIP INVITE request.

Upon receiving the INVITE request, the destination mobile responds with an *Offer Response*, which may come back from the destination mobile to the originating mobile along the same path traveled by the INVITE request. The Offer Response contains the SDP determined by the destination mobile. Once the originating mobile receives the Offer Response, it determines the offered set of media streams and issues a *Response Confirmation* (Response Conf). The Response Confirmation may also contain an SDP, which may be the same as or different from

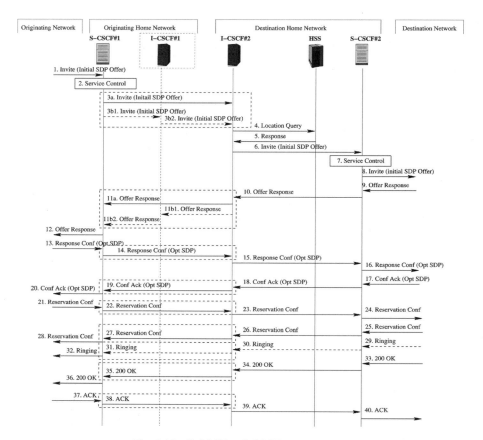

Fig. 3.16 *S-CSCF to S-CSCF signaling flow*

the SDP received from the destination mobile. Finally the destination mobile responds to the confirmation with an acknowledgment (Conf ACK). After that, the mobile issues a *Reservation Confirmation* (Reservation Conf) message, which completes the resource reservation. Please refer to Chapter 6 for issues related to QoS (Quality of Service) and resource reservation indicated in Figures 3.15–3.17. The destination mobile is alerted by the incoming request by a ringing indication. The originating mobile may also hear the ring-back tone and is alerted that the destination is ringing. Once the destination mobile user answers the request, a SIP 200 OK message is returned to the originating user, which then responds with a SIP ACK message.

In Figures 3.15–3.17, the dotted-rectangular represents two optional flows. That is, the flow may be relayed by the I-CSCF or it may bypass the I-CSCF. Although we explain the end-to-end session setup by Figures 3.15–3.17 altogether, each of the figures is independent and can be combined with other procedures. For instance, the S-CSCF to S-CSCF signaling flow of Figures 3.16 could be used with other

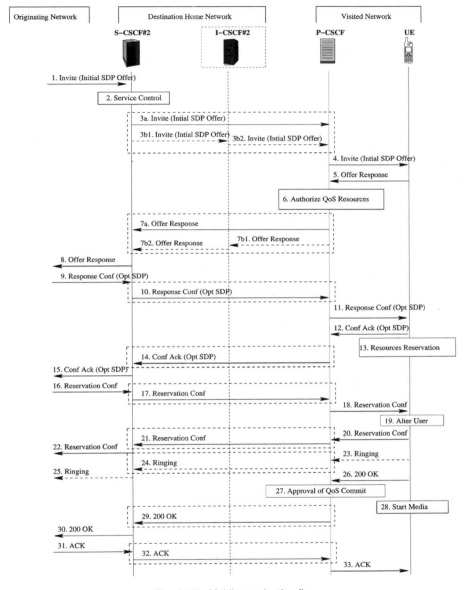

Fig. 3.17 *Mobile termination flow*

mobile origination flows and mobile termination flows, in which the originating or destination mobile user is in its own home network. It is also possible that the flow is initiated or terminated from PSTN or PLMN.

The release of a session is normally initiated by a mobile. During the session release process, the related bearers are deleted and necessary billing information is collected by the network. The session release may also be initiated by the network due to loss of radio connection, loss of IP bearer, operator intervention, etc. Figure 3.18 depicts a normal termination of SIP session initiated by mobile user. Once a mobile user hangs up, the mobile station generates a SIP BYE message, which is delivered all the way to the other mobile party via various components such as P-CSCF, and S-CSCF. Related resources are removed, and service control logics are executed when receiving the SIP BYE message. Once the destination mobile receives the BYE message, it responds with a 200 OK message back to the originating mobile. Although network-initiated session termination is not discussed in this section, it is similar to mobile-initiated session termination. The major difference is who initiates the SIP BYE message.

3.3 3GPP2 IP MULTIMEDIA SUBSYSTEM (IMS)

3GPP2 is currently defining an IP Multimedia Subsystem (IMS), which is a subsystem of 3GPP2 IP Multimedia Domain (MMD). Most of the 3GPP2 IMS specifications are still in draft status. Many of them are based on the specifications of 3GPP IMS. Because the signaling flows in 3GPP2 IMS essentially are the same as those in 3GPP IMS, this section only reviews the 3GPP2 IMS architecture.

The system requirements of 3GPP2 MMD are specified in [3]. The MMD system consists of a mobile station, radio access network, and core network. It provides end-to-end IP connectivity, services, and features through the core network to subscribers. To enable independent evolution of core network and radio access network, the core network should be able to connect with various types of radio access networks by standard protocols. The MMD system is backward compatible with the legacy packet system specified in 3GPP2 P.S0001 [1], although the MMD system could be built without the legacy packet system.

Figure 3.19 depicts the MMD core network architecture that is capable of providing Packet Data Subsystem (PDS) and IMS [4]. The collection of components that supports general packet data services is called PDS. The IMS comprises the entities that provide multimedia session capabilities. The IMS is further illustrated in Figure 3.20. Most components and interfaces in Figure 3.20 are basically the same as those defined in 3GPP IMS. In Figure 3.20, the combination of the AAA and various databases provides the functionality of the HSS (Home Subscriber Server). Please be advised that some network entities shown in Figure 3.19 are common to both PDS and IMS.

Figure 3.21 illustrates the service control of 3GPP2 MMD. Same as that in 3GPP, the CSCF (Call Session Control Function) is a SIP server and the ISC (IMS Service Control) interface is based on SIP. Compared with Figure 3.7 in 3GPP, the Application Server A and the Application Server B in Figure 3.21 essentially are the OSA Application Server and the SIP Application Server in Figure 3.7, respectively. There is no CAMEL service in 3GPP2. The Application Server C in Figure 3.21 represents generic applications that only utilize bearer resources. The Position Server could provide geographic position information. The combination of

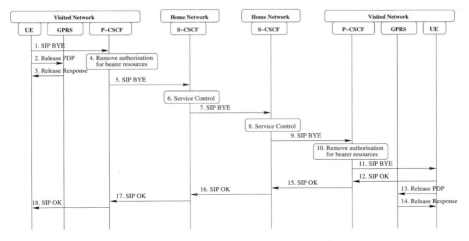

Fig. 3.18   *Release flow: mobile initiated*

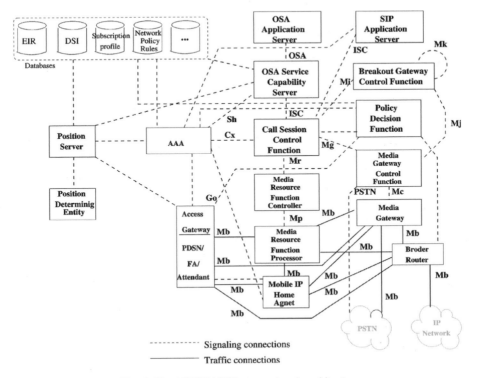

Fig. 3.19   *3GPP2 MMD core network architecture*

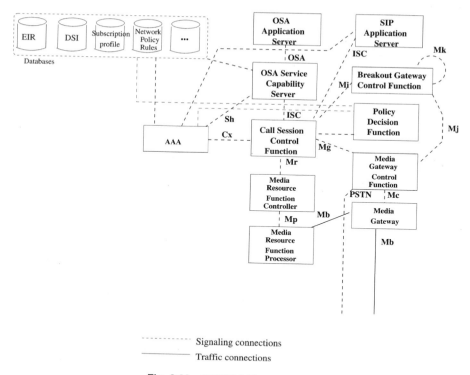

Fig. 3.20 3GPP2 IMS architecture

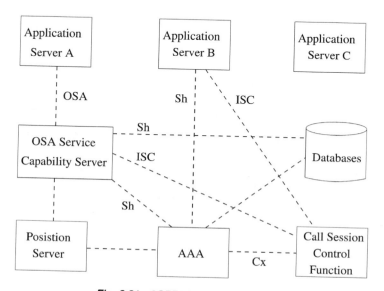

Fig. 3.21 3GPP2 IMS service platforms

the position server with the AAA and OSA service capability server is responsible for ensuring proper authorization for access request from every position. Details were still under development by 3GPP2 at the time this book was completed.

Figure 3.22 shows a functional architecture providing SIP-based multimedia services. A mobile station should perform SIP registration with the P-CSCF in the visited network before it can access any service provided by the IMS. The necessary authentication would be carried out by the local AAA server. Once authorized by the visited network, the mobile station could further connect to the S-CSCF in the home network. Same as that in 3GPP, the home network may leverage the I-CSCF to hide internal configuration. Session control such as registration, initiation, and termination is based on SIP. Details of signaling flows are presented in 3GPP2 X.P0013.2 [2], which is based on 3GPP TS 23.228 [10]. This section does not discuss the detailed flows because most of them are the same as those presented in Sections 3.2.5–3.2.7.

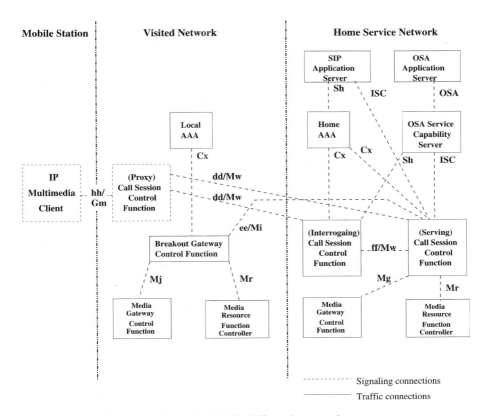

Fig. 3.22   *3GPP2 IMS service control*

To conclude this chapter, we point out that 3GPP and 3GPP2 are harmonizing their IMSs. In addition, 3GPP is integrating WLAN as well [7], [9]. It is expected that a common IMS would work over cdma2000, WCDMA, and WLAN.

REFERENCES

1. 3rd Generation Partnership Project 2 (3GPP2). Wireless IP network standard. 3GPP2 P.S0001, Version 1.0, December 1999.

2. 3rd Generation Partnership Project 2 (3GPP2). All-IP multi-media domain; IP multimedia subsystem; stage 2. 3GPP2 X.P0013.2, Version 1.6.0, March 2003.

3. 3rd Generation Partnership Project 2 (3GPP2). IP Multimedia Domain; System Requirements. 3GPP2 S.P0058, Version 0.5.4, February 2003.

4. 3rd Generation Partnership Project 2 (3GPP2). IP network for cdma2000 spread spectrum systems; 3GPP2 all-IP core network; enhancements for multimedia domain (MMD); overview (part-00). 3GPP2 X.P0013.0, Revision 0.22, March 2003.

5. 3rd Generation Partnership Project (3GPP), Technical Specification Group Core Network. Application programming interface (API); part 1: overview, release 5. 3GPP TS 29.198-1, Version 5.1.1, March 2003.

6. 3rd Generation Partnership Project (3GPP), Technical Specification Group Core Network. CAMEL application part (CAP) specification, release 5. 3GPP TS 29.078, Version 5.3.0, March 2003.

7. 3rd Generation Partnership Project (3GPP), Technical Specification Group Services and System Aspect. Feasibility study on 3GPP system to wireless local area network (WLAN) interworking, release 6. 3GPP TR 22.934, Version 6.1.0, December 2002.

8. 3rd Generation Partnership Project (3GPP), Technical Specification Group Services and System Aspect. Service requirements for the IP multimedia core network subsystem, release 5. 3GPP TS 22.228, Version 5.6.0, June 2002.

9. 3rd Generation Partnership Project (3GPP), Technical Specification Group, Services and System Aspects. 3GPP system to wireless local area network (WLAN) interworking; functional and architectural definition. 3GPP TR 23.934,Version 1.0.0, August 2002.

10. 3rd Generation Partnership Project (3GPP), Technical Specification Group, Services and System Aspects. IP multimedia subsystem (IMS) stage 2, release 5. 3GPP TS 23.228, Version 5.7.0, December 2002.

11. 3rd Generation Partnership Project (3GPP), Technical Specification Group, Services and System Aspects. Network architecture, release 5. 3GPP TS 23.002, Version 5.7.0, June 2002.

12. H. Schulzrinne. RTP profile for audio and video conferences with minimal control. IETF RFC 1890, January 1996.

13. H. Schulzrinne, S. Casner, R. Frederick, and V. Jacobson. RTP: a transport protocol for real-time applications. IETF RFC 1889, January 1996.

14. T. Berners-Lee, R. Fielding, and L. Masinter. Uniform resource identifiers (URI): generic syntax. IETF RFC 2396, August 1998.

15. B. Campbell, J. Rosenberg, H. Schulzrinne, C. Huitema, and D. Gurle. Session initiation protocol (SIP) extension for instant messaging. IETF RFC 3428, December 2002.

16. D.L. Mills. Network time protocol (version 3): specification, implementation and analysis. IETF RFC 1305, March 1992.

17. T. Dierks and C. Allen. The TLS protocol. IETF RFC 2246, January 1999.

18. R. Droms. Dynamic host configuration protocol. IETF RFC 2131, March 1997.

19. M. Handley and V. Jacobson. SDP: session description protocol. IETF RFC 2327, April 1998.

20. ITU-T Rec. H.248. Gateway control protocol, June 2000.

21. ITU-T Rec. H.323. Packet-based multimedia communications systems, November 2000.

22. R. Stewart, Q. Xie, K. Morneault, C. Sharp, H. Schwarzbauer, T. Taylor, I. Rytina, M. Kalla, L. Zhang, and V. Paxson. Stream control transmission protocol. IETF RFC 2960, October 2000.

23. A.B. Roach. Session initiation protocol (SIP)—specific event notification. IETF RFC 3265, June 2002.

24. J. Rosenberg. The session initiation protocol (SIP) UPDATE method. IETF RFC 3311, September 2002.

25. J. Rosenberg and H. Schulzrinne. An offer/answer model with the session description protocol (SDP). IETF RFC 3264, June 2002.

26. J. Rosenberg and H. Schulzrinne. Reliability of provisional responses in the session initiation protocol (SIP). IETF RFC 3262, June 2002.

27. J. Rosenberg, H. Schulzrinne, G. Camarillo, A. Johnston, J. Peterson, R. Sparks, M. Handley, and E. Schooler. SIP: session initiation protocol. IETF RFC 3261, June 2002.

28. J. Rosenberg, H. Schulzrinne, G. Camarillo, A. Johnston, J. Peterson, R. Sparks, M. Handley, and E. Schooler. SIP: session initiation protocol. IETF RFC 3261, June 2002.

29. S. Donovan. The SIP INFO method. IETF RFC 2976, October 2000.

30. H. Schulzrinne and J. Rosenberg. A comparison of SIP and H.323 for Internet telephony. In *Proc. of Network and Operating Systems Support for Digital Audio and Video (NOSSDAV)*, Cambridge, England, July 1998.

31. R. Sparks. The SIP refer method. IETF Internet Draft, ⟨draft-ietf-sip-refer-07.txt⟩, work in progress, November 2002.

32. A. Vaha-Sipila. URLs for telephone calls. IETF RFC 2806, April 2000.

4

Mobility Management

This chapter examines methodologies for supporting mobility in wireless IP networks. We begin by discussing the basic issues in mobility management, including the impact of naming and addressing on mobility management, location management, and handoffs. Then, we will focus on mobility management methodologies for IP networks, 3GPP packet networks, 3GPP2 packet networks, and MWIF networks.

4.1 BASIC ISSUES IN MOBILITY MANAGEMENT

Mobility can take different forms such as follows:

- *Terminal mobility*: Terminal mobility is the ability for a user terminal to continue to access the network when the terminal moves.
- *User mobility*: User mobility is the ability for a user to continue to access network services under the same user identity when the user moves. This includes the ability for a user to access network services from different terminals under the same user identity.
- *Service mobility*: Service mobility is the ability for a user to access the same services regardless of where the user is.

Mobility can be discrete or continuous. Take terminal mobility, for example; discrete terminal mobility is the ability for a terminal to move to a new location,

IP-Based Next-Generation Wireless Networks: Systems, Architectures, and Protocols,
By Jyh-Cheng Chen and Tao Zhang. ISBN 0-471-23526-1   © 2004 John Wiley & Sons, Inc.

connect to the network, and then continue to access the network. This is often referred to as portability. Continuous terminal mobility, on the other hand, is the ability for a terminal to remain connected to the network continuously (i.e., without user-noticeable interruptions of network access) while the terminal is on the move.

In some cases, a terminal or a user may be considered by a network to have "moved" even if the terminal or the user has not changed its physical position. This may occur, for example, when the terminal switched from one type of radio system to another (e.g., from WLAN to a cellular system).

A future wireless IP network should meet several basic mobility management requirements:

- *Support all forms of mobility*: A future wireless IP network should support all forms of mobility.
- *Support mobility for all types of applications*: A future wireless IP network should be able to support the mobility for both real-time and non-real-time data, voice, and multimedia applications.
- *Support mobility across heterogeneous radio systems*: Future wireless IP networks should allow users to move seamlessly across different radio systems in the same or different administrative domains.
- *Support session (service) continuity*: Session (or service) continuity is the ability to allow an on-going user application session to continue without significant interruptions as the user moves about. Session continuity should be maintained when a mobile changes its network attachment points or even moves from one type of radio system to another.
- *Global roaming*: An important goal of a future wireless IP network is to support global roaming. Roaming is the ability for a user to move into and use different operators' networks.

A mobility management system needs to have several basic functional components (capabilities):

- *Location management*: Location management is a process that enables the network to determine a mobile's current location, i.e., the mobile's current network attachment point where the mobile can receive traffic from the network.
- *Packet delivery to mobiles*: A process whereby a network node, mobile terminal, or end-user application uses location information to deliver packets to a mobile terminal.
- *Handoff and roaming*: Handoff (or handover) is a process in which a mobile terminal changes its network attachment point. For example, a mobile may be handed off from one wireless base station (or access point) to another, or from one router or switch to another. Roaming is the ability for a user to move into and use different operators' networks.

- *Network access control*: Network access control is a process used by a network provider to determine whether a user is permitted to use a network and/or a specific service provided by the network. Network access control typically consists of the following main steps:
  - *Authentication*: Authentication is to verify the identity of user.
  - *Authorization*: Authorization is to determine whether a user should be permitted to use a network or a network service.
  - *Accounting*: Accounting is a process to collect information on the resources used by a user.

Next, we discuss location management, packet delivery to mobiles, handoff, and roaming in greater detail. Network access control will be discussed more in Chapter 5, "Security."

4.1.1 Impact of Naming and Addressing on Mobility Management

A name identifies a network entity, such as a user, a user terminal, a network node, or a service. An address is a special identifier used by the network to determine where traffic should be routed.

How terminals are addressed at the network layer plays a critical role in how the network can handle terminal mobility. In today's networks, a terminal's address typically identifies a network attachment point from which the terminal can access the network. For example:

- A telephone number in a PSTN network identifies a port on a PSTN switch rather than the telephone set itself. Consequently, moving a telephone set from a telephone line connected to one switch to a telephone line connected to a different switch will require the telephone set to use a different telephone number. To allow a user to keep an old telephone number when the user moves from one PSTN switch to another, new technologies, such as Local Number Portability, have been developed for the network to redirect calls addressed to the user's old telephone number to the new PSTN switch to which the user is currently connected.
- An IP terminal's IP address identifies an attachment point to an IP network. As a result, when an IP terminal moves to a new attachment point to the IP network, it will have to use a new IP address to receive packets from the new network attachment point. IP-layer mobility management protocols, such as Mobile IP (Section 4.2.2), had to be used in order to allow a mobile to maintain a permanent IP address and to receive packets addressed to this permanent IP address regardless of the mobile's current location.

The ways in which terminals are named also has a significant impact on mobility management. In today's networks, the name of a terminal is often tied with the terminal's address. For example, an IP terminal has traditionally been named by the

Internet Domain Name associated with the terminal's IP address. A terminal name dependent on the terminal's address is not suitable for future mobile networks. Mobile terminals that use multiple network addresses are becoming increasingly popular. For example, a mobile terminal may have multiple radio interfaces. Each radio interface may use a different type of radio technology. Each radio interface may need to have its own IP address. Which domain name should be used as the terminal's name in this case? Notice that the mobile's radio interfaces may not all be connected to the network at any given time. Solutions have been designed to make the IP terminal names independent of the terminal's addresses. For example, the IETF has defined the Network Access Identifier (NAI) [12], [15] that allows a terminal to be identified by a single globally unique NAI regardless of how many IP addresses this terminal may have.

Now, let's consider user names and how they may impact mobility management. A user's name distinguishes the user from other users. A user's name is also needed by the network to authenticate the user and to identify the network services and resources consumed by the user for billing purpose.

Traditional circuit-switched networks, such as the PSTN, typically do not support user names. Therefore, these networks can only identify terminals but not users. They assume a static mapping between a terminal and the user responsible to pay for the services used by the terminal. For example, the PSTN cannot distinguish which user called from a telephone but simply the fact that a phone call is made from a particular telephone number. Static mapping of users to terminals could lead to a range of problems in a mobile network. Mobile users often have to, or like to, use different types of terminals in different locations depending on what types of terminals are available or best fit their needs. This suggests that a mobile user's name should not be statically tied to a mobile terminal.

Terminal-independent user names have become increasingly common in recent mobile networks. For example, in GSM, each subscriber is identified by a globally unique International Mobile Subscriber Identity (IMSI) that is independent of the terminal used by the user. A Subscriber Identity Module (SIM) carries a mobile's IMSI and can be ported from one mobile terminal to another to allow a user to use different terminals and still be recognized by the network as the same user.

In today's IP Networks, applications provide their own naming schemes for users. For example, e-mail users are identified by their e-mail addresses. SIP users are identified by their SIP URIs. If a service provider requires users to register with the service provider before they are allowed to access the services, a user may typically register with an arbitrary name. The NAI may serve as a user's globally unique and terminal-independent user name.

4.1.2 Location Management

The term *location* in the context of mobility management refers to where a mobile is, or can be, attached to the network. In other words, a mobile's location refers to the mobile's precise or potential network attachment point or attachment points, and it does not necessarily indicate the mobile's geographical position.

Location management is a process that enables the network to maintain up-to-date information regarding the mobiles' locations. Location management typically requires the following main capabilities:

- *Location Update*: A process whereby mobiles notify the network of their locations.
- *Location Discovery*: A process for the network to determine a mobile's precise current location. This process is commonly referred to as terminal paging or paging for simplicity.

4.1.2.1 Location Update Strategies

A location update strategy determines when a mobile should perform location updates and what location-related information the mobile should send to the network.

A straightforward location update strategy is to update the mobile's precise location *every time* the mobile changes its network attachment points. This, for example, is the strategy used in Mobile IP (Section 4.2.2).

Knowing a mobile's precise location allows the network to deliver traffic to the mobile via unicast. However, when mobiles change their network attachment points frequently, maintaining precise locations of all mobiles could lead to heavy location update traffic, which wastes limited radio bandwidth and scarce power resources on the mobiles.

To save scarce resources on the mobile and in the wireless network, a network can group network attachment points into *location areas* and only keep track of which location area each mobile is likely in when the mobile and the network have no traffic to send to each other. A mobile does not have to perform a location update when it remains inside the same location area. The network tries to determine a mobile's precise location only when it needs to deliver user traffic to the mobile.

A network may use multiple types of location areas simultaneously. The location areas used in a radio access network can be different from the location areas used for location management in the core network. For example, location areas inside a radio access network could be radio cells, whereas location areas for the core IP network could be IP subnets or other IP-layer location areas.

Many location update strategies exist today to determine when to perform location updates. They can be classified into the following categories [11], [54], [51]:

- *Time-based update* [11], [54]: A mobile performs location update periodically at a constant interval (called the update interval). Time-based location update is often used as a backup to other location update strategies.
- *Movement-based update* [11], [54], [13]: A mobile performs a location update whenever it traverses a predefined number of location areas. This pre-determined number of location areas is called the *movement threshold*. Most existing wireless networks (e.g., GSM, GPRS, 3GPP, 3GPP2) use a special case of a movement-based location update strategy in which the movement

threshold is one; i.e., a mobile updates its location every time it moves into a new location area.

- *Distance-based update* [11], [54], [13], [33]: A mobile performs a location update whenever it has traveled a predefined distance threshold from the location area in which it performed its last location update. Distance may be measured in many different ways, such as physical distance, or *cell distance* (i.e., distance measured in number of radio cells or location areas). The physical distance-based strategy is used, for example, as an option in 3GPP2.

- *Parameter-based update*: A mobile performs location update whenever the value of any preselected parameter changes. These strategies are sometimes referred to as profile-based strategies. This strategy is used, for example, as an option in 3GPP2.

- *Implicit update*: A mobile does not send any message explicitly for the purpose of location update. Instead, the network derives the mobile's location when the network receives other signaling or user data from the mobile. This approach is used, for example, by 3GPP2 and some micromobility management protocols designed for IP networks (e.g., Cellular IP [17] and HAWAII [42], to be discussed in Sections 4.2.7 and 4.2.8).

- *Probabilistic update*: A mobile performs location update based on a probability distribution function. A probabilistic version of time-based, movement-based, or distance-based location update strategies may be created. Consider a time-based location update, for example. The new update time interval after each update may be dynamically adjusted based on the probability distribution of the call arrival times [29].

The main difference between movement-based and distance-based location update strategies is illustrated in Figure 4.1 when distance-based strategies use "cell distance" rather than physical distance. Suppose that the mobile last performed a location update in the center location area shown in Figure 4.1. The arrowed lines indicate the mobile's movements. The number on each arrowed line indicates the number of times the mobile has crossed a cell boundary since its last location update in the center cell. Assume that the movement threshold used by a movement-based update strategy is three cell boundary crossings and the distance threshold used by the distance-based update strategy is three cells. In the example shown in Figure 4.1, the mobile will perform location update at the third, sixth, and the ninth times it crosses a cell boundary if it uses the movement-based update strategy. On the other hand, the mobile will only perform location update once, i.e., at the ninth time it crosses a cell boundary, if it uses distance-based update strategy.

Each location update strategy brings its unique advantages, but also it has its limitations. Movement-based strategies with a movement threshold of one and time-based strategies are the most commonly used strategies in existing wireless networks and in 3G wireless networks (e.g., 3GPP and 3GPP2). Selection of location update strategies should be considered in concert with paging strategies (Section 4.1.2.2) as they are closely related to each other.

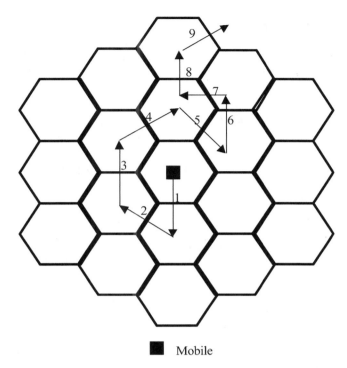

■ Mobile

Fig. 4.1 *Movement-based vs. distance-based location update strategies*

4.1.2.2 *Location Discovery (Paging)*

Location discovery or paging is necessary when a network does not maintain mobiles' precise locations at all times. The network performs paging to determine the precise location of a mobile and to inform the mobile of incoming traffic.

Typically, a network performs paging by sending one or multiple *paging messages* to a *paging area* where the mobile is likely to be located currently. A paging message is used to inform the mobile terminal that the network has traffic to deliver to the mobile. A paging area is a set of network attachment points. Paging areas do not have to be identical to location areas (Section 4.1.2.1).

Upon receiving a paging message, a mobile needs to update its precise current location with the network. The mobile may also need to establish the necessary connectivity with the network for carrying user traffic to and from the network. For example, in a circuit-switched radio access network, the physical radio channels and logical connections over these physical radio channels required for carrying user traffic need to be established when a mobile receives a paging message. Updating its precise location or establishing the necessary network connectivity can often serve as an implicit acknowledgment to the network that the mobile has received a paging message.

A key consideration in paging is that paging should be done within a reasonable time constraint [54], [44], [27]. If paging takes too long, the call setup latency could become intolerable to end users and call attempts may be dropped as a result. Other critical issues that need to be addressed in the design of paging strategies include the following:

- How to construct paging areas?
- How to search a paging area to locate a mobile?

Paging areas can be static or dynamic. A static paging area does not change unless reconfigured by the network operator manually or via a network management system. Existing second-generation wireless networks and the third-generation wireless standards (e.g., 3GPP and 3GPP2) typically use fixed paging areas. Dynamic paging areas have been proposed in the literature to reduce paging overhead. The idea is to dynamically adjust the paging area configurations in response to changing network dynamics (e.g., distribution of mobile user population and mobility patterns) so that the combined location update and paging signaling overhead can be reduced. Supporting dynamic paging areas, however, typically requires a much more complex signaling protocol than supporting static paging areas.

Once the paging area is determined for a mobile, many strategies are available for delivering paging messages to the paging area to search for the mobile. These paging strategies can be classified into the following categories:

- *Blanket paging*: A paging message is broadcast simultaneously to all radio cells inside the paging area where the mobile is located. Blanket paging is deployed in most of today's wireless networks. Its main advantages are simplicity and low paging latency. The drawback, however, is that broadcasting paging messages to a large number of radio cells could consume a significant amount of scarce resources, including radio bandwidth and power on all the mobiles in the paging area.
- *Sequential paging*: With this strategy, a large paging area is divided into small paging sub-areas (e.g., radio cells). Paging messages are first sent to a subset of the paging sub-areas where the network believes the mobile is most likely to be located. If the mobile is not in this sub-area, subsequent paging messages will be sent to another paging sub-area. This process continues until either the mobile is found or the entire paging area (or the entire network) is searched. Any technique may be used to determine how to divide a large paging area into smaller paging sub-areas and which sub-areas should be searched first.
- *Other paging strategies*: Other paging strategies also exist that cannot be clearly classified into the categories mentioned above. For example:
  - *Geographic paging*: The network uses the geographical position of a mobile to determine where a paging message should be sent [30].

- *Group paging*: When the network wants to locate a mobile, it pages a group of mobiles together instead of paging only the mobile to be located [45].
- *Individualized paging:* The network maintains an individualized paging area for each individual mobile [18].

4.1.2.3 Interactions between Location Update and Paging

Location update strategy, location area design, and paging strategy have a close interdependency. For example, if a precise location update is used every time a mobile changes its network attachment points, no paging will be required. If sequential paging is used, the network could enlarge its search area gradually if it cannot find a mobile in one location area. Therefore, a mobile may not necessarily have to update its location every time it moves into a new location area and the location update messages do not necessarily have to be delivered reliably.

Therefore, a key issue in the design of location update strategies, paging strategies, and the location update and paging protocols is how to achieve a proper balance among the following:

- *Overhead*: Network resources consumed by location updates and paging.
- *Performance*: e.g., paging latency.
- *Complexity*: The complexities of the location update and paging strategies as well as the protocols needed to support these strategies. High complexity often translates into high network costs and high level of difficulty in operating the network.

Consider the tradeoff between overhead and performance. Small location areas and frequent location updates could enable a network to locate a mobile quickly when it has traffic to deliver to the mobile (i.e., high paging performance), but they could lead to high location update overhead. On the other hand, large location areas or infrequent location updates could reduce location update overhead but increase paging overhead and paging delay.

4.1.3 Packet Delivery to Mobile Destinations

Packet delivery to mobile destinations is the process whereby a *packet originator* and the network use location information to deliver packets to a mobile destination. A packet originator may be a fixed or mobile terminal, a network node, or a user application. For example, when a network node S inside a network N receives a packet from another network, network node S may be the packet originator for delivering this packet inside network N.

Packet delivery strategies can be classified into two basic categories, as illustrated in Figure 4.2:

- *Direct Delivery strategies*: With Direct Delivery strategies, a packet originator first obtains the destination mobile's current location and then addresses and

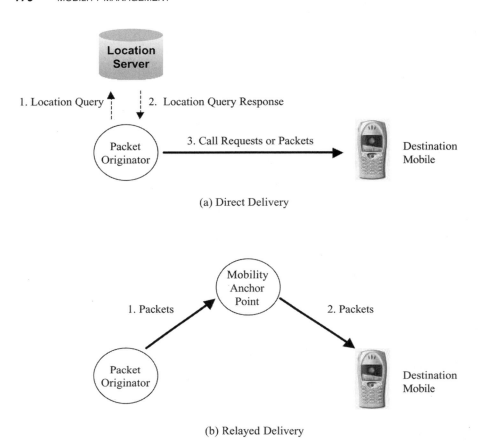

Fig. 4.2 *Strategies for delivering packets to mobiles*

sends the packets directly to the current location of the destination mobile. A packet originator may maintain mobiles' locations by itself or obtain location information from location servers.

- *Relayed Delivery strategies*: With Relayed Delivery strategies, a packet will be sent first to a *mobility anchor point*, which then relays the packet toward its final destination. The packet originator does not need to know a destination mobile's current location. In fact, it does not even need to know whether a destination is a mobile or a fixed node. Furthermore, the packet originator may not necessarily need to be aware of the existence of any mobility anchor point, nor the fact that it is sending call requests or packets to a mobility anchor point.

Direct Delivery strategies have the potential ability to route packets along the most direct paths to their destinations. However, they have several characteristics that need to be carefully considered.

- Direct Delivery strategies require a packet originator to determine whether the destination of a packet is a mobile or a fixed host in order to decide whether a location query should be performed in order to deliver the packet. This is because location query is only necessary for mobile destinations, and the network may support both mobile and fixed hosts. Performing location query for every destination could incur heavy overheads, such as consuming heavy processing power on the packet initiator and creating heavy location query traffic to the network.

- Direct Delivery strategies require every packet originator to implement protocols for determining a destination host's current location. For example, they may have to implement protocols for querying location servers or for obtaining location information by other means. When location servers are used, packet originators need to be able to discover the IP addresses of the location servers.

Most IP hosts and routers in today's Internet do not maintain sufficient information to allow them to determine whether a destination IP host is a mobile host or a fixed host. For example, most IP hosts and routers only know a destination host by its IP address or NAI, which does not tell whether the destination is mobile or fixed. Furthermore, most IP hosts and routers in today's Internet do not have the ability to query location servers. Therefore, modifications have to be made to an IP host or router in order to allow it to use Direct Delivery strategies to send packets to a mobile destination.

Relayed Delivery strategies typically do not require changes to the packet originators. Instead, the mobility anchor points are responsible for determining the mobiles' locations and relay packets to these mobiles. However, Relayed Delivery strategies have their own limitations too:

- Relayed Delivery strategies may cause packets to take longer paths than Direct Delivery strategies.

- The mobility anchor points could become traffic and performance bottlenecks.

Mobile IPv4, Mobile IPv6, and the mobility management approaches in 3GPP and 3GPP2 packet data networks all use the Relayed Delivery strategies as their basic packet delivery strategy.

Relayed Delivery and Direct Delivery strategies can be combined to take advantages of the strengths of both strategies and to overcome each other's weaknesses. This is illustrated in Figure 4.3. Initially, a packet originator does not have to know the destination's current location. Packets destined to the destination will be routed first toward a mobility anchor point. Upon receiving the packets, the mobility anchor point relays them to the mobile's current location. The mobility anchor point or the destination can then inform the packet originator of the destination's current location. Then, the packet originator can address the packets directly to the mobile's current location. Such a combined Delayed Delivery and

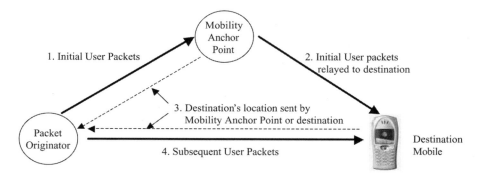

Fig. 4.3 *Integrated Relayed Delivery and Direct Delivery strategies*

Direct Delivery strategy has been used, for example, in SIP as well as in the route optimization extensions to Mobile IPv4.

4.1.4 Handoffs

Handoff is a process whereby a mobile changes from one network attachment point to another within the same network administrative domain. For example, a mobile can change its radio channels from one base station to another or from one radio frequency band to another on the same base station. Handoff in an IP-based wireless network is a much broader issue than the changing of a mobile's radio channels from one base station.

First, handoffs in an IP-based wireless network may occur at different protocol layers:

- *Physical Layer*: During a physical layer handoff, a mobile changes its network attachment point at the physical layer. For example, the mobile may change from one radio channel to another, from one wireless base station to another.
- *Logical Link Layer*: During a logical link layer handoff, a mobile changes its logical link layer over which the mobile exchanges user IP packets with the network.
- *IP Layer*: With an IP-layer handoff, the mobile changes its IP address or moves to a different IP access router.

A handoff at one protocol layer does not necessarily result in a handoff at a different protocol layer. For example, when a mobile changes its radio channels or moves from one base station to another, it does not necessarily have to change its logical link layer connection with the network, and does not necessarily have to change its IP address or move to a different IP access router.

Similarly, a mobile may change its IP address while using the same physical connectivity and the same link layer connection with the network.

Therefore, an important concept in mobility management in an IP-based wireless network is that mobility at different protocol layers can be managed by different protocols. Furthermore, mobility management at the IP layer may be independent of mobility management at the lower protocol layers.

Second, handoffs at each protocol layer may occur in different scopes. Take handoffs on the IP layer, for example; there can be

- *Intra-subnet handoff*: A mobile remains on the same IP subnet after it changes its IP address or moves from one base station to another.
- *Inter-subnet handoff*: A mobile moves into a new IP subnet and changes its IP address as a result of the handoff.
- *Inter-router handoff*: A mobile moves to a new IP access router as a result of the handoff.

Different capabilities may be required to support different handoff scopes. For example, IP-layer procedures may not be needed to support intra-subnet handoffs if the mobile does not change its IP address as a result of the handoff. When a mobile moves within the same IP access router, the mobile typically does not need to repeat some of the potentially time-consuming network access control procedures, such as authentication and authorization. However, when the mobile moves to a new IP address router, it may have to be reauthenticated and reauthorized by the network.

Third, handoffs can be hard or soft depending on how the mobile receives user data from the network during the handoff process. During a hard handoff, a mobile can receive user data from only one base station at any time. With a soft handoff, a mobile receives copies of the same user data from two or more base stations simultaneously. The mobile uses signal processing techniques to determine the most likely correct value of the data from its multiple copies. This way, even when the mobile's radio channel to one base station is experiencing low signal quality, the mobile may still be able to receive data. Soft handoff has been proven to be an effective way for increasing the capacity, reliability, and coverage range of CDMA (Code Division Multiple Access) systems.

There are two basic ways to implement a hard handoff:

- *Make-before-Break*: The mobile sets up its network connectivity via the new network attachment before it tears down the network connectivity via the old network attachment.
- *Break-before-Make*: The mobile tears down its network connectivity via the old network attachment point and then establishes its network connectivity via the new network attachment point.

Realizing soft handoff requires the following capabilities:

- *Data distribution and selection*: Separate copies of the same data need to be sent via multiple base stations to the same mobile. The mobile should be able to use the multiple copies of the data to construct a single copy and only pass the single copy to upper layer protocols or applications. Similarly, multiple copies of the same user data originated from a mobile will also be sent to the network via different base stations. The radio access network or the edge devices connecting the radio access networks to the core network should be able to select one copy of the data to send to the destination.

- *Data content synchronization*: Pieces of data arriving from multiple base stations to a mobile at the same time should be copies of the same data in order for the mobile's radio system to combine these copies into a correct single copy.

In today's circuit-switched CDMA networks such as IS-95 [5], a Selection and Distribution Unit (SDU) is responsible for data distribution in the forward direction (i.e., from network to mobile). A SDU may be located on a base station or an MSC. It creates and distributes multiple streams of the same data over layer-2 circuits to multiple base stations that relay the data to the mobile. The mobile's radio system collaborates with the base stations to synchronize the radio channel frames and combine the radio signals received from different base stations to generate a single final copy of received data. The SDU helps ensure data content synchronization at the mobile by ensuring that the matching layer-2 frames sent to different BSs contain copies of the same data. In the reverse direction, the mobile helps ensure data content synchronization by ensuring that the matching layer-2 frames sent to different base stations contain copies of the same data. The SDU then selects one of the frames received from different base stations as the final copy of the data.

4.1.5 Roaming

To discuss roaming, we first need to define *home domains* and *visited domains* for a user.

- *Home domain*: A user's home domain is the domain where the mobile maintains a service subscription account, or account for convenience. A user's account contains information regarding the subscriber's identity, billing address, service profile, and security information needed to authenticate the user. The user's service profile describes which network services are subscribed by the user, including which networks the user is allowed to use. The home domain uses the information described above to determine how to provide services to a mobile and how to charge for the services used by the mobile.

- *Visited domain*: When a user moves into a domain with which it does not have an account, this domain will be called the mobile's visited domain.

Roaming is the process whereby a user moves into a visited domain. Roaming is similar to handoff in the sense that a mobile changes its network attachment point as a result of roaming. However, supporting roaming requires networking capabilities that are not necessary for supporting handoffs within the same administrative domain. In particular, the following extra capabilities are needed to support roaming:

- Network access control for visiting mobiles.
- Roaming agreement between the mobile's home domain and the visited domains.
- Session continuity while a user crosses domain boundaries.

When a user wishes to use a visited domain, the visited domain needs to determine whether this user should be allowed to use the visited domain. To make such a decision, the visited domain needs to know, for example, who this user is, whether the user or its home domain agrees to pay for its use of the visited domain, and where to send the bill for this user. However, the visited domain itself may not have sufficient information to make this decision because it does not have an account for this user.

Therefore, to allow the user to use a visited domain, the visited domain needs to have a *roaming agreement* with the user's home domain. A roaming agreement should decide how a visiting mobile should be authenticated, authorized, and billed.

As illustrated in Figure 4.4, when a user tries to gain access to a visited domain, the visited domain may ask the user's home domain to authenticate the user and to confirm how to charge for the user's use of the visited domain. Upon successful authentication and authorization by the user's home domain, the home domain

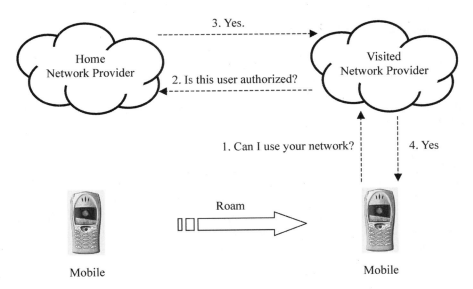

Fig. 4.4 *Roaming*

replies to the requests from the visited domain. The home domain may also send information regarding the user's service profile to the visited domain to help the visited domain to determine how to provide services to the user. Such information may include, for example, the user's QoS requirements. This basic approach shown in Figure 4.4 is used in second-generation cellular networks and is the basic framework used for supporting roaming in IP networks.

There may be many public network providers in a country. This is increasingly the case as the deployment of public WLANs continues to grow because public WLANs in different locations may be owned and/or operated by different network providers. Even when only a very small number of public network providers exist in a country, users may roam outside the countries into different network providers in other countries. It is difficult for a network provider to establish a roaming agreement with every other network provider.

One alternative to solve the problem described above is to establish a *Roaming Broker*, as shown in Figure 4.5. Each network provider only needs to establish a roaming agreement with the Roaming Broker. When a user roams into a new visited network, this visited network will ask the Roaming Broker to authenticate and authorize the user. The Roaming Broker can relay the authentication and authorization requests received from a network provider to the mobile's home network and then relay the responses back to the mobile's current visited network.

Alternatively, a user could have a single service subscription account with the Roaming Broker and the Roaming Broker will then ensure that the user can roam into any network connected to the Roaming Broker. In this case, the user's home network becomes the Roaming Broker.

4.2 MOBILITY MANAGEMENT IN IP NETWORKS

This section discusses the fundamental issues of mobility management in wireless IP networks and examines existing IP mobility management protocols.

The current standard protocol defined by the IETF for mobility management in IPv4 networks is commonly referred to as Mobile IPv4 or MIPv4 (Section 4.2.2). MIPv4 enables an IP terminal to maintain a permanent IP address and receive packets addressed to this permanent address regardless of the mobile's current attachment point to the Internet. It first became an IETF RFC in 1996 and was later revised in 2002. Currently, the IETF is leveraging MIPv4 to define an IP-layer mobility management protocol for IPv6 [20] networks—Mobile IPv6 (MIPv6).

The IETF has also been working on IP-layer mobility protocols that are aimed at providing enhanced mobility support (e.g., reduced handoff delay) within a limited geographical region, e.g., a building, campus, or a metropolitan area network. These protocols are often referred to as "micromobility" management protocols. Examples of micromobility management protocols include MIPv4 Regional Registration (Section 4.2.3), Cellular IP (Section 4.2.7), and HAWAII (Section 4.2.8). Currently, no micromobility management protocol has become an IETF standard.

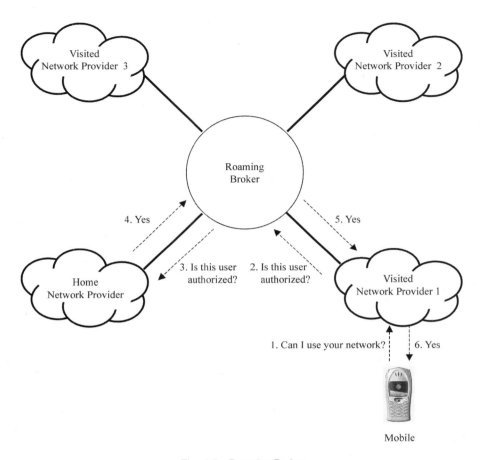

Fig. 4.5   *Roaming Broker*

Existing IP-layer mobility management protocols alone typically cannot support user mobility. Application-layer mobility management protocols have been proposed to support both terminal and user mobility. The most widely studied application-layer mobility protocol is SIP-based mobility management (Section 4.2.6). As discussed in previous chapters, SIP has been widely accepted as the session management protocol for wireline and wireless IP networks.

4.2.1   Naming and Addressing of IP Terminals

A terminal's IP address uniquely identifies the node's point of attachment to the IP network, rather than to the terminal itself. Therefore, a terminal must be attached to the network indicated by its IP address in order to receive packets addressed to its IP address. Traditionally, an IP terminal's IP address serves as the terminal's identifier.

Although this has worked well for fixed networks, it has several limitations when used in a mobile network.

- With regular IP routing protocols, when a terminal moves to a new IP network or IP subnet (commonly referred to as a *visited network* or a *foreign network*), the terminal will have to use a new IP address that is part of the IP address space of the new IP network in order to receive packets from the visited network. This means that if the mobile terminal uses its IP address as its identifier, the identifier will change as the mobile moves from one IP network to another.
- A mobile may have multiple radio interfaces, each with a different IP address. A mobile's radio interfaces may not all be reachable by the network at any given time, depending on which radio systems are available at the mobile's current location or which radio system the mobile user wishes to use if multiple radio systems are available. This makes it difficult to determine which IP address configured on the mobile should be used as the mobile's identifier.

To help resolve the issues described above, the IETF has defined the Network Access Identifier (NAI) [12], [15] that can be used to identify a mobile terminal (or user) regardless of the terminal's current location and regardless of how many IP addresses the terminal may have.

The NAI takes the form of *username@realm* [12]. The *username* identifies the terminal. The *realm* identifies the Internet domain name of a Network Access Server (NAS).

4.2.2 Mobile IPv4

With regular IP routing protocols, when a mobile terminal moves to a visited network, the mobile will have to use an IP address that is part of the IP address space of the visited network to receive IP packets. This leads to the following issues:

- A correspondent node (a node that communicates with a mobile terminal) will have to use the mobile terminal's new IP address to address the packets to be sent to the mobile. However, it is difficult to force every possible correspondent node to keep track of when a mobile terminal may change its IP address and what the mobile's new address will be.
- Changing an IP address will cause on-going TCP sessions to break. Today, a large percentage of Internet traffic is TCP (Transmission Control Protocol) traffic. For example, the WWW uses TCP for data transport. In order to maintain an on-going TCP connection, the IP addresses used by the two endpoints of the TCP connection need to remain the same. However, a mobile terminal may have to change its IP address when it moves into a different IP subnet. Therefore, mobility management should ensure that an on-going TCP application can continue as the mobile terminal moves about. This may be achieved by ensuring that the on-going TCP does not break as the mobile

moves about. Alternatively, if an on-going TCP connection breaks as a result of mobility, it can be restored quickly so that the temporary connection breakdown does not cause noticeable interruptions to the application.

Mobile IP [37] solves the problems described above by allowing a mobile terminal to maintain a permanent *home address* and to receive packets addressed to its home address regardless of its current location. A mobile's home address is a globally unique and routable IP address that does not have to change regardless of the mobile's current point of attachment to the Internet. The home address can be either preconfigured or dynamically assigned, as described in Section 4.2.2.4. Any correspondent node can always use a destination terminal's home address to address the packets to be sent to the mobile terminal, without having to know whether the destination is mobile and where the destination node is currently attached to the network. Mobile IP ensures that packets addressed to a mobile's home address be delivered to the mobile regardless of where the mobile terminal is.

Each mobile has a *home network*—the network whose network address prefix matches that of the mobile terminal's home address. Packets addressed to a mobile's home address will be routed by the regular IP routing protocol to the mobile's home network. When a mobile is inside its home network, it receives and sends packets as a fixed terminal without using Mobile IP. When a mobile is in a foreign network, a router on the mobile's home network will act as a *Home Agent* (HA) for the mobile. The HA will maintain up-to-date location information for the mobile, intercept packets addressed to the mobile's home address, and *tunnel* these packets to the mobile's current location. A mobile's location in a foreign network is identified by a temporary *Care-of Address* (CoA) assigned to the mobile by the foreign network. A mobile uses its CoA to receive IP packets in the foreign network. Each time the mobile obtains a new CoA, it will register the new CoA with its HA ("home registration").

Each foreign network may have a *Foreign Agent* (FA). An FA is a router that provides routing services to the visiting mobile terminals. These routing services include the following:

- Provides CoAs and other necessary configuration information (e.g., address of default IP router) to visiting mobiles.
- De-tunnels packets arriving from the tunnel from a visiting mobile's home agent and then delivers the packets to the visiting mobile.
- Acts as the IP default router for packets sent by visiting mobile terminals.
- Helps visiting mobiles to determine whether they have moved into a different network.

MIPv4 allows two types of CoAs:

- *Foreign Agent CoA*: An FA CoA is an IP address of a Foreign Agent. Each FA is responsible for providing FA CoAs to visiting mobiles. When FA CoA is used, the mobile's home agent tunnels the packets addressed to the mobile's

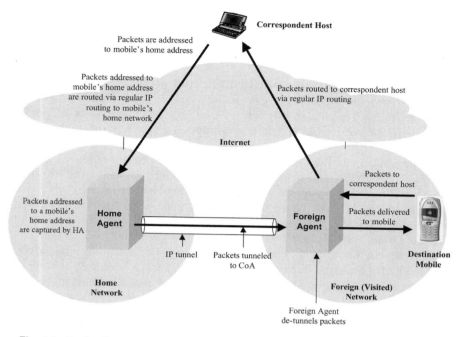

Fig. 4.6 *Packet flows between a correspondent host and a mobile: mobile uses FA CoA*

home address to the mobile's current FA. The foreign agent will then de-tunnel the packets and deliver them to the mobile.

Delivery of packets from an FA to a visiting mobile can use any link layer techniques available on that particular network and does not have to use standard IP routing protocol.

- *Co-located CoA*: A co-located CoA is a CoA acquired by a mobile terminal through any method external to Mobile IP. For example, a mobile may use the Dynamic Host Configuration Protocol (DHCP) [22] to obtain a temporary address dynamically. When a co-located CoA is used, the mobile terminal's home agent tunnels the packets addressed to the mobile's home address directly to the mobile itself; these packets do not have to go through any FA.

Figure 4.6 illustrates the packet flows between a correspondent host and a mobile when the mobile uses FA CoA, assuming that the FA is also the default router in the foreign network for the mobile. When the mobile uses co-located CoA, the packet flows are shown in Figure 4.7.

Next, we discuss the main phases of MIPv4 operation:

- Agent Discovery
- Movement detection

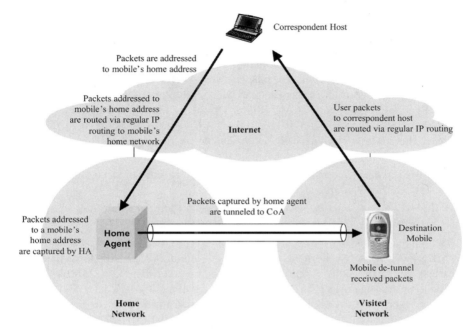

Fig. 4.7 *Packet flows between a correspondent host and a mobile: mobile uses co-located CoA*

- Leaving the home network
- Entering and staying in a visited network
- Returning to the home network

4.2.2.1 Agent Discovery In order for a mobile terminal to communicate with
a MIPv4 mobility agent (i.e., HA or FA), the mobile terminal will first need to know
the agent's IP address. Typically, each mobile terminal is configured with the IP
address of its home agent. This is because a mobile's home agent typically does not
change. However, to allow a mobile's home agent to change while the mobile is
away from home, a mobile may also discover a home agent dynamically using the
procedures described in Section 4.2.2.4. Mobiles always have to dynamically
discover the existence and the IP addresses of foreign agents because it is difficult to
predict which foreign networks the mobile may move into in the future; hence, it is
impossible for the mobile terminal to be preconfigured with the addresses of all the
foreign agents it may use.

The process for a mobile terminal to discover the mobility agents and receive
information from these agents is called MIPv4 *Agent Discovery*. Agent Discovery is
achieved by the mobility agents advertising their services and system information to
the mobiles via *Agent Advertisement* messages. The Agent Advertisement messages
may be periodically broadcast to all mobiles or delivered in any way a network
provider deems appropriate. A mobile terminal does not have to wait passively for

the Agent Advertisement messages to arrive. It may solicit an Agent Advertisement message from the mobility agents by sending an *Agent Solicitation* message to the Mobile-Agents Multicast Group address 224.0.0.11. All mobility agents are required to respond to any received Agent Solicitation message.

Instead of defining its own new messages for MIPv4 Agent Discovery, MIPv4 uses the Internet Control Message Protocol (ICMP) Router Discovery Messages [19] for MIPv4 Agent Discovery. ICMP Router Discovery Messages are standard messages defined by the IETF for IPv4 terminals to learn the IP addresses of access routers. There are two main ICMP Router Discovery messages:

- *ICMP Router Advertisement Message*: Sent by a router to terminals to inform them of the IP addresses of the router.
- *ICMP Router Solicitation Message*: Sent by a terminal to a router to ask the router to send ICMP Router Advertisement Messages.

MIPv4 Agent Advertisement message uses the ICMP Router Advertisement message with extensions to carry MIPv4-specific information. A MIPv4 Agent Solicitation message uses a format that is identical to the ICMP Router Solicitation message, except that its IP Time-to-Live (TTL) field must be set to 1 so that the Agent Solicitation message will not propagate beyond the local IP subnet.

MIPv4 defines the following two extensions to the ICMP Router Advertisement message:

- *Mobility Agent Advertisement Extension*: This extension is used to
  - Indicate that an ICMP Router Advertisement message is also a MIPv4 Agent Advertisement message.
  - Carry information specific to a MIPv4 mobility agent.
- *Prefix-Lengths Extension*: This is an optional extension used to indicate the network prefix length (in bits) of each Router Address advertised in the ICMP Router Advertisement portion of the Agent Advertisement. The network prefix lengths may be used by a mobile to determine whether it has moved into a new IP network.

The structure of a MIPv4 Agent Advertisement message is illustrated in Figure 4.8. The first portion of the message is the ICMP Router Advertisement message, which is followed by the Mobility Agent Advertisement Extension portion defined by MIPv4. The optional Prefix-Lengths Extension defined by MIPv4, if present, will follow the Mobility Agent Advertisement Extension portion.

Figure 4.9 illustrates the format of the Mobility Agent Advertisement Extension. The fields and flags are defined as follows:

- *Type*: The value of this field should be 16 to indicate that this is the Mobility Agent Advertisement Extension.

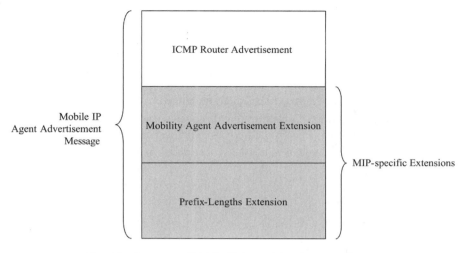

Fig. 4.8 *Structure of Mobile IP Agent Advertisement message*

- *Length*: Contains the length in octets of the extension from the beginning of the Sequence Number field to the end of the extension.
- *Sequence Number*: Contains the number of Agent Advertisement messages sent since the agent was initiated.
- *Registration Lifetime*: Contains the longest lifetime in seconds that this agent is willing to accept for any Registration Request.
- *R (Registration required)*: Indicates that Mobile IP registration through this foreign agent (or another foreign agent on this link) is required even when using a co-located CoA. How a mobile registers through a foreign agent will be described in Section 4.2.2.4.
- *B (Busy)*: Indicates that this foreign agent will not accept registrations from additional mobile terminals.

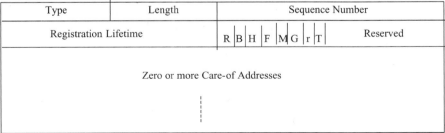

Fig. 4.9 *MIPv4 Mobility Agent Advertisement Extension to ICMP Router Advertisement message*

- *H (Home agent)*: Indicates that this agent offers service as a home agent on the link on which this Agent Advertisement message is sent.
- *F (Foreign agent)*: Indicates that this agent offers service as a foreign agent on the link on which this Agent Advertisement message is sent.
- *M (Minimal encapsulation)*: Indicates that this agent can accept tunneled messages that use Minimal Encapsulation defined in [14].
- *G (GRE encapsulation)*: Indicates that this agent accepts tunneled packets that use Generic Routing Encapsulation (GRE) defined in [26].
- *r (Reserved)*: This field is not used. It must be set to zero and ignored on reception.
- *T (Reverse tunneling)*: Indicates that this foreign agent supports reverse tunneling [35]. Reverse tunneling is discussed in Section 4.2.2.8.
- *Reserved*: This field is not currently used and shall be ignored by the mobiles.
- *Foreign Agent Care-of Addresses*: Contains the Foreign Agent Care-of Addresses, if any, provided by this foreign agent.

Figure 4.10 illustrates the Prefix-Length Extension to the ICMP Router Advertisement message. The fields are defined as follows:

- *Type*: The value of this field should be 19 to indicate that this is the Prefix-Length Extension.
- *Length*: The value of the "Num Addrs" field in the ICMP Router Advertisement portion of the Agent Advertisement. The Num Addrs field in the ICMP Router Advertisement portion of the Agent Advertisement indicates the number of Router Addresses advertised in this message.
- *Prefix Lengths*: The number of leading bits that define the network prefix of the corresponding Router Address listed in the ICMP Router Advertisement portion of the message. The prefix length of each Router Address is encoded as a separate byte, in the order that the Router Addresses are listed in the ICMP Router Advertisement portion of the message.

4.2.2.2 Movement Detection
A mobile terminal needs to know when it enters a new IP subnet, i.e., when it may need to change its care-of address. The process for a mobile to detect when it enters a new IP subnet is referred to as *movement detection* in MIPv4. Mobile terminals can use information in the received Agent Advertisement messages to detect movement. Two such methods are described in [37]:

```
0 1 2 3 4 5 6 7 8 9 0 1 2 3 4 5 6 7 8 9 0 1 2 3 4 5 6 7 8 9 0 1
```

Type	Length	Prefix Length	. . . . . .

Fig. 4.10 *MIPv4 Prefix-Length Extension to ICMP Router Advertisement message*

- *Approach 1—Use the Lifetime field in Agent Advertisement messages.* Each Agent Advertisement message has a Lifetime field (in the main body of the ICMP Router Advertisement portion of the Agent Advertisement) that indicates the length of time that this Agent Advertisement is valid. For each mobility agent from which a mobile received Agent Advertisement messages, the mobile caches the Lifetime carried in the last received Agent Advertisement message until the Lifetime expires. If the mobile does not receive any new Agent Advertisement from the same mobility agent within the remaining Lifetime, it will assume that it has lost contact with that mobility agent. In other words, the mobile will assume that it has moved into a new network. If, by this time, the mobile has already received Agent Advertisement messages with positive Lifetime from other mobility agents, it may use one of these mobility agents. Otherwise, the mobile should start searching for a new mobility agent by issuing Agent Solicitation messages.

- *Approach 2—Use network prefixes.* A mobile may detect whether it has moved into a new IP subnet by comparing the network prefix of the old network with the network prefix of the new IP subnet: If the two network prefixes differ, the mobile has just entered a new IP subnet. Using the network prefixes for movement detection requires the mobile to know the network prefix lengths of the old and the new networks. An FA may advertise the network prefix length of the local subnet (more precisely, the network prefix length of the address of the local access router) in the Agent Advertisement message with the Prefix-Lengths Extension.

It is important to note that a mobile does not have to wait until it loses contact with its current mobility agent before it starts to use a different network. Instead, a mobile may actively select the network it wants to use if multiple networks are available simultaneously. For example, if a mobile has received Agent Advertisement messages with positive Lifetimes from multiple foreign agents, the mobile may determine which network to use by selecting one of the foreign agents to use. Changing to a new foreign agent moves the mobile's network attachment point to a network served by the new foreign agent.

In addition to the methods described above, a mobile may use any other method for movement detection. For example, a mobile may use indications of changes in lower layer characteristics (e.g., change of type or identity of radio systems, change of radio access points or radio channels) for movement detection.

4.2.2.3 Leaving the Home Network

Before a mobile leaves its home network, it needs to ensure that, after the mobile is outside its home network, the home agent will be able to capture the packets addressed to the mobile's home address. This section discusses how this can be achieved in a home network that uses the Address Resolution Protocol (ARP) [39].

ARP is used in most broadcast media networks (e.g., Ethernet and IEEE 802.11 wireless LANs) by nodes to determine the hardware address associated with a target

IP address. A hardware address is an address that identifies a node at the link layer and is used by the link layer protocol to forward link-layer frames or packets. A typical hardware address is the Medium Access Control (MAC) address defined by the IEEE and used in many local area networks such as IEEE 802.11 and Ethernet.

When a node wants to send an IP packet to a target node and does not know the target node's hardware address, it broadcasts an ARP REQUEST message over the local IP network to ask all the nodes on the local network for the target node's hardware address. The ARP REQUEST message carries the following main fields:

- *Sender Protocol Address*: The IP address of the sender.
- *Target Protocol Address*: The target IP address for which the sender wishes to determine its associated hardware address.
- *Sender Hardware Address*: The hardware address of the sender.

The node that has the target IP address as its own IP address will reply with an ARP REPLY message that will carry its IP address and its hardware address. Once a node learns the mapping from an IP address to a hardware address, the node typically caches the mapping in its *ARP Cache* for later use so that it does not have to invoke the ARP protocol for every packet it has to send.

Two issues need to be resolved in order to support MIPv4 when a mobile is leaving its home network:

- After a mobile terminal leaves its home network, other nodes on the home network may still have the mapping of the mobile's IP address to its hardware address cached in their ARP caches. As a result, they will continue to send packets to the mobile's hardware address rather than to the home agent. As a result, these packets will be lost.

  MIPv4 uses *Gratuitous ARP* [40] to solve the problem described above. A Gratuitous ARP packet is a standard ARP packet sent by a node to trigger other nodes to update their ARP caches. The standard ARP protocol requires a node to use the information carried in *all* received ARP packets to update the existing entries in its ARP cache. Therefore, a Gratuitous ARP packet can be, for example, an ARP REQUEST packet.

  Before a mobile leaves its home network, it broadcasts a Gratuitous ARP packet to all other nodes (including mobility agents) on the local IP subnet. The mobile will set the ARP Sender Protocol Address and the ARP Target Protocol Address in this Gratuitous ARP packet to the mobile's home address. However, the mobile will set the ARP Sender Hardware Address in the Gratuitous ARP packet to its home agent's hardware address. Any node that receives such a Gratuitous ARP packet will update its ARP cache to map the sending mobile's home address to the home agent's hardware address. This will cause the node to forward future packets addressed to the mobile's home address to the mobile's home agent. A gratuitous ARP packet will need to be broadcast over the local subnet because the mobile does not know which nodes need to be updated.

- If a node on a mobile's home network does not have the mobile's hardware address in its ARP cache when it wants to send a packet to the mobile, this node will have to use ARP to find the mobile's hardware address. However, when the mobile is away from the home network, the mobile will not be able to reply to the ARP REQUESTs sent by nodes on the home network.

MIPv4 solves this problem using *Proxy ARP* [52]. A Proxy ARP packet is an ARP REPLY message sent by one node on behalf of another node in response to an ARP REQUEST. When the home agent receives an ARP REQUEST asking for the hardware address of the mobile that is away from the home network, the home agent will reply to this ARP REQUEST on behalf of the mobile. The home agent will set the Sender Protocol Address field and the Sender Hardware Address field of its ARP REPLY message to the home agent's own IP and hardware addresses, respectively. This will cause the nodes that receive the ARP REPLY message to forward packets addressed to the mobile's home address to the home agent.

4.2.2.4 *Entering and Staying in a Visited Network* Upon entering a visited network, a mobile terminal will have to acquire a temporary CoA from the visited network in order to receive IP packets from the visited network. The mobile will then have to register its new CoA with its Home Agent. This registration serves as a location update and will cause the Home Agent to tunnel the packets addressed to the mobile's home address to this new CoA.

MIPv4 defines two messages for the registration operation:

- *Registration Request*
- *Registration Reply*

Registration Request and Registration Reply are the only new messages defined by MIPv4. A mobile sends a Registration Request message to its home agent in order to register its current CoA with the HA. Upon receiving a Registration Request message, the home agent authenticates the mobile that originated this Registration Request. If the authentication is positive, the home agent will use the CoA carried in the received Registration Request message to update the mobile's CoA. Then, the home agent will return a Registration Reply message to the mobile in response to the Registration Request message.

Registration Request and Registration Reply messages are transported over UDP to a well-known UDP port number 434.

To register with the home agent, a mobile needs to provide its identity to the home agent. A mobile may identify itself by its home address or its NAI [15]. Using the NAI as a mobile terminal's identifier has several advantages over using a home address as the mobile's identifier.

- A mobile may have multiple radio interfaces, each using a different radio technology and having a different home address. In this case, the mobile could use its single unique NAI as its identifier to be authenticated by the network.
- A mobile may wish its home network to dynamically assign a home address to it. But, before it can be dynamically assigned a home address, the mobile needs to be identified and authenticated by the home agent.

A mobile may register its current CoA with its home agent directly or through a foreign agent. Registering with the home agent directly is to send Registration Request messages directly to the home agent without having to go through a foreign agent. Registering through a foreign agent is to send Registration Request messages first to a foreign agent, which will process the messages and then forward them to the mobile's home agent. Registering through a foreign agent allows the visited network to enforce policies on network access and accounting.

When a mobile is registering a Foreign Agent CoA with its home agent, the registration must be done via the foreign agent. When a mobile is registering a co-located CoA, the registration may be done directly with the home agent, unless the foreign agent requires registration via the foreign agent. The foreign agent can force a mobile to register through the foreign agent by setting the "R" flag in the Agent Advertisement it sends to the mobiles. When a mobile receives an Agent Advertisement with the "R" flag set, the mobile is required to register through the foreign agent even if the mobile can acquire its own co-located CoA without the assistance of the foreign agent.

Figure 4.11 shows the signaling message flows for direct registration with the home agent and registration through a foreign agent. Figure 4.11(a) is straightforward and self-explanatory. Now we take a closer look at how registration through a foreign agent works. Upon receiving a Registration Request from a mobile, the foreign agent can inspect the content of the Registration Request to determine whether the mobile is allowed to use the visited network. If the foreign agent determines that the mobile is allowed to use the visited network, it will relay the Registration Request to the mobile's home agent. In this case, the foreign agent will also relay the Registration Reply from the home agent back to the visiting mobile.

On the other hand, if the foreign agent wants to deny network access to the mobile, the foreign agent will generate a Registration Reply and send it to the mobile, indicating that the mobile's Registration Request has been denied. The foreign agent will discard the Registration Request received from the mobile without relaying it to the mobile's home agent.

When a mobile sends a Registration Request message, the IP source address of the IP packet carrying the Registration Request message will be the mobile's home address unless

- The mobile is registering a co-located CoA through a foreign agent. In this case, the IP source address will be the mobile's CoA.

(a) Registration with home agent directly.

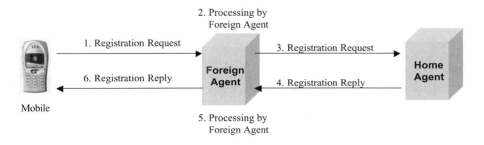

(b) Registration through foreign agent

Fig. 4.11 *MIPv4 registration message flows*

- The mobile is not registering a co-located CoA through a foreign agent and the mobile does not have a home address. In this case, the IP source address will be 0.0.0.0.

MIPv4 requires that a home agent authenticate *all* Registration Requests it receives and that a mobile terminal authenticate all Registration Reply messages it receives. This mutual authentication is designed to protect against a range of security attacks. For example, it can protect against malicious users from sending Registration Requests to a home agent to cause packets to another mobile user to be redirected (a security attack commonly known as redirection attack). Mutual authentication can also protect a malicious user from pretending to be a home agent to conduct "denial of service" attacks (i.e., to deny services to a mobile by rejecting its Registration Requests).

In addition to registering a CoA, a mobile terminal can also use Registration Request messages to

- Discover the address of a home agent.
- Discover the mobile's home address, if the mobile is not configured with a home address.
- Renew a registration that is due to expire.
- Deregister with the HA when the mobile returns home.

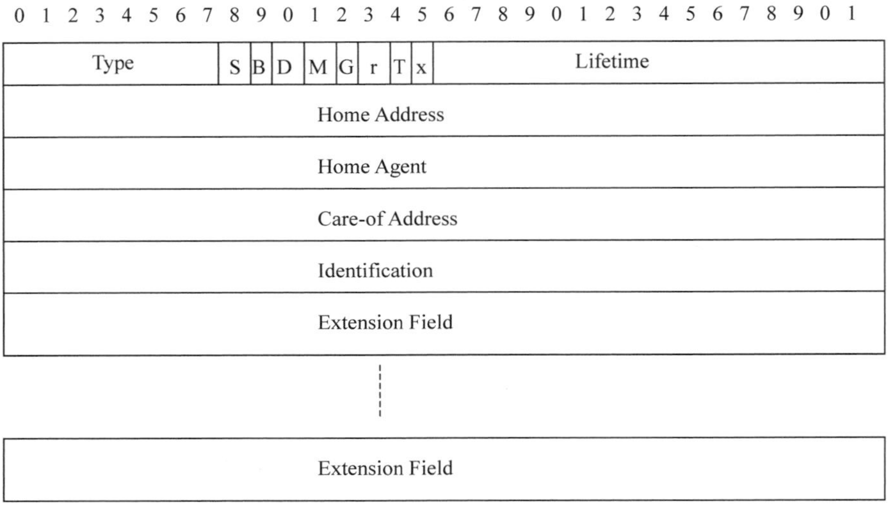

```
0  1  2  3  4  5  6  7  8  9  0  1  2  3  4  5  6  7  8  9  0  1  2  3  4  5  6  7  8  9  0  1
```

Type	S	B	D	M	G	r	T	x	Lifetime

Home Address

Home Agent

Care-of Address

Identification

Extension Field

Extension Field

Fig. 4.12 *MIPv4 Registration Request message format*

Figure 4.12 shows the format of the MIPv4 Registration Request message. The Type field indicates whether this is a Registration Request or Registration Reply. The flags S, B, D, M, G, r, T, x are defined as:

- *S: Simultaneous bindings*: If the "S" bit is set, the mobile node is requesting that the home agent retain its prior mobility bindings. In other words, a mobile may request its home agent to maintain multiple care-of addresses for the mobile at the same time. When the home agent intercepts a packet addressed to the mobile's home address, it will tunnel a copy of the packet to each currently registered care-of address.

- *B: Broadcast datagrams*: If the "B" bit is set, the mobile node requests that the home agent tunnel to it any broadcast datagrams that it receives on the home network.

- *D: Decapsulation by mobile terminal*: If the "D" bit is set, the mobile terminal will itself decapsulate datagrams that are sent to the care-of address. That is, the mobile terminal is using a co-located care-of address.

- *M: Minimal encapsulation*: If the "M" bit is set, the mobile terminal requests that its home agent use Minimal Encapsulation defined in [14] for datagrams tunneled to the mobile node.

- *G: GRE encapsulation*: If the "G" bit is set, the mobile terminal requests that its home agent use GRE encapsulation for datagrams tunneled to the mobile node.

- *r*: This field will always be zero and ignored on reception. This field should not be used for any other purpose.

- *T*: Reverse Tunneling requested.
- *x*: This field will always be zero and ignored on reception. This field should not be used for any other purpose.

The Lifetime field contains the number of seconds remaining before the registration is considered expired. A zero lifetime indicates a request for de-registration.

If a mobile has a preconfigured home address, it may put its home address in the Home Address field of the Registration Request. This allows the home network to identify the mobile by its home address. If the mobile does not have a preconfigured home address or wishes its home agent to dynamically assign a home address to it, the mobile can set the Home Address field to 0.0.0.0. In this case, the mobile should use its NAI [12] to identify itself. The mobile can specify its NAI in the Registration Request message using the Mobile Node NAI Extension [15]. When a home agent receives a Registration Request with the Mobile Node NAI Extension but without a valid home address, the home agent will assign a home address to the mobile and send the assigned home address back to the mobile in the Registration Reply message.

The Home Agent field contains the IP address of the mobile's home agent if the mobile knows its home agent's address. In case the mobile does not know the address of its home agent, it may use *Dynamic Home Agent Address Resolution* to discover the address of its home agent. In this case, the mobile terminal should set the Home Agent field to the subnet-directed broadcast address of the mobile terminal's home network. The mobile will send the Registration Request to the subnet-directed broadcast address for the mobile's home network so that the Registration Request will reach all IP nodes on the mobile's home network. Each home agent receiving such a Registration Request with a broadcast destination address will reject the mobile terminal's registration and return a Registration Reply message with a result code indicating the rejection. As every Registration Reply message will carry the IP address of the home agent sending the message, the mobile can therefore learn the IP address of a home agent by examining the Registration Reply messages it receives.

The Care-of Address field contains the mobile's CoA. The Identification field is a 64-bit number constructed by the mobile terminal used for matching Registration Requests with Registration Replies, and for protecting against *replay attacks* of registration messages. In a replay attack, a malicious user may capture packets from a user and resend ("replay") these packets while pretending to be the legitimate source of the packets. To prevent replay attacks, the mobile and its home agent can agree on a way to generate and interpret the value in the Identification field so that the packets replayed by a malicious user will be identified.

To support future enhancement, MIPv4 defines a general extension mechanism to allow optional information to be carried by the Registration Request. The fixed portion of a MIPv4 Registration Request message can be followed by one or more extension fields.

Since MIPv4 requires that a home agent authenticate all Registration Request messages it receives, a mandatory extension field in every Registration Request message is the Mobile-Home Authentication Extension (Section 4.2.2.6), which carries information needed for a home agent to authenticate the Registration Request.

Figure 4.13 shows the format of the MIPv4 Registration Reply message. The fields in the MIPv4 Registration Reply message are defined as follows:

- *Type*: Indicates whether this is a MIPv4 Registration Request or Registration Reply message.
- *Code*: Contains a value indicating the result of the corresponding Registration Request.
- *Lifetime*: For successful registration, the Lifetime field contains the number of seconds remaining before the registration is considered expired. A value of zero indicates that the mobile terminal has been deregistered. For failed registration requests, the Lifetime field should be ignored.
- *Home Address*: Contains the mobile's home address.
- *Home Agent*: Contains the IP address of the mobile terminal's home agent.
- *Identification*: Contains a 64-bit number used for matching Registration Requests with Registration Replies, and for protecting against replay attacks of registration messages. The value is based on the Identification field in the Registration Request message received from the mobile terminal, and on the style of replay protection used in the security context between the mobile terminal and its home agent.

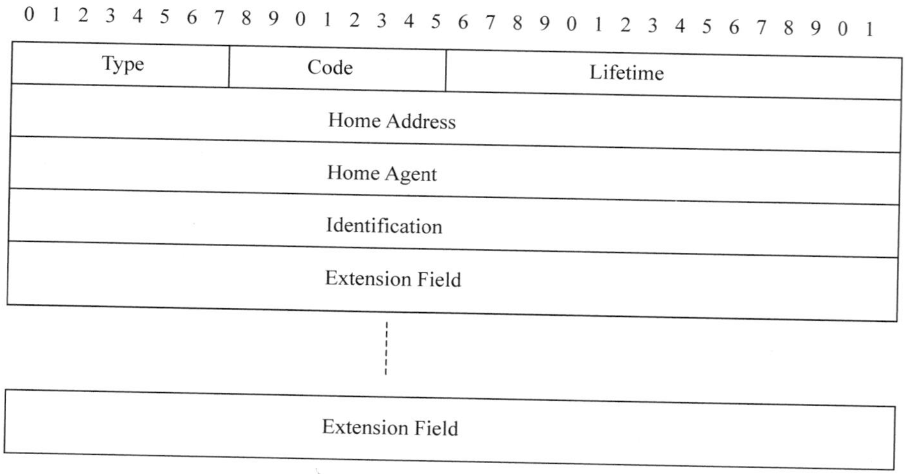

```
0 1 2 3 4 5 6 7 8 9 0 1 2 3 4 5 6 7 8 9 0 1 2 3 4 5 6 7 8 9 0 1
+---------------+---------------+-------------------------------+
|     Type      |     Code      |            Lifetime           |
+---------------+---------------+-------------------------------+
|                        Home Address                           |
+---------------------------------------------------------------+
|                        Home Agent                             |
+---------------------------------------------------------------+
|                      Identification                           |
+---------------------------------------------------------------+
|                      Extension Field                          |
                              :
                              :
+---------------------------------------------------------------+
|                      Extension Field                          |
+---------------------------------------------------------------+
```

Fig. 4.13 *MIPv4 Registration Reply message format*

Like a Registration Request message, a Registration Reply message can also have Extensions fields after the fixed portion of the Registration Reply message. A mandatory Extensions field to be carried in every Registration Reply message sent by a home agent is the Mobile-Home Authentication Extension (Section 4.2.2.6), which carries information to be used by a mobile to authenticate the Registration Reply message.

4.2.2.5 Returning to the Home Network

When a mobile terminal returns to its home network, it needs to ensure that the packets addressed to its home address will now be forwarded to itself directly, rather than to its HA. In particular, two steps need to be taken:

- The IP-to-hardware address binding cached by the nodes on the home network for the returning mobile needs to be updated so that these nodes will start to send packets directly to the mobile rather than to the home agent.
- The returning mobile should inform its HA that it is back in the home network so that the HA can remove the states it maintains for the mobile.

We now illustrate how the two steps described above can be implemented in a home network that uses the ARP protocol for address resolution. To trigger the nodes on the home network to update their ARP caches, both the returning mobile and the home agent may need to broadcast Gratuitous ARP packets over the home network. This, for example, is necessary if both the mobile and the home agent use radio transmissions and their radio coverage areas are different. Upon moving back to the home network, the mobile will first broadcast a Gratuitous ARP on its home network so that the other nodes, including the home agent, on the home network will start to forward packets destined to the mobile directly. The mobile will then send a MIPv4 Deregistration Request to the home agent to inform the home agent that the mobile is now back in the home network. The MIPv4 Deregistration Request is simply a MIPv4 Registration Request message with its Lifetime field set to zero. When the home agent receives and accepts the Deregistration Request, the home agent will also broadcast a gratuitous ARP on the home network in an attempt to trigger any node that the returning mobile cannot reach to update its ARP cache.

4.2.2.6 Mobile-Home Authentication Extension

Authentication of MIPv4 Registration Request and Registration Reply messages are achieved using a Mobile-Home Authentication Extension to the MIPv4 Registration Request and the Registration Reply messages.

The format of the Mobile-Home Authentication Extension is illustrated in Figure 4.14. The Type field should carry a value 32, indicating that this Extension is a Mobile-Home Authentication Extension. The Length field contains the length in octets of the Mobile-Home Authentication Extension from the beginning of the SPI field to the end of this Extension.

0 1 2 3 4 5 6 7 8 9 0 1 2 3 4 5 6 7 8 9 0 1 2 3 4 5 6 7 8 9 0 1

Type	Length	Security Parameter Index (SPI)
SPI (continued)		Authenticator . . . . . .

Fig. 4.14 *Mobile-Home Authentication Extensions to Mobile IP messages*

A Security Parameter Index (SPI) is a four-octet identifier used to identify a security context between a mobile and its home agent. The SPI identifies the authentication algorithm and the *secret* used by the mobile and its home agent to compute the Authenticator.

The Authenticator is a number calculated by applying an authentication algorithm on the message that needs to be protected. The default authentication algorithm uses HMAC-MD5 [32] to compute a 128-bit "message digest" (Chapter 5, "Security") and uses it as the Authenticator.

The Authenticator in the Mobile-Home Authentication Extension protects the following fields of a MIPv4 Registration Request message or a Registration Reply message, as illustrated in Figure 4.15:

- The Registration Request or the Registration Reply data.
- All other Extensions to the Registration Request or the Registration Reply message prior to the Mobile-Home Authentication Extension.
- The Type, Length, and SPI fields of this Mobile-Home Authentication Extension.

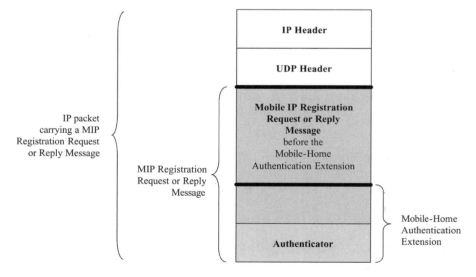

Fig. 4.15 *Fields protected by MIP Mobile-Home Authentication Extension*

The Authenticator field as well as the IP header and the UDP header in the IP packet carrying the MIPv4 Registration Request or the Registration Reply message are not included in the computation of the Authenticator value by default.

4.2.2.7 Vendor/Organization Specific Extensions to Mobile IP Messages

Vendor/Organization Specific Extensions [21] allow network equipment vendors and other organizations (e.g., network operators) to add their specific information to the Mobile IP signaling messages (i.e., Registration Request, Registration Reply, Agent Advertisement messages). These extensions enable network equipment vendors and network operators to implement creative mobility control capabilities in addition to the basic mobility control capabilities provided by the basic Mobile IP protocol. For example, a 3GPP2 packet data network uses Mobile IP signaling messages for mobility control and route management inside a 3GPP2 packet data network. There, the Vendor/Organization Specific Extensions are used to carry 3GPP2-specific control information (Section 4.4.3.4).

Two Vendor/Organization Specific Extensions have been defined in IETF RFC 3115 [21]: Critical Vendor/Organization Specific Extensions (CVSE) and Normal Vendor/Organization Specific Extensions (NVSE). The main differences between the Critical and Normal Extensions are as follows:

- *Handling by a Mobile IP entity*: When a Mobile IP entity (Mobile IP terminal, home agent, foreign agent) encounters a CVSE but does not recognize the extension, it must silently discard the entire message containing the CVSE. However, when an NVSE is encountered but not recognized, the NVSE should be ignored, but the rest of the message containing the NVSE must be processed.
- *Format of the Extensions*: The CVSE has a Length field of two octets, and the NVSE has a Length field of only one octet.

The formats of the CVSE and NVSE are shown in Figure 4.16 (a) and (b) respectively. The CVSE contains the following fields, listed in the order of their appearance in the CVSE:

- *Type*: The Type field is set to the CVSE-TYPE-NUMBER, 38, to indicate that this is a CVSE.
- *Reserved*: This field is reserved for future use and must be set to 0 by the sender and must be ignored on reception.
- *Length*: This field contains the length in bytes of this extension, not including the Type and Length bytes.
- *Vendor/Org-ID*: This field contains the identifier of the vendor or organization that is using this extension.
- *Vendor-CVSE-Type*: This field indicates the particular type of this CVSE. A vendor may assign and use different types of CVSEs at its discretion.

- *Vendor-CVSE-Value*: This field contains vendor/organization-specific data. It may contain zero or more octets.

The NVSE contains the following fields, listed in the order of appearance in the NVSE:

- *Type*: The Type field is set to the NVSE-TYPE-NUMBER, 134, to indicate that this is an NVSE.
- *Length*: This field contains the length in bytes of this extension, not including the Type and Length bytes.
- *Reserved*: This field is reserved for future use and must be set to 0 by the sender and must be ignored on reception.
- *Vendor/Org-ID*: This field contains the identifier of the vendor or organization that is using this extension.
- *Vendor-NVSE-Type*: This field indicates the particular type of this NVSE. A vendor may assign and use different types of NVSEs at its discretion.
- *Vendor-NVSE-Value*: This field contains vendor/organization-specific data. It may contain zero or more octets.

```
0 1 2 3 4 5 6 7 8 9 0 1 2 3 4 5 6 7 8 9 0 1 2 3 4 5 6 7 8 9 0 1
```

Type	Reserved	Length
Vendor/Org-ID		
Vendor-CVSE-Type	Vendor-CVSE Value • • • • • •	

(a) Critical Vendor/Origination Specific Extension (CVSE)

```
0 1 2 3 4 5 6 7 8 9 0 1 2 3 4 5 6 7 8 9 0 1 2 3 4 5 6 7 8 9 0 1
```

Type	Length	Reserved
Vendor/Org-ID		
Vendor-NVSE-Type	Vendor-NVSE Value • • • • • •	

(b) Normal Vendor/Origination Specific Extension (NVSE)

Fig. 4.16 *Vendor/Organization Specific Extensions to Mobile IP messages*

4.2.2.8 Reverse Tunneling When a mobile sends IP packets in a visited network, the IP source addresses in these outgoing packets may not belong to the IP addressing space used in the visited network. For example, the IP source address may be the mobile's home address. Today, an increasing number of routers on the Internet use information in addition to the destination IP address to make routing decisions. For example, an IP access router in a visited network may reject any packet whose source IP address is not part of the IP addressing space of the visited network (a technique commonly referred to as "ingress filtering"). As a result, outgoing packets from a visiting mobile may not be able to go through the IP access router in the visited network that implements ingress filtering.

Reverse tunneling [35] provides a solution to the problem described above. Reverse tunneling is to tunnel a mobile's outgoing packets from the mobile's CoA back to the mobile's home agent. The home agent will then decapsulate the packets and route the original packets to their final destinations.

IETF RFC 3024 [35] specifies how reverse tunneling works when a mobile uses Foreign Agent CoA. In this case, the reverse tunnel goes from a foreign agent to the mobile's home agent. A mobile arrives at a visited network, listens for Agent Advertisement messages, and selects a foreign agent that supports reverse tunnels. A foreign agent informs visiting mobiles that it supports reverse tunneling by setting the "T" flag in the Agent Advertisement messages it sends to the mobiles. The mobile requests the reverse tunneling service when it registers through the selected foreign agent, by setting the "T" flag in the MIPv4 Registration Request. This will cause the foreign agent to establish a reverse tunnel to the mobile's home agent.

After the MIPv4 registration via the foreign agent, the visiting mobile may use one of the following ways to deliver its packets to the foreign agent:

- *Direct Delivery Style*: The mobile designates the foreign agent as its default router and proceeds to send packets directly to the foreign agent, that is, without encapsulation. The foreign agent intercepts these packets and tunnels them over the reverse tunnel to the mobile's home agent.

- *Encapsulating Delivery Style*: The mobile encapsulates all its outgoing packets and sends the encapsulated packets to the foreign agent. The foreign agent de-encapsulates and tunnels them over the reverse tunnel to the mobile's home agent.

The mobile specifies the delivery style it wishes to use in the Registration Request message it sends to the foreign agent.

When reverse tunneling is used, user packets from a mobile to a correspondent host follow the path illustrated in Figure 4.17.

4.2.2.9 Limitations of MIPv4 MIPv4 in its basic form has several well-known limitations.

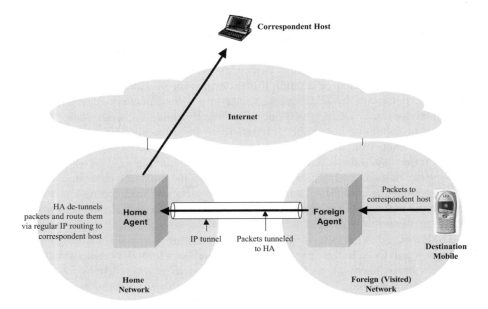

Fig. 4.17   *Mobile IPv4 reverse tunneling*

- *Triangular routing*: Triangular routing refers to the fact that, with MIPv4, packets addressed to a mobile's home address will have to be routed to the mobile's home agent first, then be forwarded by the mobile's home agent to the mobile's current care-of address. Triangular routing could introduce long end-to-end packet delays and lead to inefficient use of network resource. A technique—Route Optimization—has been proposed to reduce the number of packets that have to experience triangular routing. Route optimization will be discussed in Section 4.2.2.10.

- A home agent may become a traffic and performance bottleneck: All user traffic destined to a mobile outside its home network will have to go through the mobile's home agent. This makes a home agent a potential traffic and performance bottleneck as the number of mobile terminals and/or the traffic volume destined to these mobile terminals grow.

- *Potential long handoff delay*: When a mobile changes its CoA (e.g., when it handoffs from one IP subnet to another), it has to register its new CoA with its home agent. If the foreign network is far away from the mobile's home network, this registration process could introduce a long delay that may be unacceptable to on-going real-time sessions of voice or multimedia applications. To reduce handoff delay, "micromobility" management protocols have been proposed. These will be discussed in Sections 4.2.3, 4.2.7, and 4.2.8.

- *Potential insufficient deregistration capability*: After a mobile is registered through a foreign agent, the mobile may move away from this foreign agent into a new network. Using the basic MIPv4, the mobile does not explicitly deregister with the foreign agent in the old network. Instead, a mobile's registration with the old foreign agent expires only when the registration lifetime expires. This makes it difficult for a visited network to determine when a mobile has left the visited network, making it difficult for the visited network to release the network resources allocated to the mobile in a timely manner after the mobile left the visited network. It also makes it difficult for a visited network to determine how much time a visiting mobile has spent in the visited network.

- *Insufficient capabilities to support other mobility management requirements*: For example, current MIPv4 does not support *dormant* mobiles. A dormant mobile exchanges limited information infrequently with the network in order to save scarce resources (e.g., power on the mobile), and, therefore, the network may not know the precise location of a dormant mobile. The network will need to perform *paging* to determine the mobile's precise location when it has packets to send to a dormant mobile. To support dormant mobile terminals, IP paging protocols are being designed [31], [55]. One approach is to add paging capability to MIPv4 (Section 4.2.4).

4.2.2.10 *MIPv4 Route Optimization*

Route optimization [38] is a technique that enables a correspondent node to address packets to a mobile's current CoA directly so that these packets do not have to be first routed to the mobile's home agent. Route optimization reduces the number of packets that have to experience triangular routing.

Figure 4.18 illustrates the operation of route optimization assuming that the mobile is using a foreign agent care-of address. The basic idea is to allow a correspondent node to be aware of a mobile's current CoA and then tunnel packets

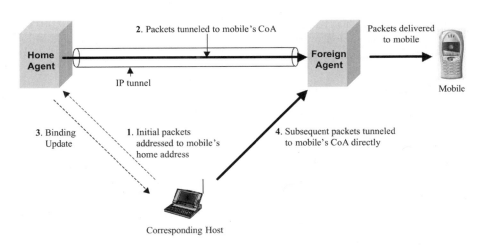

Fig. 4.18 *MIPv4 route optimization*

to the destination mobile's CoA directly. A correspondent host may maintain a *Binding Cache* that maps the mobiles' home addresses to their CoAs. When a packet is to be sent, the correspondent host will first search its Binding Cache for the mobile's CoA. If a binding cache entry is found, the correspondent host will tunnel the packets to the mobile's CoA directly. Otherwise, it will send the packet to the mobile's home address as in the basic MIPv4.

A mobile's home agent is responsible for informing correspondent hosts of the mobile's current CoA. The home agent does so by sending a Binding Update message to a correspondent host. The home agent deduces that a correspondent host does not have a binding cache entry for a mobile if the home agent intercepts a packet that is addressed to the mobile's home agent and is originated from the correspondent host. The Binding Update message will contain the mobile's home address and current CoA and the lifetime associated with the CoA.

A correspondent host does not need to acknowledge a Binding Update message received from the home agent. This is because a correspondent host will continue to send packets destined to a mobile to the mobile's home agent if the correspondent host fails to receive the Binding Update message. These future packets, when intercepted by the home agent, will trigger the home agent to send new Binding Update messages to the correspondent host.

For a correspondent host to accept Binding Updates from a mobile's home agent, a security association between the correspondent host and the home agent needs to be established. In particular, a correspondent host needs to be able to authenticate the received Binding Updates to determine whether they are from nodes that are allowed to send such Binding Updates. This is necessary to prevent malicious users from sending Binding Updates to a correspondent node to cause the correspondent host to send its packets to the wrong places. However, the requirement of a security association between the home agent and a correspondent host becomes a critical limitation of MIPv4 route optimization. Establishing a security association between a home agent and every possible correspondent host on a large network such as the Internet is difficult. This is a major cause of limited scalability of the existing MIPv4 route optimization approach and a key reason of the slow adoption of MIPv4 route optimization by the industry.

4.2.3   MIPv4 Regional Registration

As discussed in Section 4.2.2, a mobile using the basic MIPv4 protocol has to register with its home agent every time it changes its care-of address. This could introduce long handoff delay when the visited network is far away from the mobile's home network. MIPv4 Regional Registration [24] has been proposed to reduce handoff delay. It extends the basic MIPv4 protocol to allow a mobile to register its new care-of addresses locally with its visited network domain. A network domain, or domain for short, is a collection of networks sharing a common network administration.

Figure 4.19 illustrates the operation of MIPv4 Regional Registration. Each network domain consists of two or more hierarchical levels of foreign agents. We

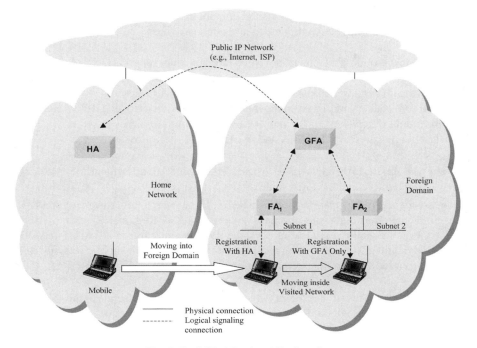

Fig. 4.19 *MIPv4 Regional Registration*

will use a two-level hierarchy of foreign agents to illustrate the principles and operation of MIPv4 Regional Registration. At the top level of the hierarchy are the Gateway Foreign Agents (GFAs). Each domain will have at least one GFA in order to support MIPv4 Regional Registration. GFAs are the foreign agents that directly interact with visiting mobiles' home agents outside the domain. Therefore, a GFA must have a publicly routable IP address. At the lower level of the hierarchy are any number of FAs.

A mobile inside a visited domain will have two CoAs:

- *GFA Address*: The mobile will register the address of a GFA in the visited domain as its CoA with its home agent.
- *Local CoA*: A local CoA is an address used by the mobile to receive packets over a network inside the visited domain. The local CoA can be shared or co-located. A shared local CoA is an address of an FA that is at the lowest level of the FA hierarchy in the visited network and that can deliver packets to the mobile. A co-located local CoA is a local IP address that is co-located on the mobile.

To support MIPv4 Regional Registration, the MIPv4 Agent Advertisement message is extended to include a flag "I" to indicate whether the domain supports

MIPv4 Regional Registration. If a visited domain does not support MIPv4 Regional Registration, the mobile continues to use standard basic MIPv4 in the visited domain.

When the mobile first enters a visited domain that supports MIPv4 Regional Registration, it needs to learn the address of a GFA in the visited domain and registers the GFA address as its CoA with its HA. This will cause the mobile's HA to tunnel future packets addressed to the mobile's home address to the GFA, which will in turn tunnel the packets to the mobile's Local CoA.

The mobile can learn the GFA address in one of the following ways:

- *From Agent Advertisement messages*: The Agent Advertisement messages are extended to carry the GFA address.

- *Dynamically assigned by visited network*: If the Agent Advertisement message indicates that the visited domain supports MIPv4 Regional Registration but does not contain any GFA address, the mobile can require the visited network to dynamically assign it with a GFA address. To do so, the mobile sets the CoA field in its Registration Request to zero.

If an FA advertises (in the Agent Advertisement messages it sends to the mobiles) support for MIPv4 Regional Registration, the FA will process Registration Requests messages in the following way. When the FA receives a Registration Request message from a mobile, it extracts the CoA from the Registration Request message. If this CoA is neither zero nor the address of the FA, the CoA must be the address of a GFA and the FA will forward the Registration Request message to the GFA. If the CoA is zero, the FA will assign a GFA to the mobile. The FA will add the following extensions to the received Registration Request message and then relay the Registration Request message with the added extensions to the GFA:

- A GFA IP Address Extension, which contains the address of the assigned GFA.
- A Hierarchical Foreign Agent Extension, which contains the address of the FA.

When a mobile moves between FAs connected to the same GFA, there will be no need for the mobile to perform MIP registration with its home agent. Instead, the mobile only needs to perform regional registration, i.e., to register its new local CoA with the GFA so that the GFA knows where to deliver packets destined to the mobile. When the mobile moves to a new GFA inside a visited domain, it needs to perform a home registration to inform its home agent of the address of the new GFA.

MIPv4 Regional Registration introduces two new messages for supporting the regional registration operation described above:

- *Regional Registration Request*: Sent by a mobile to a GFA via the FA to initiate regional registration.
- *Regional Registration Reply*: Sent by a GFA to a mobile in response to a Regional Registration Request.

4.2.4 Paging Extensions to Mobile IPv4

Mobile IP can be extended to support paging. One set of paging extensions to Mobile IPv4 is the P-MIP (Paging in Mobile IP) [56]. Here, we will use P-MIP as an example to illustrate how Mobile IPv4 may be extended to support paging.

With P-MIP, a mobile can be in active or idle state. An active mobile operates in exactly the same manner as in standard Mobile IP without P-MIP. A mobile in idle state, however, may not perform MIP registration.

A mobile uses an *Active Timer* to determine whether it should be in active or idle state. It stays in active state for an Active Timer period and changes into idle state when its Active Timer expires. Each time a mobile sends or receives a packet, it restarts its Active Timer. An idle mobile transitions into active state whenever it receives or sends any packet.

The FA through which a mobile performed its last Mobile IP registration, which is referred to as the mobile's *Registered FA*, is responsible for keeping track of whether the mobile is active or idle. The FA also uses an *Active Timer* to determine whether a mobile is active or idle. The FA considers a mobile to be in active state for an Active Timer period and assumes the mobile is in idle state when the Active Timer for the mobile expires. Each time the mobile's Registered FA sends a packet to or receives a packet from the mobile, it restarts the Active Timer for the mobile.

Since FAs are used to track the mobiles' active/idle states, P-MIP requires that

- An FA is required on each IP subnet.
- Mobiles can only use FA CoAs and have to perform Mobile IP registration through FAs.

FAs are grouped into *Paging Areas*. An idle mobile does not have to perform MIP registration when moving from one IP subnet to another inside the same paging area; it only needs to perform MIP registration when it moves into a new paging area.

Figure 4.20 illustrates how P-MIP delivers packets to idle mobiles. Packets addressed to a mobile's home address will be tunneled by the mobile's home agent to the mobile's CoA, which is the mobile's Registered FA. Upon receiving packets destined to a mobile, the mobile's Registered FA checks if the mobile is active or idle. If the FA believes that the mobile is active, it will forward the packets over its own local network directly to the mobile.

If the mobile's Registered FA believes that the mobile is idle, it will broadcast a Paging Request over its own local network and will unicast a Paging Request to every FA in the same Paging Area.

The FA that sends a Paging Request is referred to as a Paging FA. When an FA receives a Paging Request from a Paging FA, it authenticates the Paging FA to ensure that the Paging FA is authorized to send Paging Requests and then broadcasts a Paging Request over its local network if the authentication is successful.

When an idle mobile receives a Paging Request, it will transition into active mode. If it detects that it is now in a new IP subnet that is different from the subnet where it performed its last Mobile IP registration, it will acquire a new CoA and

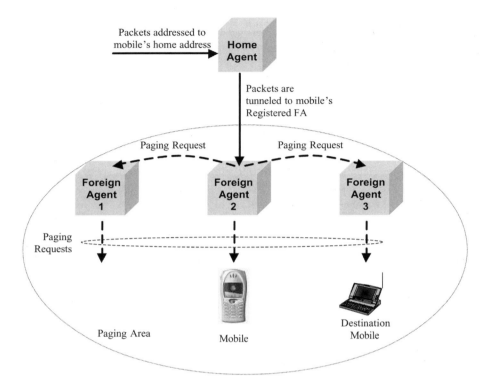

Fig. 4.20   *Paging Extensions to Mobile IPv4*

perform Mobile IP registration through the FA in the new IP subnet. This will cause the mobile's HA to tunnel the mobile's future packets to the FA in the new subnet.

To help the mobiles to determine whether they have changed paging areas, each paging area is identified by a unique *Paging Area Identifier* (PAI). The FAs are responsible for informing the mobiles which paging areas they are currently in. This is accomplished by extending the Mobile IP Agent Advertisement message to carry the PAI as well as a flag indicating whether the FA supports paging. A mobile compares the PAIs received from different FAs to determine whether it has moved into a new Paging Area.

The use of Active Timers to determine when a mobile is in active or idle state avoids the need for mobiles to use explicit signaling messages to inform an FA when the mobile will be entering idle mode, which simplifies protocol design. It, however, has some limitations.

- The value of the Active Timer depends on the nature of the application traffic. For example, when a mobile is sending or receiving a stream of packets, the value of the Active Timer should be longer than the inter-packet arrival times so that no extra paging will be needed before the last packet of the packet stream is received by the mobile. Otherwise, paging could introduce significant packet delay and delay jitters.

Different applications generate different types of traffic with widely varying interpacket arrival times. Therefore, mobiles should be able to dynamically adjust the value of its Active Timer. However, adjusting the Active Timer value dynamically will require the mobile to send signaling messages to inform its Registered FA of the new Active Timer value. This defeats the purpose of using Active Timers, i.e., to avoid the need for mobiles to use explicit signaling messages to inform an FA when the mobile will be entering idle mode.

• The value of the Active Timer maintained on the mobile should be the same as (or at least not significantly different from) the value of the Active Timer used by the mobile's Registered FA for the mobile. This requires an FA to know the value of the Active Timer for each mobile that may register with it. Pre-configuring such Active Timer values on all the FAs for every mobile does not seem to be a scalable approach. A mobile may inform the FA of its Active Timer value at the time it performs Mobile IP registration. This requires further extension to the MIP Registration message to carry the Active Timer value.

4.2.5 Mobile IPv6

Mobile IPv6, as Mobile IPv4, makes a mobile's movement (i.e., change of IPv6 address) transparent to the upper layer protocols and applications on the mobile as well as on correspondent nodes. MIPv6 uses the same concepts of home networks and home addresses as in MIPv4. Each MIPv6 mobile has a home network and an IPv6 home address assigned to the mobile within the network prefix of its home network. The mobile's IPv6 home address does not have to change regardless of where the mobile is. A correspondent node can always address packets to a mobile's IPv6 home address. Mobile IPv6 ensures that a mobile can receive the packets addressed to its home address regardless of where the mobile is.

When a mobile moves into a foreign network, it will acquire an IPv6 care-of address from the foreign network and use it to receive packets from the foreign network. To ensure that a mobile can continue to receive packets addressed to its IPv6 home address, the mobile will register its current care-of address with its home agent. The association between a mobile's home address and its care-of address is referred to as a *binding*.

As illustrated in Figure 4.21, each time a mobile changes its care-of address, it will send a Binding Update (BU) message to its home agent to register its current care-of address with the home agent. The home agent will return a Binding Acknowledgment (BA) message to inform the mobile of the status of the Binding Update. The formats of BU and BA messages are described in Section 4.2.5.4.

As in MIPv4, MIPv6 also requires that a home agent authenticate every BU message it receives and that a mobile authenticate every BA it receives. Authentication of BU and BA messages is achieved using IPsec (Chapter 5,

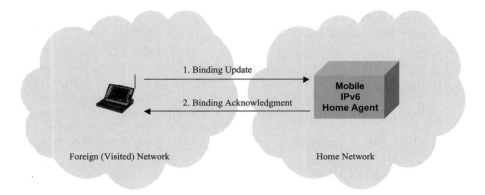

Fig. 4.21 MIPv6 address binding with home agent

"Security"). In particular, the IPsec Encapsulating Security Payload (ESP) header in transport mode should be used for the mutual authentication between a mobile and its home agent.

Unlike MIPv4, MIPv6 does not use foreign agents. Recall that foreign agents in Mobile IPv4 provide two main functions: provide care-of addresses to visiting mobiles and help the mobiles detect whether they have moved into a new network and hence have to change its care-of address (i.e., movement detection). In an IPv6 network, mobiles use only co-located care-of addresses. Therefore, there is no need for a foreign agent to provide care-of addresses. Furthermore, standard IPv6 facilities of IPv6 Neighbor Discovery [50] can be used to help IPv6 mobiles to detect movement. Movement detection is discussed further in Section 4.2.5.1.

Based on the ways packets are delivered to a mobile outside its home network, MIPv6 supports two modes of operation:

- *Bi-directional tunneling* mode
- *Route optimization* mode

The bi-directional tunneling mode of operation is similar to how MIPv4 works when an IPv4 mobile uses a co-located care-of address. As illustrated in Figure 4.22, a correspondent host does not have to use MIPv6. It treats a mobile destination in exactly the same way it treats a fixed destination. When it wants to send a packet to a mobile, it always uses the mobile's home address as the destination address in the IPv6 header of the packet (we say that these packets are addressed to the mobile's home address).

The packets addressed to a mobile's home address will be routed via regular IPv6 routing to the mobile's home network. If the mobile is inside its home network, these packets will be delivered to the mobile via regular IPv6 routing and/or the specific lower layer protocols used inside the mobile's home network, without the involvement of MIPv6. If the mobile is outside its home network, its home agent

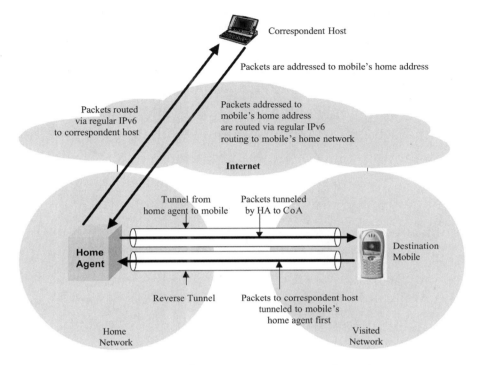

Fig. 4.22   *MIPv6 bi-directional tunneling mode of operation*

will intercept the packets addressed to its home address and then tunnel these packets to the mobile at its current location.

While a mobile is away from its home network, packets originated from the mobile will be tunneled to the mobile's home agent first. This is similar to reverse tunneling in MIPv4 (Section 4.2.2.8). The home agent will then use regular IPv6 routing to route these packets toward their final destinations. In the route optimization mode of operation, a mobile will register its binding not only with its home agent but also with its correspondent hosts. Packets from a correspondent host can then be routed directly to the care-of address of the distination mobile.

As illustrated in Figure 4.23, before a correspondent host has the binding for a mobile, it will address packets to the mobile's home address. These initial packets will be tunneled by the home agent to the mobile. The mobile can then send its binding to the correspondent host so that the correspondent host will be able to sent future packets directly to the mobile.

Route optimization is designed to be an integral part of MIPv6. To support route optimization, MIPv6 requires each IPv6 host and MIPv6 home agent to use a binding cache to maintain the binding information received from the mobiles. When an IPv6 terminal wishes to send a packet to another IPv6 terminal, it first checks its

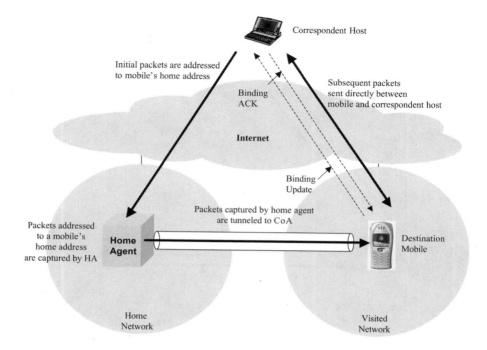

Initial packets are addressed
to mobile's home address

Binding
ACK

Subsequent packets
sent directly between
mobile and correspondent host

Correspondent Host

Internet

Binding
Update

Packets captured by home agent
are tunneled to CoA

Packets addressed
to a mobile's
home address
are captured by HA

**Home
Agent**

Destination
Mobile

Home
Network

Visited
Network

Fig. 4.23 *MIPv6 route optimization*

binding cache to see if it has a binding for the destination. If it does, it can address the packet to the destination's CoA directly. If it does not have any binding for the destination, it will address the packet to the destination's home address.

Recall that a main objective of MIPv6 is to make the change of IP addresses transparent to the protocols and applications above the IPv6 and MIPv6 layers. How can this be achieved when a correspondent host or home agent is allowed to address packets directly to the mobile's care-of address, which can change any time? This will be discussed in greater detail in Section 4.2.5.2.

When the mobile away from its home network wants to send a packet to a correspondent host or the mobile's home agent, the mobile may use its care-of address as the source IPv6 address in the IPv6 header of the packet. This allows the packet to go through access routers without having to use reverse tunneling (Section 4.2.2.8). This requires MIPv6 to solve the following problem: How can MIPv6 make the change of care-of address transparent to the protocols and applications above the IPv6 layer on the correspondent host? The solution is described in Section 4.2.5.3.

When a mobile's binding is about to expire on a correspondent node, the correspondent node may ask the mobile to refresh its binding by sending a Binding Refresh Request message to the mobile.

MIPv6 does not require a mobile and a correspondent node to have a static security association in order for the correspondent node to accept a mobile's BU.

Instead, a method called *return routability* is designed for a correspondent node to ensure dynamically that the right mobile terminal is sending a Binding Update message.

4.2.5.1 Movement Detection

The basic approach used by an IPv6 mobile for movement detection is IPv6 Neighbor Discovery [50]. IPv6 Neighbor Discovery enables an IPv6 terminal to discover new IPv6 routers and determine if a router is reachable (i.e., if the terminal and the router can receive packets from each other).

Using IPv6 Neighbor Discovery, an IPv6 router on each local network will broadcast Router Advertisement messages to mobiles on that network. These Router Advertisement messages carry, among other information, the IPv6 addresses of the router and network prefixes that can be used by mobiles to configure their care-of addresses. The information in the Router Advertisement message allows a mobile to discover new IPv6 routers. It also helps a mobile to detect whether an IPv6 router is still reachable, hence, helping the mobile to detect whether it has moved out of a network and whether it has moved into a new network. A mobile also uses other information to help determine whether it is still reachable from a router. For example, the fact that a mobile just received any packet from a router can be used as an indication that the mobile is still reachable from the router.

A mobile can also proactively probe the network to see if there are reachable routers. A mobile may do so by broadcasting Neighbor Solicitation messages over the local network. Upon receiving such a Neighbor Solicitation message, a router will send Router Advertisement messages to the mobile.

A mobile may also use any other means available to supplement the capabilities provided by IPv6 Neighbor Discovery to help perform movement detection. For example, a mobile may use indications from lower protocol layers to help detect its movement. For example, a handoff at the lower layer (e.g., change of radio channels, radio cells, or radio interfaces on the mobile) can be used as an indication that the mobile may have moved into a new IP network.

A mobile can acquire an IPv6 care-of address by using IPv6 Stateless Address Auto-configuration [48] to combine a network prefix received in the Router Advertisement messages with the mobile's own hardware address. The hardware address identifies the mobile terminal uniquely. The network prefix identifies the network to which the mobile is currently attached. A mobile may also use stateful protocols, such as DHCPv6, to acquire new care-of addresses.

4.2.5.2 Sending Packets Directly to Mobile's Care-of Address

When a correspondent host has a binding for a mobile, the correspondent host can address IPv6 packets directly to the mobile's care-of address. A mobile's care-of address can change any time. Mobile IPv6 wants to make these address changes transparent to the protocols and applications above the IP and Mobile IP layers.

This is achieved using an IPv6 *routing header* defined by MIPv6. In IPv6, a routing header is used by an IPv6 source node to list one or more nodes that should process the IPv6 packet, in addition to the node identified by the destination IPv6

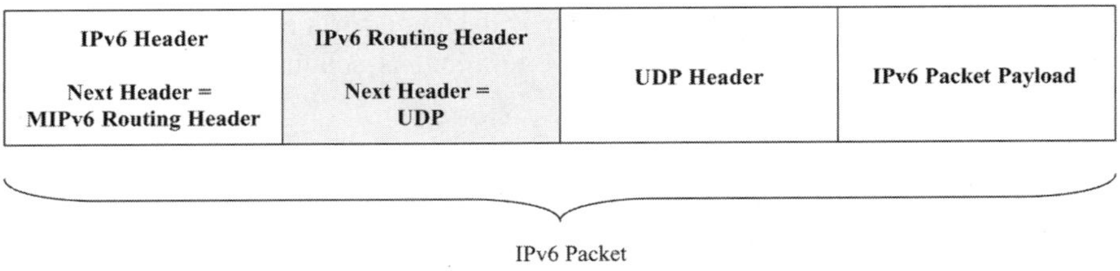

IPv6 Header Next Header = MIPv6 Routing Header	IPv6 Routing Header Next Header = UDP	UDP Header	IPv6 Packet Payload

IPv6 Packet

Fig. 4.24   IPv6 routing header

address in the IPv6 header of the IPv6 packet. When a packet is processed by a node, we say that the packet *visited* the node.

A routing header is inserted between the IPv6 header and the header of the upper layer protocol (e.g., UDP or TCP). An IPv6 packet carrying a routing header is illustrated in Figure 4.24, assuming that upper layer protocol used to transport user data is UDP.

The routing header will not be examined or processed by any node along a packet's path until the packet reaches the node identified by the destination address in the IPv6 header.

When a correspondent host sends a packet directly to a mobile, it will use the mobile's care-of address as the destination address in the IPv6 header of the packet. The mobile's home address will be carried in a routing header defined by MIPv6. When the packet arrives at the destination mobile's care-of address, the mobile will process the routing header carried in the packet. This will allow the mobile to know that the packet should be routed to the address in the routing header, i.e., to the mobile's home address. The mobile replaces the IPv6 destination address in the IPv6 header of the packet with the mobile's home address, decrements the Segments Left field in the routing header by one (i.e., the Segments Left will become 0, indicating that the mobile's home address is the final destination of the packet), and resubmits the packet to the IPv6 for processing. As the mobile's home address and the final destination of the packet is the mobile itself, the IPv6 layer on the mobile will deliver the packet to the upper layer protocol. Hence, the change of care-of address on the mobile is transparent to the upper layer protocols and applications on the mobile because the packet delivered to the upper layer carries the mobile's home address as the destination address in its IPv6 header.

The format of the routing header defined by MIPv6 is shown in Figure 4.25. The fields in the routing header are as follows:

- *Next Header*: An 8-bit code that identifies the type of header immediately following the routing header.
- *Header Extension Length*: An 8-bit unsigned integer that indicates the length of the routing header in eight-octect units, not including the first eight octets.
- *Routing Type*: The type of the routing header.

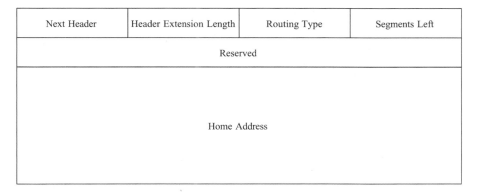

Next Header	Header Extension Length	Routing Type	Segments Left
Reserved			
Home Address			

Fig. 4.25 *MIPv6 routing header format*

- *Segments left*: An 8-bit unsigned integer that indicates the number of nodes listed in this routing header that are still to be visited. This field must be set to 1 because this MIPv6 routing header will carry only a single home address.
- *Reserved*: A 32-bit field reserved for future use.
- *Home Address*: The home address of the destination mobile.

4.2.5.3 Sending Packets While Away From Home

When a mobile away from its home network wants to send a packet to a correspondent host or the mobile's home agent, the mobile may use its current care-of address as the source IPv6 address in the IPv6 header of the packet in order to pass the access routers in a visited network without having to use reverse tunneling. However, the mobile's care-of address may change as the mobile moves around and MIPv6 seeks to make such a change of the mobile's care-of address transparent to the protocols and applications above the IPv6 and MIPv6 layers on the correspondent host.

To achieve the goal described above, MIPv6 makes use of the IPv6 Destination Options Header. The Destination Options Header is used to carry optional information that needs to be examined only by a packet's destination node. A Destination Options Header is placed between the IPv6 header and the header of the upper layer protocols (e.g., UPD). MIPv6 defines a Home Address Option that will be carried inside an IPv6 Destination Option Header. When a mobile away from its home network wants to send a packet, it uses the Home Address Option to inform the packet's recipient of the mobile's home address.

An IPv6 packet carrying the Home Address Option is illustrated in Figure 4.26, assuming for illustration purposes that the upper layer protocol is UDP. The highlighted portion of the IPv6 Destination Options Header is the Home Address Option carried in this header. The main fields of the Home Address Option are as follows:

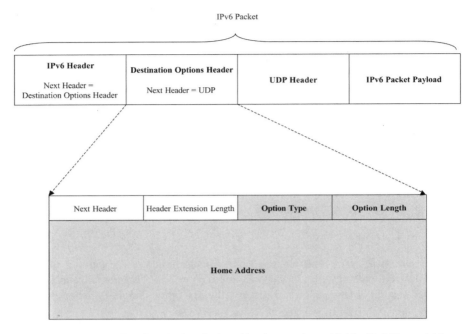

Fig. 4.26 *Format of IPv6 Destination Options Header carrying a Mobile IPv6 Home Address Option*

- *Next Header*: An 8-bit code that identifies the type of header immediately following the destination options header.
- *Header Extension Length*: An 8-bit unsigned integer that indicates the length of the destination options header in eight-octect units, not including the first eight octets.
- *Option Type*: It identifies the type of the Option carried in the IPv6 Destination Options Header. This field is defined by MIPv6 and should carry a value 201.
- *Option Length*: An 8-bit unsigned integer. It indicates the length of the Home Address Option in octets, excluding the Option Type field and the Option Length field.
- *Home Address*: The home address of the mobile sending the packet.

When a correspondent host (or a home agent) receives a packet that carries a MIPv6 Home Address Option, it processes the packet according to the following basic rules. It drops the packet if it does not have a binding entry in its binding cache for the home address carried in the Home Address Option. If the correspondent host has a binding entry for the home address, it will replace the source IPv6 address in the IPv6 header of the packet with the home address carried in the Home Address Option. It will also replace the home address carried in the Home Address Option with the source IPv6 address in the IPv6 header. This will ensure that the protocols and applications above the IPv6 and MIPv6 layers on the correspondent host will be unaware of the fact that the packet came originally from a care-of address different from the originating mobile's home address. In other words, from the perspective of upper layer protocols and applications, the packet is originated from the mobile's home address.

4.2.5.4 Formats of Binding Update and Binding Acknowledgment Messages
MIPv6 Binding Update (BU) and Binding Acknowledgment (BA) messages are transported inside a special IPv6 extension header, the *Mobility Header* defined by MIPv6. In other words, a MIPv6 BU or BA message may be piggybacked on a user IPv6 packet or transported alone without a user IPv6 packet.

As any other IPv6 extension header, the Mobility Header is placed between the IPv6 header and the upper layer protocol (e.g., UDP or TCP) header of a user IPv6 packet. The Mobility Header format is illustrated in Figure 4.27. It has the following fields:

- *Payload Protocol*: An 8-bit value that identifies the type of the header immediately following the Mobility Header.
- *Header Length*: An 8-bit unsigned integer that represents the length of the Mobility Header in units of octets, excluding the first eight octets. MIPv6

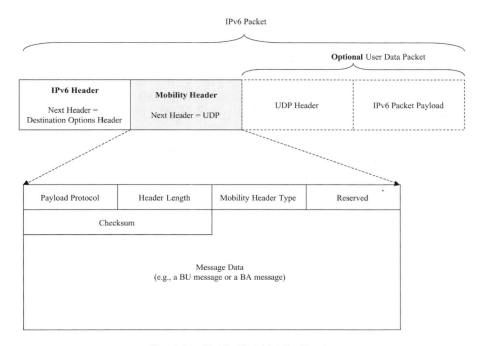

Fig. 4.27 *Mobile IPv6 Mobility Header*

specification requires that the length of the Mobility Header to be a multiple of eight octets.

- *Mobility Header Type*: An 8-bit value that identifies the type of mobility message carried in the Message Data field.
- *Reserved*: An 8-bit field reserved for future use.
- *Checksum*: An 16-bit unsigned integer that is checksum of the Mobility Header.
- *Message Data*: A variable-length field that contains a specific mobility message, such as a BU message or a BA message.

The BU or BA message is carried in the Message Data field of the Mobility Header. The format of the BU message is illustrated in Figure 4.28. It has the following fields:

- *Sequence Number*: A 16-bit unsigned integer used by the receiving node to sequence the BU messages and by the sending node to match a returned BA message with a BU message.
- *A (Acknowledge)*: A 1-bit flag, set by the sending node to request a BA message be returned by the receiving node upon receipt of the BU message.
- *H (Home Registration)*: A 1-bit flag, set by the sending node to request that the receiving node act as the sending node's home agent.
- *L (Link-Local Address Compatibility)*: A 1-bit flag that is set when the home address reported by the mobile node has the same interface identifier as the mobile node's link-local address. An interface identifier is a number used to identify a node's interface on a link. It is the remaining low-order bits in the node's IP address after the subnet prefix. A link-local address is an address that is only valid within the scope of a link, such as one Ethernet segment.

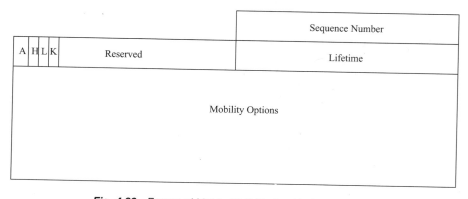

Fig. 4.28   *Format of Mobile IPv6 Binding Update message*

- *K (Key Management Mobility Capability)*: A 1-bit flag only valid in a BU message sent to a home agent. It is set by the sending node to indicate whether the protocol used for establishing the IPsec security association between a mobile and its home agent can survive movement.
- *Reserved*: Reserved for future use.
- *Lifetime*: A 16-bit unsigned integer indicating the number of time units remaining before the binding expires.
- *Mobility Options*: A variable-length field that contains one or more Mobility Options in a Type-Length-Value format.

Mobility Options in a Binding Update Message are used to carry information needed for MIPv6 mobility management, such as a mobile's care-of address or security-related information needed for a receiving node to authenticate a received message. The following Mobility Options can be included in the Mobility Options field in a BU message:

- *Alternative Care-of Address option*: An option used to carry a mobile's care-of address.
- *Binding Authorization Data option*: An option used to carry security-related information needed by the receiving node to authenticate and authorize the BU message.
- *Nonce Indices option*: A nonce is a random number used by a correspondent node to help authenticate a BU from a mobile. This option is only used when the BU message is sent to a correspondent node. The correspondent node uses the information carried in this option with the information carried in the Binding Authorization Data option to authenticate a BU message from a mobile.

The Alternative Care-of Address option is illustrated in Figure 4.29(a). The Type field carries a value 3 that identifies the Alternative Care-of Address option. The Length field contains the length in octets of the portion of the Alternative Care-of Address option starting immediately after the Length field. The Length field needs to be 16 because exactly one care-of address will be carried in the option.

The Binding Authorization Data option format is illustrated in Figure 4.29(b). The Type field carries a value 5 to indicate this is the Binding Authorization Data option. The Option Length field contains the length in octets of the Authenticator field. The Authenticator field contains a cryptographic value that can be used to determine that the message comes from a right user. The Authenticator protects the following mobility data fields:

- *Care-of address*.
- *IPv6 address of the final destination of the packet*.

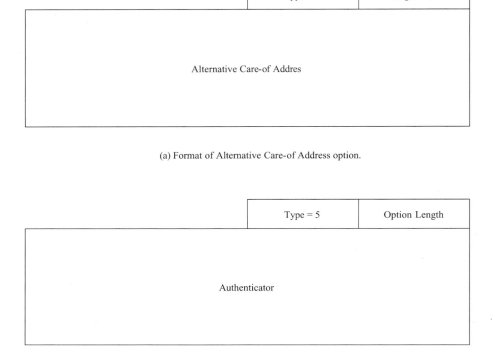

(a) Format of Alternative Care-of Address option.

(b) Format of Binding Authorization Data option.

Fig. 4.29 *Formats of Mobile IPv6 Alternative Care-of Address option and Binding Authorization Data option*

- *Mobility Header Data*: The content of the Mobility Header excluding the Authenticator field.

The Binding Acknowledgment message format is illustrated in Figure 4.30. It has the following fields:

- *Statue*: An 8-bit unsigned integer indicating the status of how the corresponding BU message is processed.
- *K*: It is used to indicate whether the protocol used by a home agent for establishing the IPsec security association between the mobile and the home agent can survive movement.
- *Reserved*: Reserved for future use.
- *Sequence Number*: The sequence number copied from the Sequence Number field of the corresponding BU message.

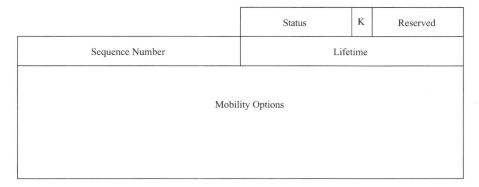

Fig. 4.30 *Format of Mobile IPv6 Binding Acknowledgment message*

- *Lifetime*: The time, in units of 4 seconds, for which the sender of this BA message will retain the binding of the receiving node of this BA message.
- *Mobility Options*: A variable-length field that contains one or more Mobility Options in a Type-Length-Value format.

A BA message may carry the following Mobility Options:

- *Binding Authorization Data option*: Used to carry the security-related information for the receiving node to authenticate the BA message.
- *Binding Refresh Advice option*: This option is used by a home agent to inform a mobile how often the mobile should send a new BU message to the home agent. Therefore, this option is only used in a BA sent by a home agent to a mobile in response to a received BU message.

4.2.5.5 *Hierarchical Mobile IPv6 Registration*   As in MIPv4, when a IPv6 mobile is far away from its HA, the process of binding update with home agent may experience a long delay. One approach to reduce binding update delay is to implement *local home agents* dynamically using the "forwarding from the previous care-of address" mechanism defined in MIPv6.

The "forwarding from the previous care-of address" mechanism is illustrated in Figure 4.31. Assume a mobile's original home network is Subnet A and its original home agent is HA A in Subnet A. Suppose that the mobile then moved from its home network first to Subnet B and then to Subnet C. While in Subnet B, the mobile acquires a care-of address CoA_B and performs a binding update with its original home agent HA A to register its care-of address CoA_B as its primary care-of address. When the mobile moves into Subnet C, it acquires a new care-of address CoA_C. But, the mobile does not have to perform address binding with its original home agent HA A. Instead, it may send a Binding Update to home agent HA B on its previous visited network Subnet B to request HA B to serve as the home agent for its previous care-of address CoA_B and use its current care-of address CoA_C as the current care-of

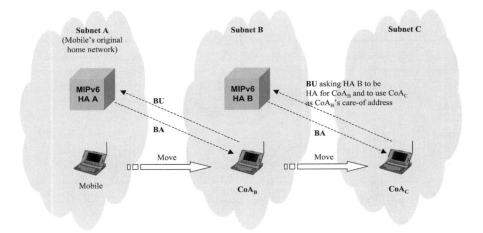

Fig. 4.31 *Mobile IPv6 "forwarding from previous care-of address" mechanisms*

address for CoA$_B$. Clearly, this will require the mobile to know the address of the HA B.

Packets addressed to the mobile's home address continue to be routed via regular IPv6 routing to the mobile's home network, where they will be captured by the mobile's home agent. The home agent continues to use CoA$_B$ as the primary care-of address for the mobile and therefore continues to tunnel the intercepted packets to CoA$_B$, i.e., to HA B. HA B will extract the original packets from the tunnel and then tunnel them to the mobile's current care-of address CoA$_C$, i.e., to the mobile itself.

The "forwarding from the previous care-of address" mechanism described above may be used to support hierarchical registration. To illustrate how this may be done, let's extend the example in Figure 4.31 to consider that the mobile subsequently moved from Subnet C to a new subnet D as illustrated in Figure 4.32.

Upon entering subnet D, the mobile will acquire a new care-of address CoA$_D$. Instead of registering the new care-of address with the mobile's original home agent HA A, the mobile can choose to make HA B its *local home agent* and register its new care-of address CoA$_D$ with this local home agent only. The mobile can do so using the "forwarding from the previous care-of address" mechanism discussed above. In particular, it sends a Binding Update message to HA B to use its current care-of address to update the care-of address for its CoA$_B$. This way, when HA B receives packets that are addressed to the CoA$_B$, it will tunnel them to the mobile's current care-of address CoA$_D$.

4.2.6 SIP-Based Mobility Management

MIPv4 and MIPv6 are IP-layer protocols. In this section, we examine how application-layer protocols may be used to support mobility over IP networks. More specifically, we discuss how the application-layer Session Initiation Protocol (SIP)

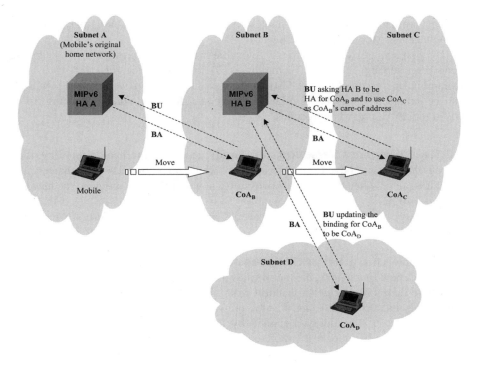

Fig. 4.32 *One approach to support hierarchical Mobile IPv6 registration*

[46] may be used to support terminal mobility. We focus on SIP-based mobility because of the following main reasons. First, SIP is currently the protocol of choice for signaling and control of real-time voice and multimedia applications over IP networks as well as over 3GPP, 3GPP2, and the MWIF network architectures. Second, significant efforts in the research community and the industry have been devoted to supporting mobility using SIP. Third, SIP appears to be the only application-layer protocol that can be readily extended to support terminal mobility today.

SIP already supports user mobility (Section 4.1), i.e., the ability for a user to originate and receive calls and access telecommunications services on any terminal and anywhere while being identified by the network as the same user. It can be extended to support terminal mobility (Section 4.1) with minor changes, i.e., by extending the existing SIP feature set only without introducing new network entities and without adding new SIP messages.

A key difference between SIP-based mobility and Mobile IP is that SIP servers may only participate in setting up the application sessions between the end users. After the application sessions are set up, user traffic may flow directly between the end users without having to go through SIP servers. This solves the triangular routing problem of MIPv4 and suggests that SIP servers will not likely become bottlenecks as Mobile IP home agents.

The first sets of extensions to SIP to enable it to support terminal mobility have been proposed by Telcordia Technologies, Toshiba America Research Inc., and Columbia University [49], [53], [23]. In the rest of this section, we describe how SIP can be extended to support terminal mobility and the limitations of existing SIP-based mobility support approaches.

4.2.6.1   *Movement Detection*

For an SIP application to handle mobility, the SIP application will have to be able to detect when the mobile terminal changes its IP address (e.g., when the mobile moves into a new IP network) and what the new IP address will be.

Detection of an IP network change and acquiring new IP addresses may be achieved using any available means to the mobile and do not have to be part of the SIP protocol.

One approach is to use DHCP. For example, to determine whether a mobile has moved to a new IP network, the mobile may ask a DHCP server for a new IP address each time the mobile detects a handoff from one radio cell to another. The mobile will supply its current IP address as the preferred address in its request sent to the DHCP server. If the address assigned by the DHCP server is the same as the mobile's current IP address, the mobile is still in the same IP subnet. Otherwise, the mobile assumes that it has just moved into a new IP network.

Once the mobile's IP address changed, the software component on the mobile that is responsible for acquiring the new address should also inform the SIP application of the change. Then, the SIP applications can use the procedures described in Sections 4.2.6.2 and 4.2.6.3 to ensure that correspondent hosts can establish SIP sessions with the mobile at its new location.

4.2.6.2   *Pre-Session Terminal Mobility*

Pre-session terminal mobility refers to the ability for correspondent hosts to establish a SIP session with a mobile regardless of where the mobile is located currently.

In this section, we describe how a SIP Redirect Server can be used to support pre-session mobility. A SIP Redirect Server in a mobile's home network tracks the mobile's current location and provides the location information to a caller so that the caller can contact the mobile at its new location directly to set up a SIP session.

As illustrated in Figure 4.33 [49], a correspondent user that needs to establish a SIP session to a mobile user will first send a SIP INVITE message to the SIP Redirect Server in the destination user's home network. This step is exactly the same as in the regular SIP session invitation procedure without considering terminal mobility. In response to the SIP INVITE message, the SIP Redirect Server in the destination user's home network will return the destination user's current location (indicated by the IP address currently used by the destination user's terminal) to the correspondent user. The correspondent user will then send a new SIP INVITE message directly to the destination user's current location to establish the SIP

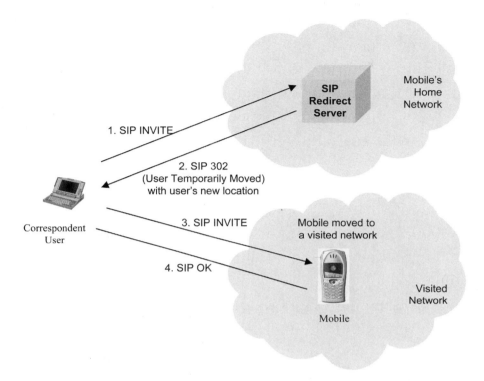

Fig. 4.33 *SIP-based pre-session terminal mobility management*

session. Once the mobile is successfully invited into the session, user data will flow between the users directly without having to traverse the SIP Redirect Server.

The SIP Redirect Server is similar to a Mobile IP home agent in that they both are responsible for tracking the current locations of the mobiles. A key difference, however, is that the SIP Redirect Server simply tells a caller where a destination is currently and will not be involved in relaying user traffic to the destination. A Mobile IPv4 home agent, however, will also be responsible for relaying user packets to the destination mobile. In essence, Mobile IPv4 uses the Relayed Delivery strategy for delivering traffic to a mobile (Section 4.1.3), whereas mobility via SIP Redirect Server uses the Direct Delivery strategy (Section 4.1.3) as the basic mechanism for delivering a call to a mobile destination.

The SIP Redirect Server in a user's home network learns about the user's current location from the SIP REGISTRATION messages received from the user. In particular, whenever a user starts to use a new IP address (e.g., when the user's mobile terminal changes its IP address or when the user starts to use a different terminal), it will register its new IP address with the SIP Redirect Server in its home network. This user registration process may be performed directly with the home

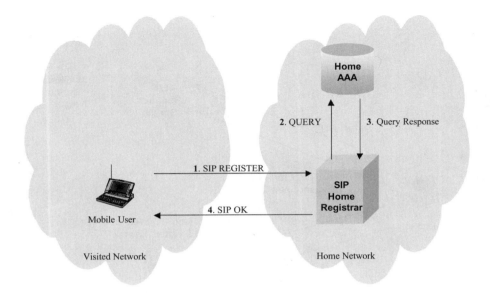

Fig. 4.34   *Location update for supporting SIP-based terminal mobility*

register or via a SIP Proxy Server in the visited network. Figure 4.34 illustrates the procedure for direct location registration with the home SIP Redirect Server.

To register its current location, the mobile user sends a SIP REGISTRATION message carrying its current location to its home SIP Redirect Server. The home SIP Redirect Server interacts with the AAA servers in the home network to authenticate the user. After positive authentication, the home SIP Redirect Server will return a positive acknowledgment to the mobile user, completing the location update process.

4.2.6.3 Mid-Session Terminal Mobility Support
Mid-session (mid-call) terminal mobility is the ability to maintain an on-going SIP session, whereas the mobile terminal moves from one IP subnet to another.

When the mobile changes its IP address in the middle of an on-going SIP session, the mobile will send a new SIP INVITE message to the correspondent host to invite the correspondent host to re-establish the SIP session to the mobile's new location. Upon receiving such update information and acknowledging the mobile's SIP INVITE request, the correspondent host will start to use the mobile's new IP address to address the packets destined to the mobile. This process is illustrated in Figure 4.35 [49].

The mobile will also update its location with its home SIP Redirect Server using the procedure described in Figure 4.34.

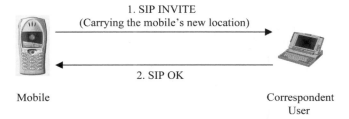

1. SIP INVITE
(Carrying the mobile's new location)

2. SIP OK

Mobile

Correspondent
User

Fig. 4.35 *SIP-based mid-session terminal mobility management*

4.2.6.4 Limitations of IP Mobility Using SIP
The basic SIP mobility approach described above has the following potential limitations:

- Similar to Mobile IP, a mobile using SIP mobility will have to register its new IP address with a SIP server (e.g., a SIP Redirect Server) in the mobile's home

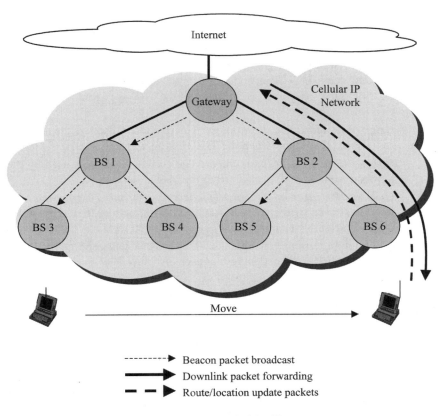

Fig. 4.36 *Cellular IP*

network every time the mobile changes its IP address. This could introduce long handoff delays when the mobile is far away from its home network. This could also create a high load on the home server.

- It is difficult for SIP-based mobility management to keep a TCP session alive while a mobile changes its IP address. With SIP-based terminal mobility, when a mobile changes its IP address, a correspondent host will have to address its outgoing packets to the mobile's new IP address. Changing the IP address on either end of a TCP session will cause the TCP session to break.

The first problem described above may be solved by similar solutions used to reduce the registration latency in Mobile IP, e.g., by hierarchical registration.

The second problem described above is more challenging technically. One solution was demonstrated by the ITSUMO (Internet Technologies Supporting Universal Mobile Operations) project—a joint research effort between Telcordia Technologies and Toshiba America Research Inc. (TARI). With this approach, a mobile terminal and a correspondent host uses a software agent called a SIP_{EYE} *agent* to hide the IP address change from the on-going TCP sessions. A SIP_{EYE} agent on a terminal operates as follows:

- It maintains a list of the on-going TCP connections on the terminal. It detects the birth and death of TCP connections by examining the headers of TCP packets. For each on-going TCP session, the SIP_{EYE} agent records the following information:
  - *Original IP address of the terminal*: The IP address used by the terminal as a source IP address when the TCP session was initiated.
  - *Current IP address of the terminal*: The current IP address used by the terminal to receive IP packets from the visited network.
  - *Original IP address of the correspondent host for this TCP session*: The IP address used by the correspondent host as its source IP address when the TCP session was initiated.
- When the mobile terminal changes its IP address, it will send a SIP INFO message [47] to the correspondent host of each on-going TCP session to inform them of the mobile's new IP address.

  The TCP application on the mobile does not need to know that the mobile has changed its IP address. The TCP application continues to use its original IP address as the source IP address in all outgoing TCP packets.
- Upon being notified that the mobile has changed its IP address, the SIP_{EYE} agent on a correspondent host will encapsulate each outgoing TCP packet with a new IP header that carries the mobile's new IP address as the destination address. These packets will be routed via regular IP routing to the mobile terminal's new location. The TCP application on the correspondent host does not need to be aware that the mobile has changed its IP address. The TCP

application continues to address its outgoing packets to the mobile's original IP address.

Upon receiving such an encapsulating packet, the SIP_{EYE} agent on the mobile terminal will strip off the encapsulating header added by the correspondent host and then deliver the payload TCP packet to the TCP process. Therefore, the TCP application can continue to use the original source and destination IP addresses throughout the on-going TCP session without any modification to the TCP protocol. This allows the TCP session to remain alive when the mobile changes its IP address.

The SIP_{EYE} approach has a potentially significant limitation: It requires a SIP_{EYE} agent to be implemented on every mobile and every correspondent host. This will be difficult over a large network such as the Internet.

4.2.7 Cellular IP

Recall that with Mobile IP, a mobile needs to change its IP address when it moves into a new IP subnet and registers the new IP address with its home agent. This could lead to long handoff delay when the mobile is far away from its home agent. Cellular IP [17] is designed to support fast handoff in a wireless network of limited size, for example, a network within the same administrative domain.

Cellular IP reduces handoff latency by eliminating the need for a mobile to change its IP address while moving inside a Cellular IP network, hence, reducing the delays caused by acquiring and registering new IP addresses.

Recall that the main reason for a mobile to change its IP address when moving into a new IP subnet is that the regular IP routing uses prefix-based routing, which divides the network into subnets and requires different subnets to use disjoint IP address spaces. To eliminate the need for a mobile to change its IP address inside a cellular network, Cellular IP does not use prefix-based routing. Instead, it uses *host-specific* routing: Network nodes perform routing and packet forwarding based on the full IP address of each individual terminal. In particular, the Cellular IP network maintains a host-specific downlink route for forwarding packets to each individual mobile, rather than maintaining a route for each IP address prefix as with regular IP routing protocols.

The configuration of a Cellular IP network is illustrated in Figure 4.36. It consists of two types of network nodes:

- *Base Stations*: Base stations are nodes internal to a Cellular IP network and do not interface directly with external networks. A base station can be a wireless access point that provides the air interface to mobiles or a router that does not have any air interface.
- *Gateway Router*: A gateway router is a router that interconnects a Cellular IP network with external IP networks.

Inside a Cellular IP network, nodes use the Cellular IP routing protocol, rather than regular IP routing protocols, to determine routes from one node to another and to determine the host-specific downlink route to each mobile. Cellular IP, however, does not change the IP packet format. After the routes are determined by Cellular IP routing, network nodes use the regular IP packet forwarding mechanism to forward IP packets along the routes.

Next, we discuss how Cellular IP routing works, how handoffs inside a Cellular IP network are supported, how handoff between two Cellular IP networks or between a Cellular IP network and a regular IP network can be supported, and how paging is supported in a Cellular IP network.

4.2.7.1 Cellular IP Routing

Inside a Cellular IP network, *uplink packets*, i.e., packets originated from mobiles inside the Cellular IP network, are first routed hop-by-hop to the gateway router, regardless of the destinations of these packets. The gateway router determines where each packet should be routed and then forwards the packet toward its final destination.

To ensure that the base stations know how to route packets to the gateway router, the gateway router periodically broadcasts a *beacon* packet throughout the Cellular IP network. Each base station records the interface on which the beacon packet is received and uses the reverse path to forward uplink packets to the gateway router.

Downlink packets, i.e., packets from the gateway router forward to a mobile inside the Cellular IP network, are sent over a host-specific downlink route from the gateway router to the mobile. The host-specific downlink routes are established and maintained by Cellular IP routing. In particular, each network node maintains a *routing cache*. An entry in a routing cache is called a *routing entry*. For each host-specific downlink route maintained by a network node, the network node maintains a routing entry that points to the next-hop network node along the host-specific route. The host-specific downlink route to a mobile is established when any packet is forwarded from the mobile toward the gateway router. As a packet from a mobile is forwarded toward the gateway router, each network node along the path of the packet will create, for this mobile, a routing entry in its routing cache that points to the base station from which the packet is received.

Network nodes maintain routes in "soft states" in the sense that they will be removed if no *route-update packet* is received during a predetermined time period. Therefore, when a mobile does not have any user packet to transmit, it may send small special route-update packets toward the gateway to refresh the route entries cached by the network for the mobile's host-specific downlink route. A router-update packet is an ICMP [19] packet and is addressed to the gateway router.

Cellular IP integrates location management with routing. Each time a mobile sends a route-update packet or any other packet, the downlink host-specific route maintained by the Cellular IP network for the mobile will also be updated. No separate location server is needed. The network maintains mobiles' locations implicitly by maintaining an up-to-date host-specific downlink route to each mobile.

The way Cellular IP base stations learn the routes to the gateway router and to the mobiles is similar to the way Transparent Bridging (defined in IEEE 802.1D) is used by most Ethernet bridges today, except that Cellular IP can work at the IP layer and the bridges can work at the link layer. An Ethernet bridge learns where a terminal attaches to the network by inspecting the source address of the link layer frames received from each terminal. No signaling is required for a terminal to inform the network of its current location. Once bridges learn the current network attachment point of a terminal, they will forward future packets destined to the terminal to its new location.

The way Cellular IP base stations learn the routes to the gateway router and to each mobile suggests that the physical configuration of a Cellular IP network has to be loop free, i.e., a tree or a string. Otherwise, routing loops may occur. For example, let's examine the physical configuration of a Cellular IP network as shown in Figure 4.36 again, and assume that there is a physical connection between base stations 3 and 4; i.e., base stations 1, 3, and 4 form a loop. When the gateway router broadcasts beacon packets, base stations 3 and 4 will receive the beacon from each other. Consequently, base station 3 will believe that the next hop for forwarding uplink packet is base station 4 and will forward uplink packets to base station 4. At the same time, base station 4 will believe that the next hop for uplink packets is base station 3 and will send the uplink packets back to base station 3, forming a routing loop. Current Cellular IP routing does not define ways to break the loop. Therefore, the physical configuration of a Cellular IP network has to be loop free.

4.2.7.2 Handoffs Inside a Cellular IP Network Cellular IP supports two types of handoffs inside a Cellular IP network: hard handoff and semi-soft handoff.

Hard handoff in a Cellular IP network is implemented using the Break-before-Make strategy (Section 4.1.4). When a mobile moves from one base station (the old base station) to another base station (the new base station), it tunes its radio to the new base station. The packets that are already on their way to the old base station may be lost. The mobile then sends a route-update packet toward the gateway router.

This route-update packet triggers the network nodes along its path to setup a host-specific downlink route for the mobile. As the physical topology of a Cellular IP network is a tree, the route-update packet will eventually reach a cross-over node. A cross-over node is a node shared by the mobile's old downlink host-specific route that goes to the old base station and the mobile's new downlink host-specific route that is being set up by this current route-update packet. For example, if the mobile in Figure 4.36 is handed off from base station 3 to base station 4, the cross-over node will be base station 1. When the mobile moves from base station 5 to base station 6, the cross-over node will be base station 2.

When the route-update packet reaches a cross-over node, this cross-over node will update the mobile's downlink host-specific route and start to forward future packets to the mobile's new base station. Packets that have already been on their way to the old base station may be lost.

Semi-soft handoff allows a mobile to receive packets from the old base station before the Cellular IP network sets up its route to the new base station. The mobile tunes its radio to the new base station, sends a semi-soft handoff packet via the new base station toward the gateway, and then tunes its radio back to the old base station immediately so that it can continue to receive user packets from the old base station while the network is setting up the mobile's downlink host-specific route to the new base station.

The semi-soft handoff packet triggers the network nodes on its path to set up a downlink host-specific route to the new base station for the mobile. When the semi-soft handoff packet reaches the first cross-over node, this cross-over node will start forwarding packets to both the old and the new base stations.

After a predetermined amount of delay when the mobile believes that the network has set up the downlink host-specific route for it to the new base station, the mobile disconnects from the old base station and tunes its radio to the new base station to receive packets from the new base station only.

4.2.7.3 Handoff Between Cellular IP Networks or Between Cellular IP and Regular IP Networks
Handoff between two Cellular IP networks interconnected via a regular IP network or between a Cellular IP network and a regular IP network are handled by a "macromobility" management protocol (e.g., Mobile IP) that is separate from Cellular IP. Here, we discuss how Mobile IP can be used to support a handoff between Cellular IP networks and between a Cellular IP network and a regular IP network.

Each mobile inside a Cellular IP network uses the IP address of the gateway router as its Mobile IP CoA, but uses its Mobile IP home address to send and receive packets over the Cellular IP network.

Upon entering a new Cellular IP network from either a regular IP network or another Cellular IP network, the mobile first sends a packet (e.g., route-update packet) toward the gateway router to trigger the new Cellular IP network to set up a downlink host-specific route for the mobile so that the Cellular IP network can deliver packets to the mobile.

The gateway router in the new Cellular IP network acts as a Mobile IP Foreign Agent. It sends Mobile IP Agent Advertisement messages to the mobile after it receives the first packet from the mobile that indicates that the mobile has just entered into the Cellular IP network. The mobile learns the IP address of the gateway router from the Mobile IP Advertisement message, uses this address as its new Mobile IP care-of address, and registers the new care-of address with its Mobile IP home agent.

After a successful Mobile IP registration, packets addressed to the mobile's home address will be tunneled by the mobile's home agent to the mobile's current care-of address, which is the IP address of the gateway router in the mobile's current Cellular IP network. The gateway router will de-tunnel the packets and forward the payload packets along the downlink host-specific route to the mobile directly without encapsulation or tunneling.

4.2.7.4 Paging
A mobile is considered by a Cellular IP network to be dormant (idle) if it has not transmitted packets for a predefined time period called *active-state-timeout*. When a mobile has not sent packets for a period of active-state-

timeout, its host-specific route maintained by the network will timeout and will be removed by the network. When a gateway router has packets to send to a mobile but does not have a valid routing entry for the mobile (i.e., the mobile is dormant), it will initiate paging to locate the mobile first.

To support paging, Cellular IP organizes the base stations into paging areas, as illustrated in Figure 4.37. When a dormant mobile crosses a paging area boundary, it updates its location with the network by sending a *paging-update packet* to the gateway router. A paging-update packet is an ICMP packet. It is addressed to the gateway router and forwarded by the base stations hop-by-hop to the gateway routers.

A network node may optionally use a *paging cache* to maintain paging routes for dormant mobiles. Each *paging entry* in a paging cache points to the next-hop network node along the paging route to a specific dormant mobile. Each paging-update packet will trigger the network nodes, which have paging caches and are along the way of the paging-update packet, to create a new paging entry or update its existing paging entry for the mobile.

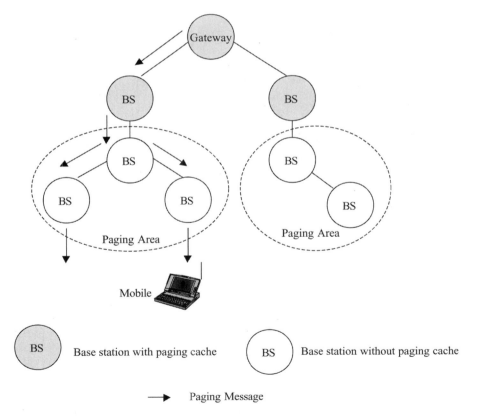

Fig. 4.37 *Paging in Cellular IP networks*

When a network node receives a downlink packet to send to a mobile but does not have a valid routing entry for the mobile, it will check if it has a valid paging entry for the mobile. If it has a valid paging entry for the mobile, it will forward a paging message along the paging route toward the mobile. If it does not have a valid paging entry, it will broadcast a *paging message* over all its interfaces except the one from which the packets destined to the mobile are received. No special paging message is defined. Instead, the first user packet destined to the mobile will serve as the paging message.

Base stations inside each paging area do not need to maintain paging caches. When a paging message reaches the first base station in the dormant mobile's current paging area, the paging message will be broadcast over the paging area to all the base stations and, hence, to all the mobiles inside the paging area.

Upon receiving a paging message or any other packet, a dormant mobile will transition into active mode and start to send route-update packets toward the gateway router to trigger the network to set up and maintain a host-specific downlink route for the mobile.

The paging entries are also maintained as "soft states." Therefore, a dormant mobile may refresh its paging route by periodically sending paging-update packets to the gateway router. However, to conserve resources, a dormant mobile sends paging-update packets at longer time intervals than it performs route updates while it is in active state. A network node can also use any type of packets received from a mobile to update the mobile's paging entry.

However, paging-update packets cannot be used to update routing caches. As a result, a network node may maintain only a paging entry for a dormant mobile, but it does not need to maintain any routing entry for the mobile. This reduces the sizes of the routing caches because large percentage of the mobiles may be dormant at any given time in a real wireless network. When a base station wants to send a packet to an active mobile, it only needs to search the routing cache, which reduces the delay incurred by table lookups at the base stations.

4.2.8 HAWAII

HAWAII (Handoff-Aware Wireless Access Internet Infrastructure) [42], [43] and Cellular IP are similar in many ways. They are both designed to support fast handoff and paging inside a wireless network under a single administrative domain. They use similar techniques. For example, they both use host-specific routes for delivering packets to mobile terminals. They both seek to reduce handoff latency by reducing the frequency at which mobiles have to change their IP addresses. However, HAWAII and Cellular IP differ significantly in routing and mobility management implementations.

HAWAII organizes a network into *domains*, as shown in Figure 4.38. A HAWAII domain is a network under the control of one administrative entity and uses HAWAII internally. A HAWAII domain consists of three types of network nodes:

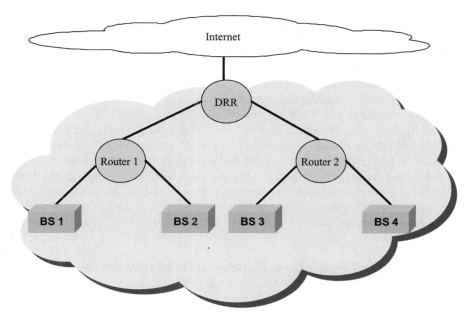

Fig. 4.38 *HAWAII*

- *Base Stations (BSs)*: A base station is a network node that supports the air interface or other radio specific functions.
- *Routers*: Routers are used to interconnect base stations.
- *Domain Root Routers (DRRs)*: A DRR is used to interconnect a HAWAII domain to external IP networks (e.g., the Internet).

Each mobile is assumed to have a home domain. A mobile's home domain belongs to the mobile's service provider. Each mobile is also required to have an IP address that is part of the IP addressing space used by the domain in which the mobile is located currently. The mobile's IP address could be statically configured when a user subscribes to network services. A mobile may also dynamically acquire an IP address from the domain it is currently in, using, for example, DHCP.

While moving inside the same HAWAII domain, a mobile does not have to change its IP address. Upon entering a new domain, the mobile has to obtain a new IP address from the new domain in order to receive IP packets in the new domain.

All user IP packets originated from or destined to any mobile inside a HAWAII domain will be routed first to the DRR in the HAWAII domain. The DRR will then forward these packets toward their final destinations. A DRR uses regular IP routing to route the packets destined to external IP networks. It uses a host-specific forwarding route established by the HAWAII Path Setup Schemes to forward packets destined to a mobile inside the HAWAII domain.

Network nodes maintain the host-specific forwarding routes in soft states. A network node needs to maintain state information about a host-specific route only when it is part of that host-specific route. This reduces the state information each router needs to maintain and as a result improves network scalability.

Unlike Cellular IP, HAWAII uses the regular IP routing protocol for routing and forwarding packets for which host-specific routes are not necessary. For example, IP packets addressed to any HAWAII network node or to external IP networks will be routed by the regular IP routing protocol. As a result, the physical configuration of each HAWAII domain can be of any topology as IP routing creates loop-free routes.

Next, we describe the procedure for the mobile's home domain to establish the initial host-specific forwarding route for a mobile when it powers up, the HAWAII Path Setup Schemes for supporting handoff from one base station to another inside the same HAWAII domain, and how Mobile IP can be used to support handoff between HAWAII domains.

4.2.8.1 Mobile Powers Up in its Home HAWAII Domain
When a mobile powers up in its home HAWAII domain, it may need to acquire an IP address from its home domain if it does not already have one. Then, the home HAWAII domain needs to establish a host-specific forwarding route from the DRR to the mobile for delivering user packets to the mobile.

After the mobile has acquired its IP address, the procedure for creating the initial host-specific forwarding route in the HAWAII domain (called the power-up procedure) is illustrated in Figure 4.39.

Upon powering up in its home HAWAII domain, the mobile sends a MIPv4 Registration Request to its current base station BS A. BS A learns that the mobile just powered up in the HAWAII domain by examining the information carried in the MIPv4 Registration Request message. It then creates a host-specific forwarding entry for the mobile to indicate that the mobile is reachable over its air interface.

Note that Mobile IP may be used to support handoff between HAWAII domains. The mobile's Mobile IP home agent may be located inside the mobile's home HAWAII domain. In this case, the mobile's IP address obtained from the home HAWAII domain can be the mobile's Mobile IP home address. Therefore, the mobile does not have to perform registration with its Mobile IP home agent. The Mobile IP Registration Request sent to the current base station is used solely for the purpose of triggering the HAWAII path setup procedure.

BS A then looks up its regular IP forwarding table to find the next-hop node toward the DRR. This next-hop node is Router 1 in the example shown in Figure 4.39. BS A sends a HAWAII power-up message (Message 2) to Router 1. This message triggers Router 1 to create a host-specific forwarding entry for the mobile. This forwarding entry points to BS A. Router 1 then sends a HAWAII message (Message 3) to its next-hop node toward the DRR. This next-hop node happens to be the DRR. Therefore, Message 3 triggers the DRR to create a host-specific forwarding entry for the mobile. This forwarding entry will point to Router 1. The DRR then sends an acknowledgment (Message 4) back to BS A. Upon receiving the

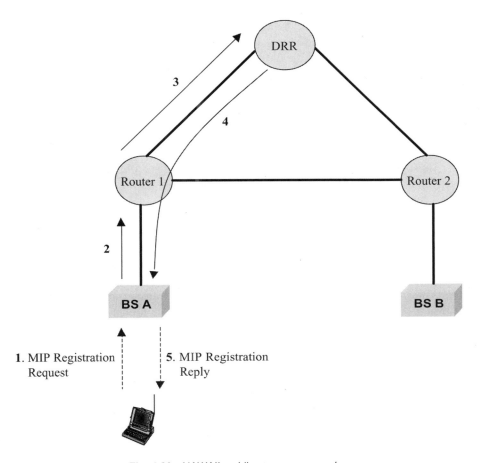

Fig. 4.39   *HAWAII mobile power-up procedure*

acknowledgment message from the DRR, BS A sends a MIP Registration Reply to the mobile.

From this point on, the DRR will use the host-specific forwarding route created during the above powering up procedure to forward user IP packets to the mobile.

As network nodes maintain the host-specific routes in soft states, a mobile has to refresh its host-specific route by sending HAWAII path refresh messages to the DRR periodically.

4.2.8.2   *Handoffs Inside a HAWAII Domain*   Now, we consider how handoff is supported when a mobile moves from one base station to another. HAWAII provides two basic path setup schemes for supporting handoff:

- *Forwarding Path Setup Scheme*: This scheme allows the old base station to forward user packets to the new base station (which in turn forwards the packets to the mobile). With this scheme, a host-specific route will be established from the old base station to the new base station so that packets that have already arrived at the old base station can be forwarded by the old base station to the new base station.

- *Nonforwarding Path Setup Scheme*: With this scheme, user packets may not be forwarded from the old base station to the new base station. When user packets destined to the mobile arrive at a cross-over router, the packets will be forwarded by the cross-over router to the new base station, which then forwards the packets to the mobile. A cross-over router is a router shared by the host-specific forwarding route from the DRR via the old base station to the mobile and the host-specific forwarding route from the DRR via the new base station to the mobile.

Figure 4.40(a) illustrates the operation of a Forwarding Path Setup Scheme. When a mobile connects to a new base station BS B, it sends a MIPv4 Registration Request message (Message 1) to the new base station. This message will inform the new base station that the mobile's previous base station was BS A. Upon receiving the MIPv4 Registration Request from the mobile, BS B initiates a HAWAII Handoff

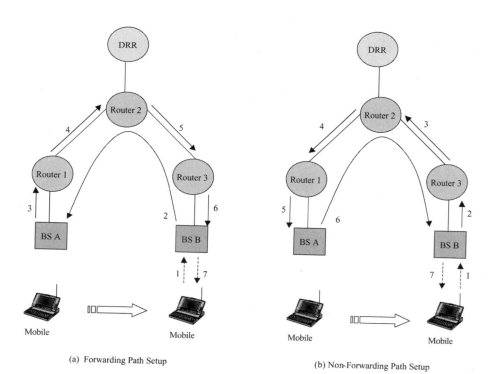

(a) Forwarding Path Setup

(b) Non-Forwarding Path Setup

Fig. 4.40 *HAWAII path setup schemes*

Message (Message 2) and sends it to BS A along the route created by regular IP routing.

Upon receiving the HAWAII Handoff Message from BS B, BS A uses the IP address of BS B and the forwarding table generated by regular IP routing to determine Router 1 as the next-hop router for forwarding packets to BS B. BS A will then set up a host-specific forwarding entry for the mobile using Router 1 as the next-hop router. This will allow BS A to forward future packets destined to the mobile to Router 1. Then, BS A will send a HAWAII message (Message 3) to Router 1 to trigger Router 1 to set up a host-specific forwarding entry for the mobile. Router 1 again uses the IP address of BS B and the forwarding table generated by regular IP routing to determine the next-hop router for forwarding packets to BS B. In this example, this next-hop router will be Router 2. Therefore, Router 1 will establish a host-specific forwarding entry for the mobile and set the next-hop for the host-specific route to Router 2. From now on, Router 1 will forward packets destined to the mobile to Router 2. Router 1 will also send HAWAII message 4 to Router 2 to trigger Router 2 to create a host-specific forwarding entry for the mobile.

Router 2 is the cross-over router. Therefore, upon receiving Message 4 and setting up the host-specific forwarding entry for the mobile, Router 2 will start to forward future user packets destined to the mobile along the new host-specific route to the mobile. That is, Router 2 will forward future user packets to the new base station BS B directly, rather than sending these packets to the old base station BS A.

As previous routers, Router 2 sends a HAWAII message to the next-hop router along the route created by regular IP routing toward BS B. The process continues until a host-specific route is established from BS A to BS B for the mobile. BS B will then send a MIP Registration Reply message to the mobile, completing the path setup process.

HAWAII defined multiple forms of Nonforwarding Path Setup Schemes. Figure 4.40 (b) illustrates one form of Nonforwarding Path Setup Scheme referred to as the Unicast Nonforwarding Scheme. As before, when a mobile moves to a new base station BS B, it sends a MIPv4 Registration Request message to the new base station to initiate the handoff procedure. This message will carry the IP address of the old base station BS A.

Upon receiving the MIPv4 Registration Request, BS B creates a host-specific forwarding entry for the mobile with the outgoing interface set to the interface on which it received the MIPv4 Registration Request. The MIPv4 Registration Request message will carry the IP address of the old base station BS A. BS B uses the IP address of BS A and looks up the IP forwarding table established by regular IP routing to determine the next-hop router toward BS A. This next-hop router is Router 3 in this example. BS B then sends HAWAII Message 2 to Router 3.

Upon receiving Message 2, Router 3 creates a host-specific forwarding entry for the mobile with the next-hop node set to BS B. Router 3 then determines the next hop toward BS A is Router 2 (based on its forwarding table created by regular IP routing) and forwards HAWAII Message 3 to Router 2.

Upon receiving Message 3, Router 2 creates a host-specific forwarding entry for the mobile with the outgoing interface set to Router 3. Router 2 is the cross-over router. Therefore, after Router 2 creates the new host-specific forwarding entry for the mobile, it will forward future user packets destined to the mobile to the new base station BS B directly, rather than forwarding these packets toward the old base station BS A.

Router 2 also sends a HAWAII message (Message 4) to the next-hop router along the route created by regular IP routing toward BS A. This process continues until BS A receives a HAWAII message (Message 5). When BS A receives the HAWAII Message 5, it will establish a host-specific forwarding entry for the mobile and then send an acknowledge message (Message 6) back to BS B. This message will trigger BS B to send a MIPv4 Registration Reply message back to the mobile to complete the path setup and handoff procedure.

4.2.8.3 Moving into Foreign HAWAII Domains

A "macromobility" management protocol such as Mobile IP can be used to support handoff between HAWAII domains to ensure that a mobile is always addressable by a permanent home address when it moves into foreign HAWAII domains. Here, we discuss how Mobile IP can be used to support handoff between HAWAII domains [42].

The interdomain handoff procedure using Mobile IP is illustrated in Figure 4.41. When a mobile enters a new HAWAII domain, it first needs to obtain a new IP address from the new HAWAII domain. This may be achieved using, for example, DHCP and is not shown in Figure 4.41.

After the mobile acquires a new IP address, the new HAWAII domain needs to set up the initial host-specific forwarding route from the DRR to the mobile. This is achieved using the power-up procedure discussed in Section 4.2.8.1.

When the mobile's current base station receives the acknowledgment message (Message 4 in Figure 4.41) from the DRR indicating that a host-specific forwarding route has been set up from the DRR to the mobile, the base station will forward the mobile's Mobile IP Registration Request message to the mobile's Mobile IP home agent.

The mobile uses the IP address it obtained from the new HAWAII domain as its co-located care-of address. Therefore, packets addressed to the mobile's home address will be tunneled by the mobile's home agent to the mobile directly. These packets will enter the mobile's current HAWAII domain via a DRR in the domain. The DRR will forward the packets along the host-specific forwarding route to the mobile. The mobile will then de-tunnel the packets to extract the original user packets.

4.2.8.4 Paging

The basic paging mechanism used in a HAWAII domain is described in [42]. More recently, an improved paging mechanism that can be used in a HAWAII domain is presented in [41].

To support paging, HAWAII groups the base stations into paging areas. Each paging area is identified by an IP Multicast Group Address (MGA) that has a scope of the administrative domain. In other words, packets sent to such a multicast group address will stay inside the administrative domain.

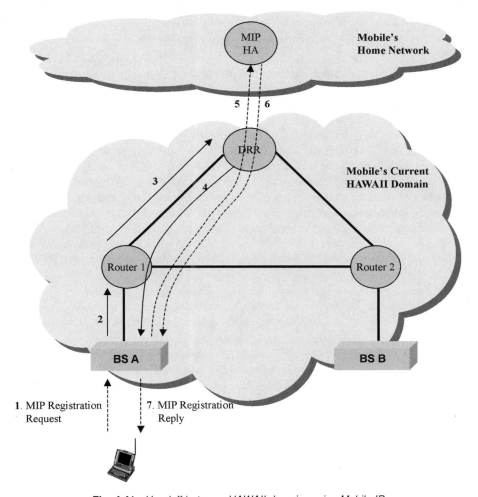

Fig. 4.41 *Handoff between HAWAII domains using Mobile IP*

While a mobile is in dormant mode (or in standby mode using the terminology in [41]), the network will only know in which paging area it is currently located but may not know to which base station the mobile is connected. A dormant mobile will send a location update message toward the DRR every time it crosses a paging area boundary. The location update message is propagated hop-by-hop from the mobile's current base station to the DRR. It triggers the base station and each router along its path to create a new or update its existing host-specific *routing entry* and *paging entry* for the mobile. A routing entry is for forwarding regular user packets to a mobile. A paging entry is used for forwarding paging messages to a mobile. Routing entries and paging entries on a network node will be maintained separately. The paging entry contains the following main information:

- The MGA that identifies the mobile's current paging area.
- The outgoing interface for sending paging messages to this mobile.

To page a mobile, a network node (router or base station) will need to act as a Paging Initiator. A paging initiator is responsible for creating a paging message and multicasting it to the MGA of the paging area in which the mobile performed its last location update, i.e., to all the base stations in the paging area. Each base station in the paging area will in turn send a paging message over the air to all the mobiles it is serving. The paging initiator is also responsible for buffering packets destined to a dormant mobile while the mobile is being paged.

To increase network reliability and scalability, paging in a HAWAII domain does not use a centralized paging initiator for all the mobiles. Instead, the paging initiator functionality is dynamically distributed to the network nodes. A network node may dynamically elect itself as a mobile's paging initiator. However, to ensure that a mobile's paging initiator knows the latest paging area of the mobile, several constraints have to be met before a network node can initiate paging for a mobile:

- *Only a network node along a mobile's latest paging route can be a paging initiator for the mobile.* A mobile's latest paging route is the host-specific paging route established by the mobile's latest location update message.
- *A paging initiator initiates paging only when it receives packets destined to the mobile from an upstream network node on the mobile's latest paging route.* Node A is node B's upstream node when both nodes are on the latest paging route from the DRR to the mobile's last-known base station, but node A is closer to the DRR. The last-known base station is the base station through which the mobile performed its last location update.

The above rules suggest that packets originated inside a dormant mobile's current HAWAII domain have to be routed first to the DRR. The DRR will forward the packets along the mobile's latest paging route toward the mobile's current paging area. Upon receiving a packet from the DRR, a router or the base station on the mobile's latest paging route may then elect itself as the mobile's paging initiator and initiates a paging message.

These rules are necessary to ensure that the mobile's most update-to-date paging area is used to page the mobile. This is because network nodes along the mobile's old paging routes from the DRR to the previous paging areas traversed by the mobile may still contain paging entries for the mobile. These paging entries are now obsolete. Therefore, if any of these network nodes is allowed to be the paging initiator for the mobile, it may not be able to send paging messages to the mobile's current paging area.

Any network node along a mobile's latest paging route may elect to be the paging initiator for this mobile. A method to dynamically determine which network node should be the paging initiator is given in [41]. Using this method, the mobile's last-known base station will be the default paging initiator. Packets destined to the dormant mobile will be forwarded by the DRR and the routers along the mobile's latest paging route toward the mobile's last-known base station. The mobile's last-known base station will then initiate a paging message and multicast it to the base stations in the mobile's current paging area.

A router along the mobile's latest paging route may take over the role of paging initiator when it detects that a potential failure along the mobile's latest paging route from itself to the mobile's last-known base station. A router assumes there is a potential failure if it does not receive any HAWAII path refresh messages from the mobile in a predetermined time period. A router may also decide to become the paging initiator in order to support load balancing (e.g., when the router's traffic load drops to a predetermined threshold).

When a dormant mobile receives a paging message, it will transition into active mode and send a *paging response* message back to the paging initiator. As the paging response message is forwarded hop-by-hop toward the paging initiator, the base station and the routers along the path will create host-specific routing entries for forwarding packets to the mobile. Upon receiving the paging response, the paging initiator will start to forward the packets along this newly established host-specific routing path to the mobile.

4.3 MOBILITY MANAGEMENT IN 3GPP PACKET NETWORKS

This section discusses mobility management in the packet-switched (PS) domain of a 3GPP network, based on Release 5 of the 3GPP Technical Specifications. When network nodes inside a RAN needs to be considered in the discussion, we assume the RAN is UTRAN unless stated explicitly otherwise.

As we have discussed in Chapter 2 "Wireless IP Network Architectures," all packet-switched user data to and from a mobile is first sent to a GGSN called the mobile's serving GGSN. The serving GGSN will in turn forward the user data toward their final destinations. The mobile and its serving GGSN use a host-specific route to exchange user data. Therefore, mobility management in 3GPP PS domain is, in essence, to manage the changes of the host-specific route between each mobile and its serving GGSN.

The host-specific route between a mobile and its serving GGSN consists of the following connections (bearers), which can be changed separately:

- *A RRC connection (Chapter 2)* between the mobile and the RAN (e.g., RNC in a UTRAN).
- *A RANAP connection (Chapter 2)* between the RAN and the SGSN.
- *CN Bearers* between the SGSN and the mobile's serving GGSN.

Furthermore, sometimes, the mobile's serving RNC may distribute downlink traffic to a mobile via another RNC. A mobile's Serving RNC is the RNC that receives from the PS CN domain the user data destined to the mobile and then distributes the data inside the RAN to the mobile.

In order for a mobile to exchange signaling messages with the PS CN (e.g., to set up and manage the traffic bearers, to perform location update), a dedicated logical signaling connection needs to be established between the mobile and the SGSN. Recall that this signaling connection consists of a signaling Radio Bearer and an I_u Signaling Bearer.

A mobile does not have to maintain all the traffic bearers in the RAN or the CN if it does not expect to send or receive user data soon. The mobile does not even need to maintain its dedicated signaling connection to the SGSN at all times. Releasing the radio resources that a mobile is unlikely to need soon allows these radio resources to be used by other mobiles and helps conserve the scarce power resources on the mobile.

Which network connections (bearers) that make up the host-specific route between a mobile and its serving GGSN need to be changed when the mobile moves around depend on the scope of mobility. Figure 4.42 illustrates the different scopes of mobility and which parts of the host-specific route between a mobile and its serving GGSN may be affected in each mobility scope.

- *Inter-Node B Handoff*: Handoff from one Node B (called the source Node B) to another Node B (called the target Node B) requires that the mobile's Radio Bearers to be changed from the source Node B to the target Node B.
- *Inter-RNC Handoff*: With an inter-RNC handoff, a mobile moves its radio bearers from one RNC (called the source RNC) to another RNC (called the

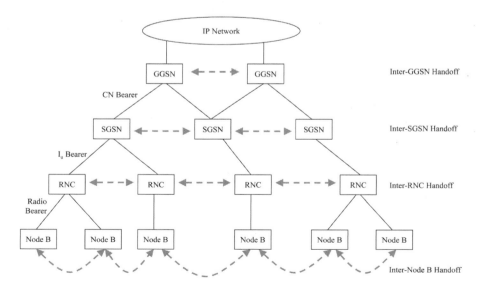

Fig. 4.42 *Scope of mobility in 3GPP packet-switched domain*

target RNC). Therefore, if the target RNC also becomes the mobile's serving RNC after the handoff (e.g., in a hard inter-RNC handoff), the mobile's I_u Bearers also need to be changed, in addition to the changes of the Radio Bearers.

- *Inter-SGSN Handoff*: With an inter-SGSN handoff, a mobile moves from one SGSN (called the source SGSN) to another SGSN (called the target SGSN) as a result of the handoff. Inter-SGSN handoff requires that the mobile's PDP context be updated and a new CN Bearer be established, in addition to the changes in the I_u Bearers and the Radio Bearers.

- *Inter-GGSN Handoff*: With an inter-GGSN handoff, a new GGSN becomes a mobile's serving GGSN as a result of the handoff. This requires the mobile's new serving GGSN to create a PDP context for the mobile. It also requires a CN Bearer to be established between the mobile's new serving GGSN and the mobile's new serving SGSN. In addition, the mobile's Radio Bearers and I_u Bearers will need to be changed.

In the rest of this section, we will focus on the following key aspects of mobility management for 3GPP packet-switched services:

- *Packet Mobility Management (PMM) contexts and states*: A mobile's PMM context is a set of information used by the network to track the mobile's location. The state of a mobile's PMM context determines which network connections (bearers) between the mobile and the SGSN should be maintained for the mobile and how the network tracks the mobile's location. PMM states are described in Section 4.3.1.

- *Location management and its interactions with the management of the host-specific route between a mobile and its serving GGSN:* Section 4.3.2

- *Changes of the I_u Bearers*: When a mobile moves around, its serving RNC may need to change from one RNC to another. As a result, the mobile's I_u Bearers need to be changed. The process for relocating the RNC side of the endpoint of an I_u bearer from one RNC to another, i.e., the Serving RNS Relocation Procedure, will be discussed in Section 4.3.4.

- *Handoffs*: Intra-RNC handoffs are managed by protocols and procedures completely inside each specific RAN and therefore will not be discussed further in this section. Inter-RNC handoffs for packet-switched services require the support of PS domain protocols and procedures and will therefore be discussed in Section 4.3.5.

4.3.1 Packet Mobility Management (PMM) Context and States

A mobile's PMM context is a set of information used by the network to track the mobile's location. The state of a mobile's PMM context determines which network connections (bearers) between the mobile and the SGSN should be maintained for the mobile and how the mobile's location should be tracked by the network.

In the 3GPP PS domain, the SGSNs are responsible for tracking the locations of mobiles that are using PS services. Therefore, the SGSNs need to maintain the PMM contexts of the mobiles. Each mobile also needs to maintain a PMM context in order to collaborate with the network for location tracking. The GGSNs, on the other hand, are not directly involved in location tracking. Therefore, a GGSN does not need to be aware of any mobile's PMM context or PMM state.

A PMM context on a mobile or on an SGSN can be in one of the following states (for UMTS) [10]:

- *PMM-DETACHED State*: In this state, there is no communication between the mobile and the SGSN. The mobile and the SGSN do not have valid location or routing information for the mobile. The mobile does not react to system information related to the SGSN. The SGSN cannot reach the mobile.

- *PMM-CONNECTED State*: In this state, the SGSN and the mobile have established a PMM context for the mobile and a dedicated *signaling connection* is established between the mobile and the SGSN. Recall that this signaling connection consists of an RRC connection between mobile and RAN and an I_u signaling connection over the I_u interface between the RAN and SGSN (Chapter 2 "Wireless IP Network Architectures"). The PS domain-related signaling and circuit-switched (CS) domain-related signaling share one common RRC connection but use different I_u signaling connections, i.e., one I_u signaling connection for the CS domain and one I_u signaling connection for the PS domain.

  In PMM-CONNECTED state, a mobile's location inside the RAN is tracked by the RNCs at an accuracy level of radio cells. In the PS CN, the SGSN tracks a mobile's location by tracking the mobile's serving RNC. In the PMM-CONNECTED state, the mobile's PDP context may or may not be activated. Recall that before a mobile's PDP context is activated, the mobile will not be able to send or receive user packets over the PS CN domain.

- *PMM-IDLE State*: In this state, the SGSN and the mobile have established the PMM contexts for the mobile. The mobile's location is tracked by the SGSN at an accuracy level of a Routing Area (Section 4.3.2). The mobile is reachable by the CN via paging. No signaling or traffic connection exists between the mobile and the SGSN. A mobile moves into PMM-IDLE state to conserve scarce resources (e.g., power off the mobile, reduce the transmissions of signaling messages to conserve radio bandwidth).

How location tracking is handled in different PMM states will be discussed in greater detail in Section 4.3.2.

Figure 4.43 illustrates the state transition machines for the PMM states on a mobile and on an SGSN (assuming RAN is UTRAN).

- *From PMM-DETACHED state to PMM-CONNECTED state*: A mobile's PMM state transitions from PMM-DETACHED to PMM-CONNECTED

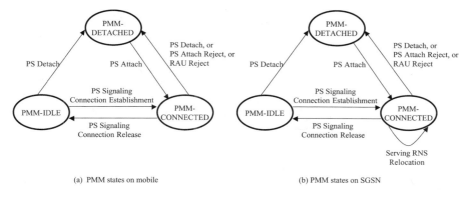

(a) PMM states on mobile   (b) PMM states on SGSN

Fig. 4.43   *3GPP PMM state transition machines*

when the mobile performs GPRS Attach, indicating that it wishes to attach to the PS domain. To support the GPRS Attach procedure, a signaling connection needs to be established between the mobile and its serving SGSN (if such a signaling connection does not already exist).

- *From PMM-CONNECTED state to PMM-IDLE state*: A mobile's PMM state changes from PMM-CONNECTED to PMM-IDLE whenever the signaling connection between the mobile and its serving SGSN is released. For example, when the GPRS Attach process is finished, this signaling connection may be released immediately, which will cause the mobile's PMM state to change from PMM-CONNECTED to PMM-IDLE.

- *From PMM-IDLE state to PMM-CONNECTED state*: A mobile's PMM state changes from PMM-IDLE to PMM-CONNECTED whenever a signaling connection is established between the mobile and a SGSN.

  A mobile in PMM-IDLE state may need to establish a signaling connection to the SGSN for various purposes from time to time. For example, a mobile needs to establish a signaling connection to the SGSN to perform routing area update (Section 4.3.3). When this signaling connection is not expected to be needed in the near future (e.g., after routing area update is completed), it may be released to allow the mobile's PMM state to change back to PMM-IDLE.

- *From PMM-CONNECTED state to PMM-DETACHED state*: A mobile's PMM state transitions from PMM-CONNECTED to the PMM-DETACHED when the GPRS Detach procedure is performed, when the mobile's GPRS Attach request is rejected by the SGSN, or when the mobile's Routing Area Update (RAU) request is rejected by the SGSN.

- *From PMM-IDLE state to PMM-DETACHED state*: The PMM state on a mobile or a SGSN may change from PMM-IDLE to PMM-DETACHED locally as a result of a local event. For example, the PMM state on a mobile may change from PMM-IDLE to PMM-DETACHED when the SIM, USIM, or battery is removed from the mobile. The PMM state on a SGSN may change from PMM-IDLE to PMM-DETACHED when the lifetime of the PMM state expires.

- *From PMM-DETACHED state to PMM-IDLE state*: The state of a PMM context cannot change from PMM-DETACHED to PMM-IDLE directly. Before a mobile's PMM context can be in PMM-IDLE state, the mobile's PMM context will have to be created first on the SGSN. To create a PDP context on the SGSN, the mobile has to perform GPRS Attach, which will cause the mobile's PMM state to change from PMM-DETACHED to PMM-CONNECTED first, before it can transition into PMM-IDLE.

While a mobile is in the PMM-CONNECTED state, the mobile's PDP context may have been created and activated. This will be the case, for example, if the mobile has sent user packets over the PS CN domain. When the mobile's PMM state transitions from PMM-CONNECTED to PMM-IDLE subsequently, the mobile's existing active PDP contexts will continue to remain in ACTIVE state on the GGSN and the SGSN. Maintaining an active PDP context in the CN does not consume much network resource, but it creates significant benefits:

- It reduces the time for a mobile to change from PMM-IDLE state back to PMM-CONNECTED state when the mobile needs to send packets to, or receive packets from, over the PS CN.
- It makes it easier for the PS CN domain to support paging. In particular, an active PDP context allows the GGSN to always know a mobile's serving SGSN. Therefore, the GGSNs do not have to be aware of the paging operations. Only the SGSNs need to support paging functions.

When a mobile's serving RNC changes, the SGSN will participate in a process called Serving RNS Relocation that will relocate the RNC side of the mobile's I_u Bearers from the old serving RNC to the new serving RNC (Section 4.3.4). The Serving RNS Relocation process can only be performed while the mobile is in PMM-CONNECTED state and it will not change the mobile's PMM state.

Sometimes, the PMM states of the mobile and the SGSN may lose synchronization. For example, the PMM state on the mobile may be PMM-IDLE while the SGSN still thinks that the mobile is in the PMM-CONNECTED state. This situation will be corrected when any one of the following events occurs:

- The mobile performs Routing Area Update, which will change the PMM state on the mobile into PMM-CONNECTED. After the Routing Area Update, the mobile's PMM state on the SGSN will continue to be in PMM-CONNECTED state. Therefore, synchronization between the PMM states on the mobile and on the SGSN is regained.
- The SGSN sends data to the mobile but receives messages from the RAN, indicating that the mobile is not known. This will trigger the paging process (Section 4.3.6). Paging will cause the mobile to establish a signaling

connection and the traffic connections to the SGSN, transferring the PMM states on both the mobile and the SGSN into PMM-CONNECTED.

4.3.2 Location Management for Packet-Switched Services

4.3.2.1 Location Concepts The RANs and the CN in a 3GPP network use different location concepts to track mobile terminal locations. The RAN uses the following location concepts [9]:

- *Cell Area (or Cell)*: A Cell is the geographical area served by one wireless base station.
- *UTRAN Registration Area (URA)*: A URA is an area covered by a set of cells.

Cells and URAs are used to track the locations of mobiles that are using CS, PS, or both CS and PS services. Cells and URAs are used only in the RAN and are invisible to CN nodes.

The CN uses the following location concepts [9]:

- *Location Area (LA)*: A Location Area is a group of Cells used by the CS CN domain to track the locations of mobiles that are using CS services.
- *Routing Area (RA)*: A Routing Area is a group of Cells used by the PS CN domain to track the locations of mobiles that are using PS services.

The relations between LAs, RAs, Cells, SGSNs, and MSCs are illustrated in Figure 4.44. An LA consists one or more Cells that belong to the RNCs that are

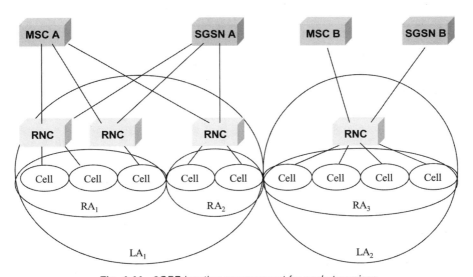

Fig. 4.44 *3GPP location management for packet services*

connected to the same MSC/VLR. All Cells in the same URA have to be served by the same MSC/VLR. In other words, one LA is handled by only one MSC/VLR. Each LA is identified by a globally unique Location Area Identifier (LAI). When a mobile moves inside an LA, it does not have to perform location update with the CN CS domain.

An RA consists of one or more Cells that belong to the RNCs that are connected to the same SGSN (or one combined SGSN and MSC). In other words, one RA is handled by only one SGSN. An RA is either the same as an LA or a subset of one and only one LA [9]. That is, one RA cannot belong to more than one LA, whereas each LA may contain multiple RAs. Each RA is identified by a globally unique Routing Area Identifier (RAI).

The structures of the LAI and the RAI are illustrated in Figure 4.45 [7]. An LAI is composed of a Mobile Country Code (MCC), a Mobile Network Code (MNC), and a Location Area Code (LAC). The MCC identifies the country in which the 3GPP network is located. The value of the MCC will be the same as the MCC in the IMSIs allocated to the mobile users in the same country. The MNC identifies a 3GPP network in that country. The MNC will have the same value as the MNC in the IMSIs allocated to mobile subscribers of this particular 3GPP network. The LAC identifies a Location Area within a 3GPP network.

An RAI consists of an LAI and a Routing Area Code (RAC). The LAI field of the RAI contains an LAI that identifies the Location Area in which the RA resides. The RAC identifies a Routing Area inside the LA identified by the LAI.

4.3.2.2 Location Tracking
3GPP uses hierarchical location tracking. The methods and the accuracy level of location tracking for each mobile terminal can vary over time depending on the activeness level of the mobile (i.e., how likely the mobile will transfer traffic soon). For location tracking purpose, a mobile's

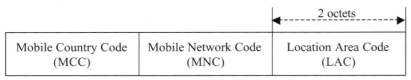

(a) Structure of Location Area Identifier (LAI)

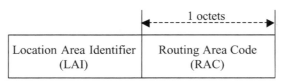

(a) Structure of Routing Area Identifier (RAI)

Fig. 4.45 *Structures of 3GPP Location Area Identifier and Routing Area Identifier*

activeness level is represented by the mode of its RRC connection. The same RRC connection is used by the mobile to transport all signaling traffic and user traffic for its CS and PS services [9], [8].

A mobile's RRC connection has two modes:

- *RRC-CONNECTED mode*: A mobile in RRC-CONNECTED mode has an established RRC connection.
- *RRC-IDLE mode*: A mobile in RRC-IDLE mode has not established any RRC connection.

The RRC connection makes up one portion of the signaling connection between the mobile and the SGSN. The other portion of this signaling connection is the I_u signaling connection between the RAN (e.g., the RNC in a UTRAN) and the SGSN. Therefore, when the RRC connection is in RRC-IDLE mode, the mobile's PMM state can only be PMM-IDLE or PMM-DETACHED because no signaling connection between the mobile and the SGSN can exist without an RRC connection. However, when the RRC connection is in RRC-CONNECTED mode, the mobile may be either in PMM-CONNECTED state or PMM-IDLE state. This is because a mobile uses a single RRC connection for both CS and PS services; hence, the mobile can be in RRC-CONNECTED mode and PMM-IDLE state at the same time because the RRC connection may be present but is currently used only for CS services; i.e., no signaling connection is established to the SGSN.

Location tracking is performed as follows depending the mode of the mobile's RRC connection [9]:

- When a mobile is in the RRC-IDLE mode (hence, also in PMM-IDLE state), the mobile's location is tracked at the RA level by the SGSNs.

  A mobile in RRC-IDLE mode will receive the Mobility Management (MM) system information *broadcast* by the RNCs at the RRC layer. The MM system information informs the mobile which Cell and RA it is in currently. A mobile will initiate RA Update (Section 4.3.3) toward the CN upon receiving MM system information, indicating that it moved into a new RA.

- When a mobile is in RRC-CONNECTED mode, its location inside the RAN is tracked at the cell level by the RNCs. To track the mobiles in RRC-CONNECTED mode, an RNC identifies a mobile by a temporary identifier, the Radio Network Temporary Identity. The Radio Network Temporary Identity is assigned to the mobile dynamically by an RNC.

  When a mobile is in RRC-CONNECTED mode, it receives the MM system information from the serving RNC over the established RRC connection. It uses the MM system information to determine if it has moved into a new Cell, RA, or LA.

  If the mobile is in RRC-CONNECTED mode and PMM-IDLE state, the SGSNs will also track the mobile's location at the RA level. Therefore, the mobile will initiate RA Update toward the CN PS domain upon receiving MM system information, indicating that it has just moved into a new RA.

If the mobile is in RRC-CONNECTED mode and PMM-CONNECTED state, the mobile's serving SGSN will know the mobile's serving RNC because the serving SGSN maintains a signaling connection through the mobile's serving RNC to the mobile. When the mobile's serving RNC function needs to be changed to a new RNC as the mobile moves about, the mobile's serving SGSN will participate in this change process (i.e., the Serving RNS Relocation procedure to be described in Section 4.3.4) to ensure that the signaling connection between the mobile and the SGSN will go through the new serving RNC.

In addition to performing RA update when crossing RA boundaries, a mobile in PMM-IDLE state may also perform periodic RA updates.

4.3.3 Routing Area Update

Routing area update in 3GPP achieves several main objectives. It allows the following:

- The mobile's serving SGSN to know which RA the mobile is currently in.
- The mobile's existing active PDP contexts to be updated. For example, if moving into a new Routing Area also means that the mobile has to use a new SGSN, a PDP context between the new SGSN and the mobile's serving GGSN will need to be established. This ensures that the mobile's serving GGSN always knows where to forward user packets destined to the mobile.

A mobile performs RA update when:

- The mobile enters a new Routing Area.
- The mobile's periodic routing area update timer expires.
- The mobile is directed by the network to re-establish its RRC connection.
- The mobile's Network Capability changes. A mobile's Network Capability is a set of information describing the mobile's non-radio-related capability. It includes, for example, information needed for performing ciphering and authentication.

An RA update may be an intra-SGSN or an inter-SGSN RA update. An intra-SGSN RA update occurs when the new RA and the old RA connect to the same SGSN. In other words, the target SGSN is the same as the source SGSN. An inter-SGSN RA update occurs when the new RA and the old RA connect to the different SGSNs.

4.3.3.1 *Intra-SGSN Routing Area Update* The intra-SGSN RA update procedure (for I_u mode CN) is illustrated in Figure 4.46.

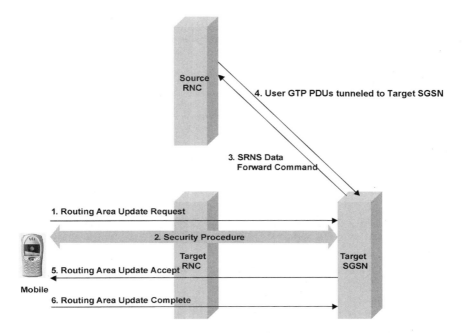

Fig. 4.46 *3GPP intra-SGSN routing area update procedure*

To send uplink signaling messages to perform an RA update, the mobile first needs to establish an RRC connection with the target RNC if such a channel does not already exist. This suggests that a mobile has to be in PMM-CONNECTED state for at least the duration of the RA Update procedure. If the mobile is in PMM-IDLE state before it starts RA Update, establishing the necessary signaling connection to the target SGSN changes the mobile's PMM state into PMM-CONNECTED.

The mobile initiates RA update by sending a Routing Area Update Request to the target SGSN. The mobile does not have to know whether the RA update is an intra-SGSN or inter-SGSN RA update; the Routing Area Update Request is the same for both intra-SGSN and inter-SGSN RA updates.

The Routing Area Update Request carries the following main information elements:

- *P-TMSI*: This field carries the P-TMSI that the mobile has been using immediately before sending the Routing Area Update Request message. This P-TMSI is likely a P-TMSI assigned to the mobile by the mobile's source SGSN.

- *Old RAI*: The Routing Area Identifier of the previous (old) Routing Area. The old RAI will be used by the target SGSN to determine whether the RA Update is intra-SGSN or inter-SGSN by examining whether it also serves the old RA.

- *Old P-TMSI Signature*: The current P-TMSI signature the mobile has for its current P-TMSI. Recall that a P-TMSI signature is used by a SGSN to authenticate a P-TMSI.

- *Update Type*: The Update Type tells the target SGSN whether the RA Update is triggered by a change of RA, a periodic RA update, or a combined RA/LA update.
- *Network Capability*.

The mobile's Routing Area Update Request arrives first at the target RNC, which in turn forwards it to the target SGSN. Forwarding the Routing Area Update Request from the target RNC to the target SGSN will trigger the establishment of an I_u signaling connection between the target RNC and the target SGSN for the mobile if such a connection does not already exist (e.g., if the mobile was in PMM-IDLE state before sending the Routing Area Update Request).

The target SGSN determines whether the RA update is intra-SGSN or inter-SGSN RA update by examining the Old RAI carried in the Routing Area Update Request received from the mobile. The RA update is intra-SGSN RA update if the target SGSN also serves the old RA.

The target SGSN needs to authenticate the mobile to determine whether the Routing Area Update Request can be accepted. As the mobile identifies itself by its P-TMSI (rather than its IMSI) in the Routing Area Update Request message, the target SGSN will try to authenticate the mobile by validating the mobile's P-TMSI first.

Only the SGSN that assigned the P-TMSI has sufficient information (i.e., the mobile's IMSI and the correct P-TMSI Signature for the P-TMSI) to validate the P-TMSI. As the target SGSN is identical to the source (and serving) SGSN with an intra-SGSN handoff, the target SGSN should be the SGSN that assigned the old P-TMSI to the mobile and therefore should be able to validate the P-TMSI locally. If the P-TMSI validation fails, the target SGSN may initiate further security procedures to authenticate the mobile.

Upon positive authentication of the mobile, the SGSN updates the mobile's RAI it maintains for the mobile. If the mobile was in PMM-CONNECTED state on the target SGSN, some user traffic destined to the mobile may have already been sent by the target SGSN to the source RNC and are still buffered at the source RNC waiting to be delivered to the mobile. As the mobile is now connected to the target RNC that is different from the source RNC, the source RNC may not be able to deliver these buffered user traffic over its own radio connections to the mobile. If these traffic belong to a Radio Access Bearer that requires in-order delivery user packets, the target SGSN may send a Serving RNS (SRNS) Data Forward Command to the source RNC to instruct the source RNC to tunnel the user traffic buffered at the source RNC to the target SGSN. The target SGSN will in turn deliver the traffic to the mobile before subsequent traffic is sent to the mobile.

The SGSN will also send a Routing Area Update Accept message to the mobile to inform the mobile that its Routing Area Update Request is accepted. The target SGSN may assign a new P-TMSI to the mobile. In this case, the new P-TMSI together with a P-TMSI Signature for the new P-TMSI will be carried in the Routing Area Update Accept message. The mobile confirms the acceptance of the new P-

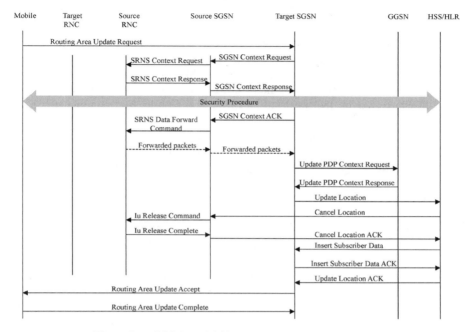

Fig. 4.47 *3GPP inter-SGSN routing area update procedure*

TMSI by returning a Routing Area Update Complete message to the SGSN, which completes the RA Update procedure.

4.3.3.2 *Inter-SGSN Routing Area Update*
The inter-SGSN RA update procedure (for I_u mode CN) is illustrated in Figure 4.47 [10]. A mobile initiates an inter-SGSN RA update by sending a Routing Area Update Request to the SGSN in exactly the same way as in initiating an intra-SGSN RA update. The Routing Area Update Request has the exact same format and information elements as in the Routing Area Update Request used in an intra-SGSN RA update.

Just as intra-SGSN RA update, the target SGSN needs to authenticate the mobile in order to determine if the Routing Area Update Request can be accepted. However, for an inter-SGSN RA update, the target SGSN is different from the source SGSN. The mobile's P-TMSI in the Routing Area Update Request was not assigned by the target SGSN, but it was instead most likely assigned by the source SGSN. Therefore, the target SGSN will ask the source SGSN to help validate the P-TMSI. To do so, the target SGSN first derives the source SGSN from the Old RAI and the P-TMSI carried in the Routing Area Update Request received from the mobile. Then, the target SGSN will send a SGSN Context Request message to the source SGSN to ask the source SGSN to validate the mobile's P-TMSI. The SGSN Context Request carries the following information elements: Old P-TMSI, Old RAI, and Old P-TMSI Signature.

The source SGSN will validate the P-TMSI and act as follows:

- *Upon positive validation of the P-TMSI*: The source SGSN will send a SGSN Context Response message back to the target SGSN. The SGSN Context Response message will carry the mobile's PMM context and PDP context. The PMM and PDP contexts contain critical information needed by the target SGSN to handle the traffic to and from the mobile. For example, the PDP contexts describe the mobile's active PDP contexts immediately before the RA update. The target SGSN will have to initiate the process to update these PDP contexts on the mobile's GGSN during the RA update process. Furthermore, if the mobile was in PMM-CONNECTED state on the source SGSN, the source SGSN could be sending packets to the mobile immediately before the RA update. Therefore, the target SGSN will also needs to know the sequence number of the next user packet the target SGSN should send to the mobile in order to ensure in-sequence delivery of user packets to the mobile.

  Some PDP context information (e.g., sequence number of the next packet to be sent to the mobile) requested by the target SGSN may be maintained by the source RNC. In this case, the source SGSN will send an SRNS Context Request to the source RNC to collect such information. After receiving this message, the source RNC will stop sending downlink data to the mobile and returns an SRNS Context Response message, which carries the information requested by the source SGSN, to the source SGSN.

- *Upon negative validation of the P-TMSI*: The source SGSN will send an appropriate error cause to the target SGSN. This will trigger the target SGSN to initiate the security procedures directly with the mobile to authenticate the mobile. If this authentication is also negative, the target SGSN will reject the mobile's Routing Area Update Request.

  If this authentication is positive, the target SGSN will send another SGSN Context Request message to the source SGSN to retrieve the mobile's PMM context and PDP context. This time, the SGSN Context Request will carry the following information: the mobile's IMSI, Old RAI, and an indicator ("MS Validated") to indicate that the mobile has been positively authenticated by the target SGSN. The source SGSN will respond with an SGSN Context Response message carrying the mobile's PMM context and PDP context if the source SGSN has these information elements requested by the target SGSN, or an appropriate error cause if the source does not have the mobile's PMM context and PDP context.

After receiving an SGSN Context Response from the source SGSN indicating a positive validation of the mobile's P-TMSI, the target SGSN responds with an SGSN Context ACK message.

If the mobile was in PMM-CONNECTED state on the source SGSN, some user traffic destined to the mobile may have already been sent by the source SGSN to the source RNC and are still buffered at the source RNC waiting to be delivered to the mobile. As the mobile is now connected to the target SGSN that is different from the source SGSN, the source RNC may not be able to deliver these buffered user traffic over its own radio connections to the mobile. If these traffic belong to a Radio Access Bearer that requires in-order delivery user packets, the source SGSN may send an SRNS Data Forward Command to the source RNC to instruct it to tunnel the user traffic buffered at the source RNC to the source SGSN, which will further tunnel the traffic to the target SGSN. The target SGSN will in turn deliver the traffic to the mobile.

After sending the SGSN Context ACK message to the source SGSN, the target SGSN will also initiate the process to update the mobile's active PDP contexts in order to ensure that the mobile's serving GGSN knows to which SGSN packets destined to the mobile should be delivered. To illustrate how the mobile's existing PDP contexts are updated during the RA Update process, we consider the case where the mobile's serving GGSN remains the same after the mobile moves into the new RA, as shown in Figure 4.47.

To update the mobile's PDP context, the target SGSN will send an Update PDP Context Request to the serving GGSN to update each existing PDP context for the mobile. This will trigger the serving GGSN to update the mobile's PDP context.

For a successful PDP context update, the target SGSN will also update the mobile's location with the HLR, which tracks each mobile's serving SGSN. When a GGSN has user packets to send to a mobile but does not have an active PDP context for the mobile, the GGSN may query the HLR to find out the address of the mobile's current serving SGSN and then use Network-requested PDP Context Activation (Chapter 2 "Wireless IP Network Architectures") to establish a PDP context for the mobile so that it can forward the packets to the mobile.

The SGSN uses the G_r interface to interact with the HLR for location update. It sends an Update Location message to the HLR. Upon receiving the Update Location message, the HLR will inform the source SGSN to cancel its location information regarding the mobile. The source SGSN will remove the location and service subscription information it has been maintaining for the mobile. The source SGSN will also release the I_u connections between the source SGSN and the source RNC used by the mobile. In the meantime, the HLR will also send the user's service subscription to the target SGSN by sending an Insert Subscriber Data message to the target SGSN. The target SGSN records the mobile's service subscription information received in this Insert Subscriber Data message and responds to the HLR with an Insert Subscriber Data ACK message. Now, the HLR will send an Update Location ACK message back to the target SGSN to indicate that location update with the HLR is complete.

Upon a successful location update with the HLR, the target SGSN will create a PMM context for the mobile. Then, the target SGSN will send a Routing Area Update Accept message to the mobile to inform the mobile that its Routing Area Update Request is accepted. The target SGSN will assign a new P-TMSI to the mobile. The new P-TMSI together with a P-TMSI Signature for the new P-TMSI will be carried in the Routing Area Update Accept message. The mobile confirms the acceptance of the new P-TMSI by returning a Routing Area Update Complete message to the SGSN, which completes the RA Update procedure.

Routing Area update in 3GPP has an important characteristic: It is integrated with GPRS routing inside the PS CN. In particular, when a mobile performs Routing Area update, the host-specific route maintained by the PS CN for forwarding user packets to and from the mobile will also be updated if necessary. For example, before an inter-SGSN Routing Area update, the mobile's host-specific route inside the PS CN is between the mobile's serving GGSN and the source SGSN. The inter-SGSN Routing Area update causes this host-specific route to be changed to between the mobile's serving GGSN and the target SGSN.

It is interesting to note that a similar integration of location update and routing is the foundation for some existing micromobility management protocols designed for IP networks, such as Cellular IP (Section 4.2.7) and HAWAII (Section 4.2.8).

4.3.4 Serving RNS Relocation

A mobile in PMM-CONNECTED state has a serving RNC. The mobile's serving RNC receives user traffic directly from the CN and distributes the traffic over the RAN to the mobile. The RNS containing a mobile's serving RNC is referred to as the mobile's serving RNS. As shown in Figure 4.48 (a), a mobile's serving RNS may forward user traffic via another RNC to the mobile.

When a mobile connects to a target RNC, the target RNC may become the mobile's new serving RNC. As an I_u connection needs to be maintained between the

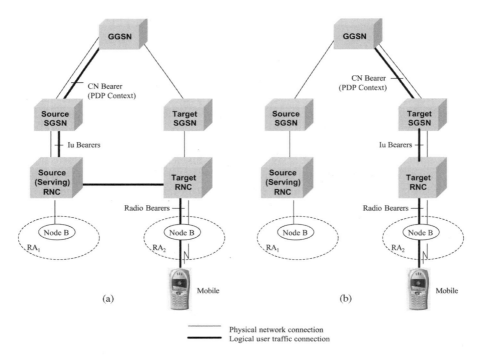

Fig. 4.48 *3GPP data path before and after Serving RNS Relocation and RA Update*

mobile's serving RNC and the mobile's serving SGSN while the mobile is in PMM-CONNECTED state, the RNC side of the mobile's I_u connections needs to be relocated from the old serving RNC to the new serving RNC. This is achieved using the Serving RNS Relocation procedure.

In this section, we describe the Serving RNS Relocation procedure under the assumption that, before the relocation, the mobile's serving RNC is using the I_{ur} interface to forward signaling and user traffic to another RNC, which in turn delivers the user traffic to the mobile. Such a scenario may occur during or after a soft inter-RNC handoff. During a soft inter-RNC handoff, the source RNC distributes copies of user traffic to one or more other RNCs, which in turn deliver the user data to the mobile simultaneously.

Under the assumption that the I_{ur} interface is used, the Serving RNS Relocation procedure can be triggered by RA Update, hard handoff, or soft handoff. When the I_{ur} interface does not exist and hence soft handoff is not possible between the RNCs, the Serving RNS Relocation procedure may be triggered by RA Update or hard handoff. Serving RNS Relocation for the case in which the I_{ur} interface is not present is described in Section 4.3.5 together with the hard handoff procedure.

Consider a mobile that moves from a source RNC to a target RNC. Assume that the source RNC is also the mobile's serving RNC before the handoff. After the mobile connects to the target RNC but before Serving RNS Relocation or RA Update is performed, the user traffic route between the GGSN and the mobile may look like the one illustrated in Figure 4.48 (a), assuming that the source RNC and the target RNC are connected to different SGSNs. In particular, user traffic destined to the mobile continues to be routed by the PS CN to the source RNC, which is still the mobile's serving RNC at this moment. The source RNC will then forward the user traffic to the target RNC. The target RNC will in turn transmit the traffic to the mobile. After the Serving RNS Relocation and RA Update procedures are completed, the user traffic route between the mobile and the GGSN will look like the one shown in Figure 4.48(b). For ease of illustration, Figure 4.48 (b) assumes that the mobile's serving GGSN does not change after the Serving RNS relocation and the RA Update procedures.

The Serving RNS Relocation procedure is illustrated in Figure 4.49 [10]. Only the source RNC can initiate Serving RNS Relocation. The source RNC decides whether to initiate Serving RNS Relocation based on measurement results of the quality of the radio channels to the mobile and based on its knowledge of the RAN topology.

When the source RNC decides to initiate Serving RNS Relocation, it sends a RANAP Relocation Required message to the source SGSN. The Relocation Required message carries the following main information elements:

- *Relocation Type*: The Relocation Type indicates whether the mobile terminal should be involved in carrying out the serving RNS relocation procedure, in particular, whether the mobile's RRC connection also needs to be relocated during the serving RNS relocation procedure.

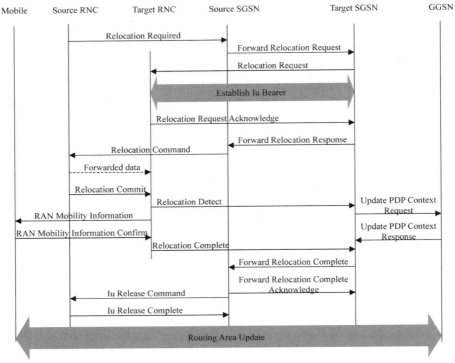

Fig. 4.49 *3GPP Serving RNS Relocation*

– *Relocation Type is "UE not Involved"*: This indicates that the network itself carries out the serving RNS relocation procedure and the mobile's RRC connection does not need to be relocated during the serving RNS relocation process. In other words, the mobile has already set up the necessary RRC connection with the target RNC; no handoff procedure needs to be performed for the RRC Connection during the serving RNS relocation procedure. The network only needs to move the RNC side of the mobile's I_u bearers to the target RNC to make it the mobile's new serving RNC.

The serving RNS relocation illustrated in Figure 4.49 is not combined with any handoff procedure. Therefore, the Relocation Type should be set to "UE not Involved."

– *Relocation Type is "UE Involved"*: This indicates that the mobile will also need to be involved in the serving RNS relocation process to relocate its RRC connection to the target RNC. The "UE Involved" Relocation Type is used, for example, in the combined handoff and serving RNS relocation procedure (Section 4.3.5).

• *Source ID*: Identifier of the source RNC.
• *Target ID*: Identifier of the target RNC.

- *Source RNC to target RNC transparent container*: This information element contains the information needed by the target RNC to perform serving RNC relocation. It includes the security information regarding the mobile, and the RRC protocol context that describes the mobile's RRC connection and the mobile's capabilities.

The source SGSN determines if the RNS relocation is intra-SGSN or inter-SGSN by inspecting the identifiers of the source and the target RNCs. For inter-SGSN relocation, the source SGSN will send a RANAP Forward Relocation Request message to the target SGSN to request the target SGSN to establish the I_u connection for the mobile. The target SGSN will then send a RANAP Relocation Request message to the target RNC to trigger it to establish the necessary RABs for the mobile. Recall that each RAB consists of I_u bearers between the SGSN and the RNC and Radio Bearers between the mobile and the RNC. To set up the RABs between the target SGSN and the mobile, the mobile's I_u bearers between the source SGSN and the source RNC need to be relocated between the target SGSN and the target RNC. There will be no need to establish new Radio Bearers for the mobile. Instead, the existing Radio Bearers between the mobile and the source RNC will be relocated between the mobile and the target RNC. After the target RNC has allocated all the necessary resources for all required RABs, it will send a RANAP Relocation Request Acknowledge message back to the target SGSN to inform the target SGSN the I_u bearers for which RABs have been successfully set up and the I_u bearers for which RABs have failed to be set up.

At this point, the resources for transporting user packets between the target RNC and the target SGSN have been allocated and the target RNC is ready to become the new serving RNC for the mobile. The target SGSN will send a RANAP Forward Relocation Response message back to the source SGSN.

Upon receiving the Forward Relocation Response from the target SGSN, the source SGSN will send a Relocation Command to the source RNC to instruct the source RNC to start to hand over the role of the serving RNS to the target RNC. This message will also inform the source RNC which RABs for the mobile should be released and which ones should be kept for a little longer so that user packets already received by the source RNC can be forwarded to the target RNC. At this moment, the source RNC may start to forward downlink user data, which it has already received, toward the target RNC.

Now, the source RNC will suspend uplink and downlink user data transfer for all the RABs that require delivery order. Then, the source RNC will transfer its serving RNC role to the target RNC. To transfer the role of serving RNC, the source RNC sends a Relocation Commit message, over the I_{ur} interface, to the target RNC. This Relocation Commit message also sends the Serving RNS (SRNS) Contexts for the mobile from the source RNC to the target RNC. These SRNS Contexts contain information regarding the RABs between the mobile and the source SGSN.

Upon receiving the Relocation Commit message from the source RNC or detecting radio connections from the mobile, the target RNC will send a RANAP Relocation Detect message to the target SGSN to request the SGSN to update the

PDP Context for the mobile if necessary (i.e., if the relocation is inter-SGSN). Immediately after sending out the Relocation Detect message, the target RNC will start to serve as the serving RNC for the mobile and will start to send RAN Mobility Information messages to the mobile. The RAN Mobility Information messages contain the identity of the mobile's new serving RNC, Location Area Identifier, and Routing Area Identifier.

The mobile may begin to send uplink traffic toward the target RNC immediately after the mobile receives the RAN Mobility Information message. The mobile will also use the information in the received RAN Mobility Information message to reconfigure itself. After the self reconfiguration, the mobile will send the RAN Mobility Information Confirm message to the target RNC. This message indicates to the target RNC that the mobile is ready to receive user traffic from the target RNC.

Upon receiving the RAN Mobility Information Confirm message, the target RNC will send Relocation Complete to the target SGSN and the target SGSN will in turn inform the source SGSN of the completion of Serving RNS relocation procedure. Upon being informed by the target SGSN that the Serving RNS relocation is completed on the target SGSN, the source SGSN will instruct the source RNC to release the I_u Bearers allocated to the mobile.

When the mobile starts communication with the target RNC, the mobile may find that it has moved into a new RA. In this case, the mobile will initiate the RA Update procedure.

4.3.5 Hard Handoffs

Handoffs between Node Bs under the same RNC (i.e., intra-RNC handoffs) in a UTRAN are handled solely by procedures inside the UTRAN. These procedures are not visible to the PS Domain.

This section focuses on inter-RNC hard handoff, assuming that no I_{ur} interface is implemented. When no I_{ur} interface is implemented, inter-RNC handoff can only be hard handoff.

Before an inter-RNC hard handoff, the source RNC is the mobile's serving RNC. During and after an inter-RNC hard handoff, the target RNC will become the mobile's new serving RNC. This requires the RNC side of the mobile's I_u Bearers to be relocated from the source RNC to the target RNC during the inter-RNC hard handoff. Therefore, an inter-RNC hard handoff is usually combined with the Serving RNC Relocation procedure. This combined procedure is illustrated in Figure 4.50. The procedure shown in Figure 4.50 also applies to handoff between GSM BSSs, and between GSM BSS and UTRAN RNS.

Only the source RNC can initiate the inter-RNC hard handoff process. The source RNC determines whether to initiate the handoff process based on the measurement results of the radio channel qualities and its knowledge of the RAN topology. It initiates the combined handoff and serving RNS relocation procedure by first initiating the serving RNS relocation procedure. The source RNC sends a RANAP Relocation Required message to the source SGSN and sets the Relocation Type in this message to "UE Involved" (Section 4.3.4). The "UE Involved" Relocation Type

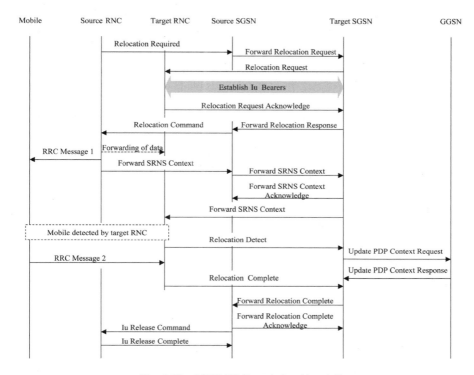

Fig. 4.50 *3GPP PS Domain hard handoff*

indicates that the mobile will also be involved in the serving RNS relocation process to relocate its RRC connection to the target RNC (i.e., to perform handoff).

After the source RNC sends out the Relocation Required message to the source SGSN, the serving RNS relocation procedure proceeds in the same manner as the Serving RNS Relocation procedure described in Section 4.3.4. In particular, if the handoff is an inter-SGSN handoff, the source SGSN will send a Forward Relocation Request to the target SGSN to ask the target SGSN to start the process to relocate the mobile's RABs between the mobile and the target SGSN.

Upon receiving the Forward Relocation Request, the target SGSN instructs the target RNC to relocate the RABs required for the mobile by sending a Relocation Request message to the target RNC. The target RNC proceeds to allocate all the necessary resources needed to set up the I_u Bearers for the required RABs. The target RNC then sends a Relocation Request Acknowledge message back to the target SGSN.

Unlike the Serving RNS Relocation procedure described in Section 4.3.4, the Relocation Request Acknowledge here will carry an extra information element— Target RNC to Source RNC Transparent Container. This information element contains all the radio-related information that the mobile will need in order to tune its radio to the radio channels of the target RNS. The Target RNC to Source RNC

Transparent Container will be passed on by the target SGSN to the source SGSN, then by the source SGSN to the source RNC, and finally by the source RNC to the mobile.

When all the I_u bearers have been established for the mobile between the target SGSN and the target RNC and the target RNC is ready to act as the serving RNC for the mobile, the target SGSN sends a Forward Relocation Response message to the source SGSN. The Forward Relocation Response message will contain, among other information, the Target RNC to Source RNC Transparent Container. The Forward Relocation Response message triggers the source SGSN to send a Relocation Command to the source RNC to instruct the source RNC to hand over the role of the serving RNS to the new RNC. The Relocation Command contains, among other information, the Target RNC to Source RNC Transparent Container received in the Forward Relocation Response message from the source SGSN.

Upon receiving the Relocation Command message, the source RNC takes the following main actions:

- *Forwarding user data to target RNC*: The source RNC begins to forward downlink user data, which have already arrived at the source RNC, toward the target RNC. Forwarding of user data from the source RNC to the target RNC is only performed for the RABs that require such data forwarding.

- *Instruct the mobile to relocate its RRC connection to the new RNC*: The source RNC will send an RRC-layer message (referred to as "RRC Message 1" in Figure 4.50) to instruct the mobile to relocate its Radio Bearers to the new RNS. This "RRC Message 1" may be different for different radio systems and different types of handoffs. For example, it may be a Physical Channel Reconfiguration message for RNS to RNS relocation in a UTRAN, a Handover Command message for BSS to BSS relocation in a GERAN, or an Intersystem to UTRAN Handover for BSS to RNS relocation.

  For the RABs that require in-sequence delivery of user data, the source RNC will suspend both uplink and downlink data transfer before instructing the mobile to relocate its Radio Bearers to the target RNS. This step helps to ensure that data will be delivered to the target RNC in order.

  After the mobile has reconfigured its Radio Bearers and established its RRC connection to the target RNC, it will send a RRC-layer message ("RRC Message 2" in Figure 4.50) to inform the target RNC that handoff on the mobile side has been completed. For example, if "RRC Message 1" is Physical Channel Relocation for RNS to RNS relocation in a UTRAN, "RRC Message 2" will be a Physical Channel Relocation Complete message.

The source RNC continues the execution of serving RNS relocation by sending a Forward SRNS Context message to the target RNC (via the source SGSN and the target SGSN). This message provides the mobile's Serving RNS Context to the target RNC. An SRNS Context contains information regarding the mobile's RABs through the source RNC and can be used by the target RNC to establish I_u bearers with the same parameters for the mobile. For each RAB that requires data delivery order, the SRNS Context also contains the sequence numbers of the next downlink and uplink GTP protocol data units to be transmitted.

When the target RNC detects that the mobile has connected to the target RNC (which may occur before the target RNC receives "RRC Message 2" from the mobile), the target RNC will inform the target SGSN by sending a Relocation Detect message to the target SGSN. For an inter-SGSN hard handoff, the Relocation Detect message will also trigger the target SGSN to initiate the PDP Context Update procedure to ensure that the GGSN will start to send user packets destined to the mobile to the target SGSN.

After the target RNC has detected that the mobile is connected to the target RNC and received the Forward SRNS Context message with the required data delivery sequence numbers for the RABs that require delivery order, the target RNC can start to exchange user data with the mobile on all RABs. However, if the target RNC receives "RRC Message 2" before it receives the required data delivery sequence numbers for the RABs that require delivery order, the target RNC can only exchange user data with the mobile over the RABs that do not require data delivery order, until it receives the required sequence numbers.

The handoff process completes when the mobile has connected to the target RNS, the mobile's PDP context on the mobile's serving GGSN has been updated, and the target RNC has received all the required Serving RNS Context information from the source RNC. The target SGSN informs the source SGSN of the handoff completion by sending a Forward Relocation Complete message to the source SGSN. This message will trigger the source SGSN to release the mobile's I_u Bearers between the source SGSN and the source RNC.

After an inter-RNC hard handoff, the mobile may perform Routing Area Update if it has moved into a new Routing Area as a result of the handoff.

4.3.6 Paging Initiated by Packet-Switched Core Network

When an SGSN wants to send user data to a mobile in PMM-IDLE state, the SGSN will have to initiate the paging process. Paging initiated by SGSN is illustrated in Figure 4.51 [10].

Upon receiving downlink user data or signaling messages destined to a mobile in PMM-IDLE state, the SGSN initiates paging by sending a RANAP Paging message to every RNC in the Routing Area in which the mobile is located. The RANAP Paging message carries the following main information:

- *Identities of the mobile to be paged*: The RANAP Paging message carries the mobile's IMSI. If the mobile is using a temporary identity, a P-TMSI, assigned by the SGSN, the RANAP Paging message will also contain the mobile's P-TMSI.
- *CN Domain Identifier*: The CN Domain Indicator indicates which CN domain (i.e., PS CN domain or CS CN domain) initiated this RANAP Paging message.
- *Area*: The Paging Area in which the mobile is to be paged.

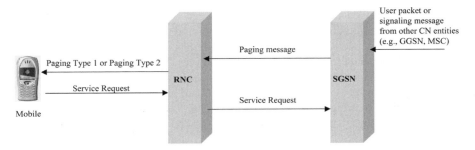

Fig. 4.51 *3GPP paging in packet switched domain*

Upon receiving a RANAP Paging message, each RNC determines how paging should be carried out over the air interface. Depending on the way a paging message is physically delivered by the RNC to the mobile, paging inside the RAN can be classified into two types:

- *Type 1 Paging*: Type 1 paging is performed when there is no dedicated RRC connection between the RNC and the mobile. In this case, the RNC will send a Paging Type 1 message to the mobile over the Paging Channel, a physical radio broadcast channel.

  Type 1 paging may also be initiated by the RAN, i.e., by an RNC in the RAN. Therefore, the Paging Type 1 message will carry a Paging Originator field that indicates whether the paging is initiated by the CN or the RAN.

- *Type 2 Paging*: Type 2 paging will be used when the mobile has a dedicated RRC connection to the RNC. In this case, the RNC will deliver a Paging Type 2 message to the mobile over this dedicated RRC connection.

Upon receiving a Paging Type 1 or 2 message, the mobile will start the Service Request procedure (Section 4.3.7) to establish the necessary signaling and traffic connections with the CN and use them to send uplink signaling messages (to, for example, respond to the received paging message and to activate the PDP Context) and user packets.

4.3.7 Service Request Procedure

The Service Request Procedure is used by a mobile in PMM-IDLE state to request the establishment of a signaling connection between the mobile and the SGSN so that the mobile can begin to exchange signaling messages with the SGSN. The Service Request Procedure is also used by a mobile in PMM-CONNECTED state to request resource reservation for the mobile's active PDP contexts.

Figure 4.52 illustrates the mobile-initiated Service Request Procedure [10] (for I_u mode operation). First, the mobile will have to establish an RRC connection with the RNC, if such a connection does not already exist. Then the mobile sends a Service

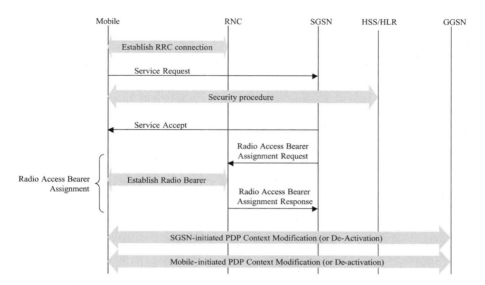

Fig. 4.52 *3GPP Mobile-initiated Service Request Procedure*

Request message to the SGSN. Upon receiving the Service Request, the SGSN will perform security procedures to authenticate the mobile and check if it is authorized to use the network. If the mobile is authorized to use the network, the SGSN takes actions based on the Service Type in the received Service Request.

- *If the Service Type indicates DATA*: This means that the mobile has user data to send over the PS CN domain or expects to receive user data from the PS CN domain. In this case, a signaling connection between the mobile and the SGSN will be established first so that the mobile can send signaling messages to the SGSN. Then, the RABs will be allocated for the mobile's existing active PDP contexts using the RAB Assignment Procedure described in Chapter 2, if such RABs do not already exist, to allow the mobile to exchange user data with the GGSN. A mobile may have an active PDP context but does not have any RAB if the mobile was in PMM-IDLE state immediately before it started the Service Request Procedure.

- *If the Service Type indicates SIGNALING*: This means that the mobile has no user data to send over the PS CN domain and does not expect to receive any user data from the PS CN domain at this moment. Instead, the mobile just wishes to exchange signaling messages with the SGSN. Therefore, only a signaling connection between the mobile and the SGSN will be established. No RAB will be allocated for any active PDP context of the mobile. Once the signaling connection between the mobile and the SGSN has been established, the mobile may use it to, for example, create and activate new PDP contexts.

The Service Request is acknowledged in different ways depending on the Service Type and the mobile's PMM state:

- If the mobile is in PMM-CONNECTED state and the Service Type indicates DATA, the SGSN will respond to the Service Request message by returning a Service Accept message to the mobile if the SGSN accepts the mobile's service request.
- If the Service Type indicates SIGNALING or the mobile is in PMM-IDLE state, the SGSN does not send any explicit signaling message to the mobile to indicate the acceptance of the mobile's Service Request. Instead, the mobile will learn that its Service Request was successfully received by the SGSN when the mobile receives certain RRC-layer signaling messages from the RNC.

After a RAB has been re-established, the QoS profile negotiated for this newly established RAB may be different from the old negotiated QoS profile maintained in the PDP contexts on the SGSN, GGSN, and the mobile. In such a case, the SGSN will trigger the SGSN-initiated PDP Modification procedure to inform the GGSN and the mobile of the new negotiated QoS profile for the RAB. In case a RAB cannot be successfully set up, the SGSN may use the SGSN-initiated PDP Modification procedure to trigger the mobile and the CN (i.e., the SGSN and the GGSN) to renegotiate the QoS profile. The SGSN may also use the SGSN-initiated PDP De-Activation procedure to delete a PDP context if the required RABs cannot be set up for this PDP context.

The mobile may also use the Modification procedure to request the CN to renegotiate the QoS profile for a RAB, or use the Mobile-initiated PDP Context De-Activation procedure to delete an active PDP context.

4.3.8   Handoff and Roaming Between 3GPP and Wireless LANs

As we discussed in Chapter 1, deployment of enterprise and public wireless LANs (WLANs) have been growing rapidly worldwide. This creates an increasing need for users to be able to handoff or roam between a cellular network such as 3GPP and an enterprise or public WLAN.

Here, we discuss how Mobile IP (v4 or v6) can be used to support handoff and roaming between 3GPP and WLAN. This same approach can also be used to support handoff/roaming between WLAN and any other cellular network (e.g., GPRS, EDGE, and 3GPP2). Other IP-based mobility management protocols (e.g., SIP-based mobility management) may also be used in a similar manner to support handoffs and roaming between WLAN and a cellular network.

Figure 4.53 illustrates a simplified network configuration for using Mobile IP to support handoff between 3GPP and WLAN. The cellular network shown in the figure can be any cellular network that can provide mobiles access to an IP network. For example, the cellular network may be 3GPP, GPRS, EDGE, or 3GPP2. The

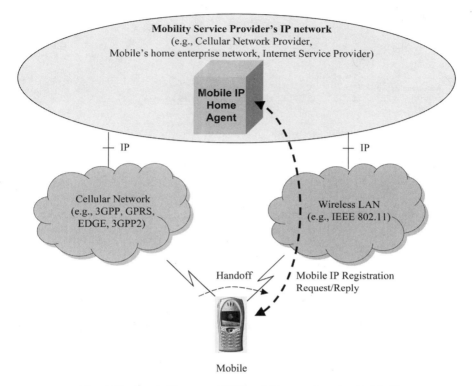

Fig. 4.53 *Handoff between 3GPP and IP networks using Mobile IP*

mobility service provider shown in Figure 4.53 refers to any network provider that provides a Mobile IP Home Agent to the mobile. A mobile's mobility service provider may be, for example, a cellular network provider, the mobile's home enterprise network, or an Internet Service Provider. The cellular network connects to the mobility service provider's IP network directly or via a standard IP network.

When a mobile moves from a cellular network to a WLAN, the mobile acquires a local IP address that it can use to receive IP packets from the new network. Then, the mobile uses this new local IP address as its new Mobile IP Care-of Address (CoA) and uses Mobile IP to register the new CoA with its Mobile IP Home Agent, as illustrated in Figure 4.53. This will allow the mobile to continue to receive IP packets addressed to its Mobile IP Home Address sent by any mobile or fixed terminal that can send packets to the mobile's mobility service provider's network. As in regular Mobile IP, user IP packets addressed to the mobile's Home Address will be routed to the mobile's home agent. The home agent will tunnel these packets to the mobile's current CoA.

Now, let's consider the scenario in which a mobile moves from a WLAN into a cellular network. Figure 4.54 illustrates a sample signaling message flow to support

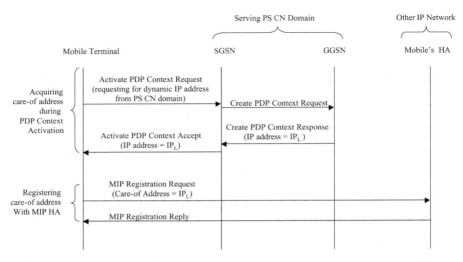

Fig. 4.54 Sample signaling flow for handoff between 3GPP and IP networks using MIPv4

handoff from WLAN to 3GPP. First, the mobile performs standard 3GPP procedures to attach to the 3GPP network (if not already attached), to activate its PDP context, and to acquire a local care-of IP address during the PDP Context Activation process. The mobile registers this care-of address with its home agent. From this point on, IP packets addressed to the mobile's MIP home address will be tunneled by the mobile's home agent to the mobile's current care-of address. If the care-of address is a co-located care-of address, the packets will be tunneled to the mobile directly. The mobile will then de-tunnel the packets to extract the original payload packets. If the mobile uses a Foreign Agent (FA) care-of address in the 3GPP network and the FA is on the mobile's serving GGSN (Chapter 2), the packets will be tunneled to the FA/ GGSN which will de-tunnel the packets and forward the resulting packets to the mobile.

It is important to note that no Mobile IP FA is necessary inside the 3GPP network. The mobile can use "Transparent Access" described in Chapter 2 to acquire a co-located local care-of address via the PDP Context Activation procedure without any assistance from a Mobile IP FA.

Many WLAN users are likely business users. The WLAN interface on a business user's mobile device will likely have a home address assigned by the user's enterprise network. When such a business user accesses the public WLAN hotspots and 3GPP networks provided by the same network operator, the public network operator may also be using Mobile IP to support handoff between its 3GPP network and its public WLANs. In such a scenario, the WLAN interface on a mobile device may have two Mobile IP home addresses: one assigned by the user's enterprise network and the other one assigned by the public carrier.

The mobile device, however, should be known to other IP terminals by only one IP home address so that other IP terminals do not have to be concerned with which

IP address on a mobile terminal is currently reachable. This can be achieved by enhancing the implementation of the Mobile IP client software on the mobile terminals to support dual home addresses. Suppose a mobile has a home address H_E assigned by its enterprise and a home address H_P assigned by a public network provider. For ease of discussion, let us refer to H_E and H_P as the mobile *enterprise home address* and *public home address*, respectively. Figure 4.55 illustrates one approach for the mobile to make H_E its only globally known home address.

As shown in Figure 4.55, when a mobile moves into the public network (cellular or public WLAN), it will obtain a local care-of address (address IP_L in the figure) that it can use to receive packets in the public network. It then registers the local care-of address with its Mobile IP home agent HA_P in the public network. It will then register its public home address H_P as its care-of address with its home agent HA_E in its enterprise network. Now, packets addressed to the mobile's home address H_E will be captured by its home agent HA_E and tunneled to its home agent HA_P in the public network. Home agent HA_P extract the original packet from the tunnel and then tunnel the original packet to the mobile local care-of address.

When the mobile is using a GPRS-based cellular network (e.g., GPRS, EDGE, 3GPP), the mobile may use its public home address H_P as its PDP Address. This way, the home agent HA_P in the public network does not have to tunnel user packets to the mobile. Instead, HA_P can use the mobile's PDP context to deliver the original packets addressed to the mobile's home address H_E directly to the mobile.

An important benefit of the dual home address approach to mobility management is that a mobile's public home address does not have to change while the mobile is

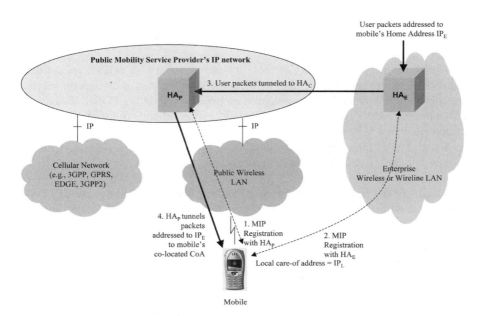

Fig. 4.55 *Mobile terminals with dual home addresses*

moving between cellular networks and public WLANs inside the same public network provider's administrative domain. This feature can be used, for example, to ensure that a VPN (Virtual Private Network) does not break while a mobile is on the move. Today's most prevailing form of IP VPN, IPsec tunnel, breaks when the IP address identifying either end of the VPN changes. When the mobile uses its public home address to identify its end of an IPsec tunnel, this IPsec tunnel will be able to stay on when the mobile changes its local care-of address.

4.4 MOBILITY MANAGEMENT IN 3GPP2 PACKET DATA NETWORKS

As we discussed in Chapter 2, "Wireless IP Network Architectures," all user IP packets to and from a mobile inside a 3GPP2 packet data network are sent first to the mobile's *serving PDSN*, which in turn forwards the packets toward their final destinations. A mobile and its serving PDSN maintains a PPP connection and uses it as the link layer for exchanging user IP packets.

The PPP connection between a mobile and its serving PDSN rides over the following concatenated connections (bearers):

- Radio Bearer (radio channel) between the mobile and a BSC
- A8 connection between BSC and a PCF
- A10 connection (i.e., R-P connection) between the PCF and the mobile's serving PDSN
- An optional P-P (PDSN-to-PDSN) connection between the mobile's serving PDSN and a target PDSN to support fast inter-PDSN handoff (Section 4.4.4).

Therefore, the primary tasks of mobility management in a 3GPP2 packet data network are to manage the changes of the underlying bearers and connections that support the PPP connection between each mobile and its serving PDSN, the change of the mobile's PDSN and the PPP connection, and the change of the mobile's care-of address when Mobile IP is used.

Some or all of the changes mentioned above may be necessary, depending on the scopes of mobility, as illustrated in Figure 4.56.

- *Intra-PDSN handoff*: With an intra-PDSN handoff, a mobile's serving PDSN remains unchanged. Therefore, the mobile's PPP connection to its serving PDSN does not need to change during and after an intra-PDSN handoff. Furthermore, the mobile does not need to change the IP address it uses to exchange user IP packets with its serving PDSN. Therefore, the mobile does not have to perform registration with its home agent if Mobile IP is used.

  However, some or all of the bearers that make up the path of the PPP connection may need to be changed depending on the scopes of "micro-mobility" within the same PDSN:

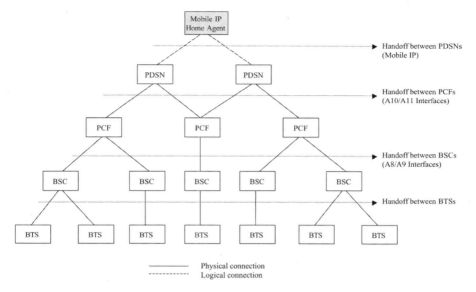

Fig. 4.56   *Scopes of mobility in a 3GPP2 packet data network*

- *Inter-BTS handoff*: With an inter-BTS handoff, a mobile moves from one BTS (called the source BTS) to another BTS (called the target BTS). An inter-BTS handoff requires that the mobile set up Radio Bearers to the target BTS.
- *Inter-BSC handoff*: With an inter-BSC handoff, a mobile moves from one BSC (called the source BSC) to another BSC (called the target BSC). An inter-BSC handoff requires that A8/A9 connections be established between the target BSC and the target PCF for the mobile, in addition to the change of Radio Bearers. The target PCF is the PCF that the target BSC connects to directly.
- *Inter-PCF handoff*: With an inter-PCF handoff, a mobile moves from one PCF (called the source PCF) to another PCF (called the target PCF). An inter-PCF handoff requires that A10/A11 connections be established between the target PCF and the target PDSN for the mobile, in addition to the changes of the A8/A9 connections and the Radio Bearers. The target PDSN is a PDSN directly connected to the target PCF and is selected by the PCF to handle user traffic for the mobile. For intra-PDSN handoff, the target PDSN will be the same as the source PDSN.

- *Inter-PDSN handoff*: 3GPP2 defines the following two basic types of inter-PDSN handoff:
  - *Regular Inter-PDSN Handoff*: With regular inter-PDSN handoff, the target PDSN becomes the mobile's new serving PDSN as a result of the handoff. To turn the target PDSN into the mobile's new serving PDSN,

the mobile will have to establish a PPP connection to the target PDSN and configure a network protocol (i.e., IPv4 or IPv6) over the PPP connection as part of the handoff process. New Radio Bearers between the mobile and the target BSC, new A8/A9 connections between the target BSC and the target PCF, and new A10/A11 connections between the target PCF and the target PDSN will also need to be established to allow the mobile to exchange user traffic with the target PDSN. Furthermore, if Mobile IP is used, the mobile will need to acquire a new care-of address and register it with the mobile's Mobile IP home agent.

– Fast Inter-PDSN Handoff: With fast inter-PDSN handoff, the mobile's serving PDSN remains unchanged during and after the handoff as long as the mobile has an active packet data session. The mobile continues to use the same PPP connection that was established before the handoff started and that terminates on the mobile's serving PDSN, to exchange user IP packets with the serving PDSN during and after a fast inter-PDSN handoff. The serving PDSN tunnels downlink PPP frames to the target PDSN, which will then tunnel them toward the mobile. As the mobile's PPP connection remains unchanged during and after the handoff, the mobile does not have to change its care-of address and therefore does not have to perform registration with its Mobile IP home agent.

For the serving PDSN and the target PDSN to tunnel user PPP frames to each other, a PDSN-to-PDSN (P-P) connection will need to be established between the serving PDSN and the target PDSN for the mobile. For the mobile and the target PDSN to exchange user traffic, new Radio Bearers between the mobile and the target BSC, new A8/A9 connections between the target BSC and the target PCF, and new A10/A11 connections between the target PCF and the target PDSN will also need to be established.

Fast inter-PDSN handoff will be discussed in Section 4.4.4.

In the rest of this section, we discuss several key aspects of mobility management in a 3GPP2 packet data network:

- *Packet Data Service States*: All the bearers that make up the path between a mobile and its serving PDSN do not have to be maintained for the mobile at all times. Some of these bearers may be released if the mobile does not expect to send or receive user traffic soon so that the resources can be used by other mobiles. A mobile's Packet Data Service State determines which bearers are maintained for the mobile.

- *Location Management*: We will discuss how users' locations are tracked.

- *Handoffs and location update procedures*: We will discuss the procedures for supporting handoffs and discuss how the host-specific route between a mobile and its serving PDSN is managed as the mobile moves around. We will also describe location update procedures together with the handoff procedures because the location update procedure is the same as the procedure for supporting handoff of dormant users.

- *Paging*: We will discuss how the 3GPP2 packet data core network locates a dormant mobile without implementing any paging protocol inside the packet data core network.

4.4.1 Packet Data Service States

3GPP2 defines three *Packet Data Service States* for managing packet data services [3].

- *ACTIVE/CONNECTED state*: In the ACTIVE (or CONNECTED) state, a bidirectional traffic radio channel is established between the mobile and the BSC. The A8/A9 and A10/A11 connections are also established for the mobile. Furthermore, the mobile and its serving PDSN maintains a PPP connection as the data link layer over which the mobile and its serving PDSN exchange user IP packets. When Mobile IP is used for mobility management, the mobile will also have already performed Mobile IP registration with its home agent. A mobile in ACTIVE state is referred to as an active mobile.
- *DORMANT state*: In the DORMANT state, no traffic radio channel exists between the mobile and the BSC. No A8 connection exists for the mobile either. However, the mobile's A10 connection is maintained between the mobile's serving PCF and the mobile's serving PDSN. Furthermore, the PPP connection between the mobile and its serving PDSN will also be maintained. A mobile in DORMANT state is referred to as a dormant mobile.
- *NULL/INACTIVE state*: In the NULL (or INACTIVE) state, there is no traffic radio channel between the mobile and the BSC. No A8/A9 or A10/A11 connection exists for the mobile. Furthermore, no PPP connection exists between the mobile and the PDSN.

Figure 4.57 illustrates the transitions between different Packet Data Service States [3].

- *From ACTIVE state to DORMANT state*: A mobile's Packet Data Service State enters DORMANT state by releasing its A8 connection and its traffic radio channel. Either the mobile or the network can initiate the process to transition the mobile's Packet Data Service State from ACTIVE to DORMANT.
- *From ACTIVE to NULL state*: A mobile's Packet Data Service State changes from ACTIVE state to NULL state when the mobile's traffic radio channel, A8 connection, A10 connection, and the PPP connection to its serving PDSN are closed. Either the mobile or the network may initiate the process to change the mobile's Packet Data Service State changes from ACTIVE to NULL state.
- *From DORMANT state to ACTIVE state*: A mobile's Packet Data Service State transitions from DORMANT state to ACTIVE state by re-establishing the

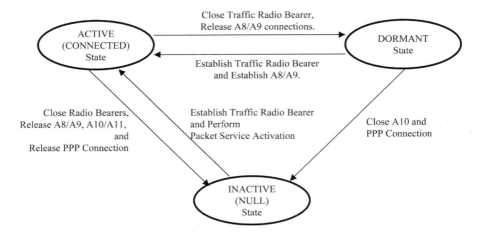

Fig. 4.57   *3GPP2 Packet Data Service State transitions*

traffic radio channel and the A8 connection. The mobile or the network can initiate the process to bring the mobile from DORMANT state to ACTIVE state. A mobile can initiate this transition using the Packet Service Activation procedure described in Chapter 2, "Wireless IP Network Architectures."

- *From DORMANT state to NULL state*: A mobile's Packet Data Service State transitions from DORMANT state to NULL state when the mobile's A10 connection and the mobile's PPP connection to its serving PDSN are closed. Either the mobile or the network may initiate the process to change a mobile's Packet Data Service State transitions from DORMANT state to NULL state.

- *From NULL state to DORMANT state*: A mobile's Packet Data Service State cannot enter the DORMANT state directly from the NULL state without first entering the ACTIVE state. Recall that a dormant mobile needs to maintain a PPP connection with the PDSN. This means that a mobile has to first enter the ACTIVE state to establish the PPP connection to the PDSN before it can enter the DORMANT state.

The Packet Data Service States are maintained in both PCF and mobile terminal. The mobile terminal's Packet Data Service States will not be communicated across the PCF-PDSN interface. In other words, the PDSN will not be aware whether a mobile is in ACTIVE or DORMANT state.

4.4.2   Location Management for Packet Data Services

3GPP2 defines Packet Zones for location management in the packet data core network. A Packet Zone is the geographical area served by a single PCF. Each Packet Zone is uniquely identified by a Packet Zone ID (PZID). Each BS periodically broadcasts, over the broadcast radio channels, the PZID of the Packet Zone it serves. A dormant mobile will be able to receive such broadcast system information and use it to determine whether it has moved into a new Packet Zone.

A dormant mobile needs to perform a Packet Zone update whenever it detects that it has just moved into a new Packet Zone. Packet Zone update is integrated with route management in the packet core network. In particular, whenever a mobile moves from one Packet Zone to another, i.e., from one PCF to another, its Packet Zone update will trigger the packet core network to establish a new A10 connection for the mobile between the new PCF and the mobile's serving PDSN. Establishing this new A10 connection is a necessary part of an inter-PCF handoff process. Therefore, 3GPP2 does not define any new protocol, message, or procedure uniquely for performing Packet Zone update. Instead, the procedure for inter-PCF dormant handoff is used to serve the purpose of Packet Zone update.

Any or multiple of the following location management strategies can be used in a cdma2000 radio system [4]:

- *Power-up and power-down location update*: A mobile performs location update whenever it powers up or down.
- *Time-based*: A mobile terminal performs location update periodically at predetermined time intervals.
- *Distance-based*: A mobile terminal performs location update whenever the physical distance between the current base station and the base station where it last updated its location exceeds a predetermined distance threshold.
- *Zone-based*: A mobile performs location update whenever it enters a new location zone (or registration zone using the terminology in [4]). A local zone is a group of BSs within a given provider's network. Each location zone will be uniquely identified by a zone number plus the SID and the NID of the zone. It is up to a network operator to determine how the BSs are grouped into location zones.
- *Parameter-based*: A mobile terminal performs location update whenever any predetermined parameter it stores changes.
- *Ordered update*: A mobile terminal performs location update whenever ordered by the base station.
- *Implicit location update*: Whenever a mobile issues an Origination Message or a Page Response Message, its location is updated implicitly because the base station can infer the mobile's location. This is the form of location update used for Packet Zone update. With implicit location update, the mobile does not need to send any separate location update message.

4.4.3 Handoffs for Supporting Packet Data Services

This section examines key procedures for supporting handoffs of packet data services. We consider both intra-PDSN and inter-PDSN handoffs.

Handoffs in a 3GPP2 network rely heavily on the circuit-switched network entities. In particular, handoffs for both circuit-switched and packet-switched services are controlled largely by the MSC. Procedures in addition to the handoff capabilities provided by the MSC are then introduced to handle the changes of the network connections that are specific to packet data services, i.e., the A8 connections, A10 connections, P-P connections, and the PPP connections.

Therefore, in this section, we will first describe the basic inter-BSC hard handoff procedure. Then, we will discuss how intra-PDSN handoff can be performed for mobiles using packet data services. We will consider intra-PDSN handoff for both active and dormant mobiles. We will also discuss procedures for inter-PDSN handoffs, including both regular and fast inter-PDSN handoff procedures.

4.4.3.1 Inter-BSC Hard Handoff within the Same PCF

Inter-BSC hard handoff is initiated by the source BSC and controlled by the MSC. The BSCs and the MSC use the A1 signaling interface (between each BSC and the MSC) to exchange signaling messages needed to perform inter-BSC hard handoff.

The inter-BSC hard handoff procedure is illustrated in Figure 4.58 assuming that the target BSC and the source BSC are connected to the same PCF. The source BSC

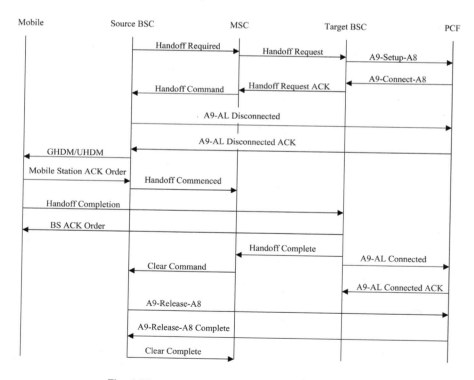

Fig. 4.58 *Inter-BSC hard handoff in cdma2000 RAN*

uses radio measurement reports received from the mobile to determine whether an inter-BSC hard handoff should be recommended. To initiate inter-BSC hard handoff, the source BSC sends a Handoff Required message over its A1 signaling interface to the MSC. This Handoff Required message will carry, among other information, one or more target radio cells for the mobile to be handed off to.

The MSC decides whether and when handoff should take place. After the MSC receives the Handoff Required message from the source BSC and decides that the handoff should occur, the MSC will construct a list of candidate target radio cells. The MSC may construct its list of candidate target radio cells based on the list of target radio cells received in the Handoff Required message received from the source BSC or based on the information it maintains. Then, the MSC selects a target radio cell on its list of candidate target radio cells and sends a Handoff Request message over its A1 interface to the BSC in the selected radio cell to instruct the BSC to allocate resources for the mobile so that the mobile can be handed off to this selected cell.

The target BSC checks whether it has the required resources to support the handoff requested by the MSC. It will initiate the process to set up the A8 connection needed to support the mobile's active data session by sending an A9-Setup-A8 message to the PCF. This message triggers the PCF to establish an A8 connection for the mobile and returns an A9-Connect-A8 message to the target BSC.

Once the necessary A8 connections are set up, the target BSC will send a Handoff Request Acknowledge message to the MSC. The Handoff Request Acknowledge message carries information regarding the characteristics of the radio channels in the target radio cell. Such information will be needed by the mobile to configure its radio channels in order to access the radio channels in the target radio cell and, therefore, will be passed along by the MSC to the mobile.

Upon receiving the Handoff Request Acknowledge message, the MSC will send a Handoff Command message to the source BSC to instruct it to order the mobile to move its radio channels to the target base station. The Handoff Command message carries the information regarding the radio channels in the target radio cell, which the MSC received in the Handoff Request Acknowledge message from the target BSC.

Upon receiving the Handoff Command message from the MSC, the source BSC sends an A9-AL (Air Link) Disconnected message to the PCF to instruct the PCF to stop transmitting packets for the mobile to the source BSC. The PCF will respond with an A9-AL Disconnected ACK message. The source BSC will also order the mobile to move its radio channels to the target base station. The source BSC does so by sending a General Handoff Direction Message (GHDM) or a Universal Handoff Direction Message (UHDM) to the mobile. The mobile confirms the receipt of the GHDM by returning a Mobile Station ACK Order message to the source BSC.

When the source BSC receives the MS ACK Order message from the mobile, it sends a Handoff Commenced message to the MSC to notify the MSC that the mobile has been ordered to move to the radio channels of the target base station. In the mean-time, the mobile will tune its radio to connect to the target base station and send a Handoff Completion message to the target BSC to inform the target BSC that the mobile can now successfully access the target BSC. The target BSC will return a BS ACK Order message to the mobile in response to the Handoff Complete message received from the mobile.

The target BSC will send a Handoff Complete message to the MSC to inform the MSC that the mobile is connected to the target BSC. The target BSC then send an A9-AL Connected message to the PCF to ask the PCF to forward future packets for the mobile to the target BSC. In response, the PCF will respond with an A9-AL-Connected ACK message.

Upon receiving the Handoff Complete message, the MSC will instruct the source BSC to release its radio resource allocated to the mobile by sending a Clear Command message to the source BSC. The source BSC releases all the resources it has allocated to the mobile and then returns a Clear Complete message to the MSC to signal the completion of inter-BSC hard handoff. For example, to release the traffic bearer A8 used by the mobile, the source will send an A9-Release-A8 message to the PCF. The PCF will return an A9-Release-A8 Complete message to inform the source BSC that the PCF's end of the A8 connection has been removed.

4.4.3.2 Inter-PCF Hard Handoff within the Same PDSN for Active Mobiles

Figure 4.59 illustrates the inter-PCF hard handoff procedure for an active mobile [2], [3], [5], [6]. It is assumed that the source PCF and the target PCF are connected to the same PDSN. The inter-PCF handoff process is initiated by the source BSC. This inter-PCF handoff procedure shown in Figure 4.59 is the inter-BSC hard handoff procedure plus additional procedures to handle the network connections specific to packet-switched services.

As in the inter-BSC hard handoff procedure shown in Figure 4.58, the source BSC determines that a hard handoff to a target BSC should take place based on the radio measurement results received from the mobile and then sends a Handoff Required message to the MSC to initiate the handoff.

The Handoff Required message carries information elements to indicate that the requested handoff is for packet-switched services, and whether the requested handoff is an inter-PCF hard handoff. For example, the Handoff Required message carries the following information elements used only for packet-switched services in the case of inter-PCF handoff:

- *PDSN IP Address*: This field will be set to the IP address of the mobile's current serving PDSN if and only if the requested handoff is an inter-PCF hard handoff for packet-switched services.
- *Protocol Type*: This field identifies the link-layer protocol used at the mobile and its serving PDSN to exchange user IP packets with each other. This field is used only for packet-switched services in the case of inter-PCF hard handoff.

The MSC selects a target BSC and sends a Handoff Request message to this target BSC to instruct it to allocate the necessary resources for the mobile so that the mobile can be handed off to a radio cell controlled by this target BSC. The Handoff

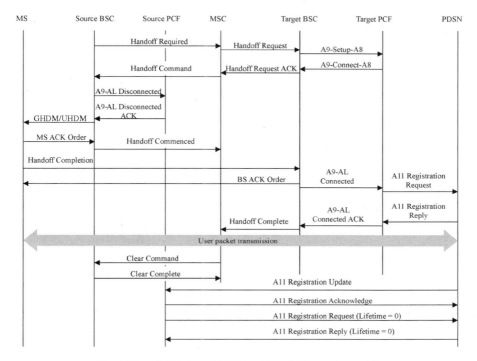

Fig. 4.59 *3GPP2 intra-PDSN hard handoff for active mobile*

Request message carries the same set of information elements as in the Handoff Required message that are specific to packet-switched services (e.g., the PDSN IP Address field and the Protocol Type).

Upon receiving the Handoff Request indicating inter-PCF hard handoff for packet switched services, the target BSC will first initiate the process to establish an A8 connection between the target BSC and the target PCF for the mobile, before it allocates radio resources for the mobile. The target BSC initiates the A8 connection establishment process by sending an A9-Setup-A8 message to the target PCF. The target PCF establishes the A8 connection for the mobile and responds to the target BSC with an A9-Connect-A8 message.

Upon establishment of the A8 connection, the target BSC will send a Handoff Request Acknowledgment message to the MSC to inform the MSC that the target BSC is ready for the handoff of the mobile's radio channel. The target BSC will then wait for the mobile to arrive on its radio channels.

Upon receiving the Handoff Request Acknowledgment message the MSC will send a Handoff Command message to the source BSC to instruct the source BSC to order the mobile to move its radio channels to the target BSC. Upon receiving the Handoff Command message, the source BSC will first send an A9-AL Disconnected message to the source PCF to ask the source PCF to stop transmitting user data to the source BSC. The source PCF will respond with an A9-AL Disconnected ACK message.

The source BSC will then instruct the mobile to transfer its traffic radio channel from the source BSC to the target BSC by sending a General Handoff Direction Message or a Universal Handoff Direction Message to the mobile. The mobile responds with a Mobile Station ACK Order message to notify the source BSC that the mobile will be connecting to the target BSC. The source BSC will then send a Handoff Commenced message to the MSC to notify the MSC that the mobile has been ordered to move to the target BSC.

Upon completion of the radio channel transfer to the target BSC, the mobile sends a Handoff Completion message to the target BSC. The target BSC responds with a BS ACK Order message. The target BSC then sends a Handoff Complete message to the MSC to notify it that the handoff inside the RAN is complete. The target BSC will also notify the target PCF of the arrival of the mobile by sending an A9-AL Connected message to the target PCF. This message will trigger the target PCF to use A11 signaling to request the PDSN to establish the A10 connection between the target PCF and the PDSN for the mobile.

To request the PDSN to set up an A10 connection, the target PCF sends an A11 Registration Request message to the PDSN. As we have discussed in Chapter 2, "Wireless IP Network Architectures," the A11 Registration Request uses the same format as the MIPv4 Registration Request (Figure 4.12). For the purpose of intra-PDSN handoff, several key fields in the A11 Registration Request message need to be set as follows:

- *Care-of Address field*: This field is set to the IP address of the target PCF.
- *Home Address field*: This field is set to the IP address of the PDSN.
- *Home Agent field*: This field is set to zero, which indicates that the requested A10 connection is for supporting an intra-PDSN handoff.

The PDSN will validate this request. Upon positive validation, the PDSN establishes the A10 connection and respond to the target PCF with an A11 Registration Reply message. The A11 Registration Reply message informs the target PCF that the A10 connection is established for the mobile. The target PCF will in turn inform the target BSC by sending an A9-AL Connected ACK message.

At this point, the mobile and its serving PDSN can begin to use the existing PPP connection between the mobile and its serving PDSN to exchange user IP packets.

As soon as an A10 connection is established between the target PCF and the PDSN for the mobile, the PDSN will tear down the A10 connection to the source PCF used by the mobile before the handoff. To do this, the PDSN sends an A11 Registration Update message to the source PCF to notify the PCF that the PDSN intends to tear down the A10 connection to the source PCF. The source PCF replies with an A11 Registration Acknowledge message to acknowledge the receipt of the A11 Registration Update message. Then, the source PCF will send an A11 Registration Request message with a zero Lifetime to the PDSN to release the mobile's A10 connection to the source PCF. This A11 Registration Request message will also carry accounting information related to the resources used by the mobile.

Upon receiving the A11 Registration Request message from the source PCF, the PDSN records the accounting information carried in the received message and releases the resources it has allocated to the mobile's A10 connection to the source PCF. Then, the PDSN returns an A11 Registration Reply message to the source PCF. This message will trigger the source PCF to close the A10 connection for the mobile and release all the resources it has allocated to the mobile.

The MSC will also send a Clear Command to the source BSC to instruct it to release the resources allocated to the mobile on the source BSC and between the source BSC and the source PCF (i.e., the mobile's A8 connection). Upon receiving this Clear Command, the source BSC will request the source PCF to close the A8 connection. After the A8 connection is closed, the source BSC will return a Clear Complete message to the MSC to notify the MSC that clearing has been completed.

4.4.3.3 *Regular Inter-PDSN Hard Handoff for Active Mobiles*   With a regular inter-PDSN hard handoff, the source PDSN is the mobile's serving PDSN before the handoff, no P-P interface is implemented between the mobile's serving PDSN and the target PDSN, and the target PDSN becomes the mobile's new serving PDSN after the handoff. Therefore, the mobile needs to establish a new PPP connection to the target PDSN during the handoff process.

Figure 4.60 illustrates the signaling message flow for regular inter-PDSN hard handoff [3], [5], [6]. The regular inter-PDSN hard handoff process is similar to the intra-PDSN hard handoff process shown in Figure 4.59 in Section 4.4.3.2 with the following main differences.

- The target PCF will have to select a target PDSN for each mobile that is performing inter-PDSN handoff. The target PCF may be physically connected to multiple PDSNs. How to determine which PDNS should be the target PDSN for a mobile is an implementation issue outside the scope of 3GPP2 specifications.

- The target PCF will have to establish an A10 connection to the selected target PDSN for the mobile. The protocol and procedure used to establish an A10 connection with any PDSN are the same. The mobile will also have to establish a new PPP connection, i.e., a PPP connection to the target PDSN, after the radio channel, the A8 connection, and the A10 connection are established. Furthermore, with a regular inter-PDSN handoff, a mobile has to use a new care-of address after it is handed off to the target PDSN. Therefore, the mobile will need to perform Mobile IP registration to register its new care-of address with its home agent.

- The A10 connection used by the mobile at the source PDSN is released in a different way. During an intra-PDSN handoff, the PDSN remains the same after the handoff. Therefore, the PDSN can instruct the source PCF to release the A10 connection between the source PDSN and the source PCF. In an inter-

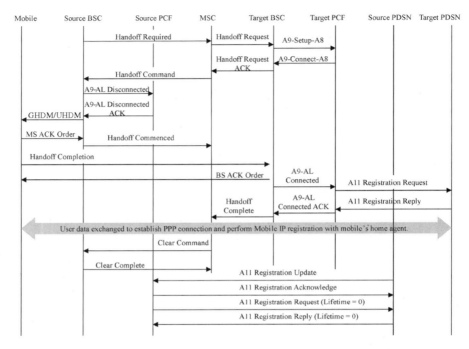

Fig. 4.60 *3GPP2 regular inter-PDSN hard handoff for active mobile*

PDSN handoff, however, the target PDSN may not be able to communicate with the source PCF. As a result, the target PDSN may not be able to directly signal the source PCF or the source PDSN to close the mobile's A10 connection between the source PCF and the source PDSN. Therefore, the mobile's A10 connection between the source PCF and the source PDSN is released when the Lifetime associated with this A10 connection expires on the source PDSN or the source PCF. When the Lifetime of this A10 connection expires on the source PDSN, the source PDSN uses the same procedure, as shown in Figure 4.59 for intra-PDSN hard handoff, to release the A10 connection.

If the source PCF is co-located on the source BSC, the source PCF could also proactively request the closure of the mobile's A10 connection between the source PCF/BSC and the source PDSN. In particular, when the source PCF/BSC receives a Clear Command from the MSC indicating that the handoff is complete, the source PCF can send an A11 Registration Request with a zero Lifetime to the source PDSN to request the source PDSN to release the mobile's A10 connection. The source PDSN will close the mobile's A10 connection and respond to the source PCF with an A11 Registration Reply message.

4.4.3.4 Inter-PCF Dormant Handoff within the Same PDSN

In this section, we illustrate the handoff of a dormant packet data session during the Dormant State of the packet data session from a source PCF to a target PCF connected to the same PDSN. Such a dormant handoff may be performed when the mobile detects a change of the Packet Zone ID (PZID), Network ID (NID), or System ID (SID). When dormant handoff is triggered by a change of PZID, it also serves the purpose of Packet Zone update.

Figure 4.61 illustrates the inter-PCF dormant handoff procedure [3], [5], [6]. The main task of intra-PDSN dormant handoff is to establish a new A10 connection for the mobile between the target PCF and the PDSN.

Dormant handoff is initiated by a mobile. The mobile initiates the dormant handoff procedure by sending an Origination Message to the target BSC (not the source BSC). This is the same Origination Message a mobile uses to initiate a circuit-switched or packet-switched call, or to initiate the process to access the 3GPP2 packet-switched networks (Chapter 2, "Wireless IP Network Architectures").

The Origination Message informs the BSC whether the mobile is requesting for packet-switched services or circuit-switched services. The Origination Message also carries a Data-Ready-to-Send (DRS) flag, which is used to inform the BSC whether

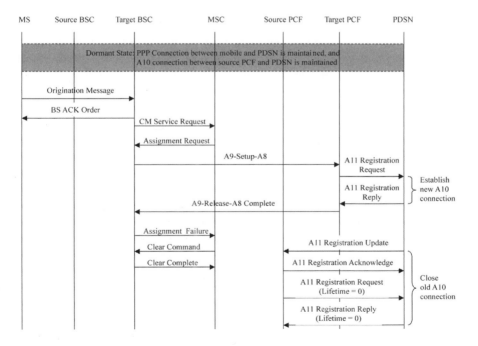

Fig. 4.61 *3GPP2 intra-PDSN dormant handoff*

the mobile has user data to send at this moment (DRS $= 1$) or it is simply performing dormant handoff or Packet Zone update (DRS $= 0$).

The target BSC acknowledges the receipt of the Origination Message immediately upon receiving the message by returning a BS ACK Order message to the mobile. The target BSC will then send a Connection Management (CM) Service Request to the MSC to ask the MSC for permission for the dormant handoff. If the MSC accepts the CM Service Request, it sends an Assignment Request message back to the target BSC to instruct the target BSC to assign radio resources for the dormant mobile.

As the mobile has indicated in its Origination Message that it is only requesting for dormant handoff (or Packet Zone update) and does not have user packets to send at this moment, the target BSC will not establish any traffic radio channel for the mobile. The target BSC will, however, send an A9-Setup-A8 message to the target PCF to trigger the target PCF to establish an A10 connection between the target PCF and the PDSN for the mobile. This A9-Setup-A8 message will carry an indication that the mobile has no data to send now. Therefore, the target PCF will not establish any A8 connection between the target BSC and the target PCF for the mobile.

Upon receiving the A9-Setup-A8 message from the target BSC, the target PCF will send an A11 Registration Request message to the PDSN to request the PDSN to set up an A10 connection for the mobile. The PDSN will validate the A11 Registration Request. Upon positive validation, the PDSN will establish the requested A10 connection for the mobile. It will then return an A11 Registration Reply message to the target PCF.

Although the mobile does not have any user data to send at this moment, the PDSN may have user data to be sent to the mobile. Therefore, the A11 Registration Reply message carries an indication to inform the target PCF whether the PDSN has user data to send to the mobile at the moment. This indication is carried in a Vendor/ Organization Specific Extension (Section 4.2.2.7) to the A11 Registration Reply message.

- *If the PDSN has no user data to send to the mobile*: The target PCF will reply to the A9-Setup-A8 message received from the target BSC with an A9-Release-A8 Complete message to inform the target BSC that no A8 connection has been established for the mobile.
- *If the PDSN has user data to send to the mobile*: The target PCF will set up an A8 connection to the target BSC for the mobile and reply to the A9-Setup-A8 message with an A9-Connect-A8 message. The A9-Connect-A8 message will inform the target BSC that the cause for setting up the A8 connection is that the target PCF has user data to send to the mobile. This A9-Connect-A8 message will trigger the target BSC to start the process to set up the traffic radio channel for the mobile.

The A9-Release-A8 Complete message (or the A9-Connect-A8 message) also implies to the target BSC that the MSC's role in the handoff procedure is complete.

Therefore, the target BSC will now acknowledge the Assignment Request received from the MSC by returning an Assignment Failure message to the MSC. This Assignment Failure message will carry a Failure Cause value indicating Packet Call Going Dormant rather than any real failure. This message will cause the MSC to send a Clear Command to the target BSC to instruct the target BSC not to notify the mobile of the results of the handoff procedure described above. As soon as an A10 connection is established between the target PCF and the PDSN for the mobile, the PDSN will tear down the mobile's A10 connection to the source PCF. The PDSN does so by sending an A11 Registration Update message to the source PCF. The source PCF replies with an A11 Registration Acknowledge message to acknowledge the receipt of the A11 Registration Update message. Then, the source PCF will send an A11 Registration Request message with a zero Lifetime to the PDSN to trigger the PDSN to remove the mobile's A10 connection to the source PCF. Upon receiving the A11 Registration Request message from the source PCF, the PDSN releases the mobile's A10 connection to the source PCF. Then, the PDSN returns an A11 Registration Reply message to the source PCF. This message will trigger the source PCF to close the A10 connection for the mobile and release all the resources it has allocated to the mobile.

4.4.4 Fast Inter-PDSN Handoff

Fast inter-PDSN handoff [1] can only be supported when a mobile is in ACTIVE state during the inter-PDSN handoff. An important concept used in fast inter-PDSN handoff is that the mobile's serving PDSN remains unchanged as long as the mobile's Packet Data Service State remains in ACTIVE state and the mobile does not renegotiate its PPP connection.

As illustrated in Figure 4.62, user IP packets destined to the mobile will continue to be routed to the mobile's serving PDSN first. The serving PDSN in turn sends the user IP packets to the mobile over the PPP connection between the serving PDSN and the mobile. In particular, the PPP frames will be tunneled by the serving PDSN to the target PDSN. The target PDSN de-tunnels the PPP frames received from the serving PDSN and then tunnels them over an A10 connection to the target PCF, which in turn tunnels the PPP frames over an A8 connection to the target BSC. The BSC de-tunnels the PPP frames and forwards them to the mobile over the Radio Bearer. Tunneling PPP frames between the mobile's serving PDSN and a mobile via a target PDSN allows the mobile to maintain its PPP connection to its serving PDSN even after the mobile's radio channels are connected to the target PDSN, as long as the mobile's Packet Data Service State remains in ACTIVE state.

Figure 4.63 illustrates a simplified flow for supporting fast inter-PDSN handoff. For ease of illustration, we assume that a PCF is co-located on each BSC. We use PCF/BSC to refer to a combined PCF and BSC node.

The target PCF/BSC realizes that the handoff is an inter-PDSN handoff by detecting, based on the information received from the mobile, that the target PCF/BSC cannot reach the mobile's serving PDSN, and therefore, a target PDSN will have to be used for the mobile. The target PCF/BSC selects a target PDSN for the

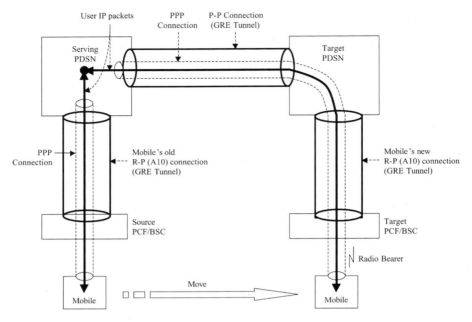

Fig. 4.62   *3GPP2 fast inter-PDSN handoff: user traffic flow*

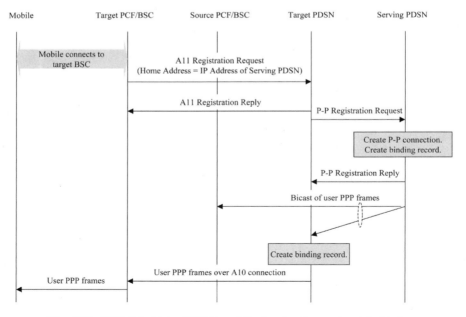

Fig. 4.63   *3GPP2 fast inter-PDSN handoff: signaling flow and user traffic flow*

mobile and sends an A11 Registration Request to the selected target PDSN to establish an A10 connection for the mobile between the target PCF/BSC and the target PDSN.

The Home Address field of this A11 Registration Request will be set to the IP address of the mobile's serving PDSN. Recall that for intra-PDSN handoff or for regular inter-PDSN handoff, the target PCF will set the Home Address field in the A11 Registration Request to zero. A nonzero Home Address field in the A11 Registration Request tells the target PDSN that a P-P connection should be set up between the target PDSN and the serving PDSN identified by the IP address in the Home Address field of the A11 Registration Request message in order to support fast inter-PDSN handoff for the mobile.

The target PDSN replies immediately to the A11 Registration Request received from the target PCF/BSC with an A11 Registration Reply message. The target PDSN then sends a P-P Registration Request message to the mobile's serving PDSN to request the serving PDSN to establish a P-P traffic connection for this mobile. If the mobile's serving PDSN accepts the P-P Registration Request, it will

- Establish the requested P-P traffic connection (i.e., a GRE tunnel).
- Update its binding record for the mobile by creating an association between the identity of the mobile, address of the target PDSN, address of itself, and the identifiers of the P-P connection for the mobile.
- Return a P-P Registration Reply message to the target PDSN.

As we have discussed in Chapter 2, "Wireless IP Network Architectures," the P-P Registration Request and the P-P Registration Reply messages use the same formats as the MIPv4 Registration Request and MIPv4 Registration Reply messages (Figures 4.12 and 4.13). The Care-of Address, Home Address, and Home Agent fields in the P-P Registration Request will be set as follows:

- Care-of Address = IP address of the target PDSN
- Home Address = 0.0.0.0
- Home Agent = IP address of the mobile's serving PDSN

The target PDSN sets the "Simultaneous Bindings" flag (i.e., the S flag) in the P-P Registration Request message to 1. This is to request the mobile's serving PDSN to maintain the mobile's old A10 connection between the serving PDSN and the source PCF/BSC after the P-P connection is established for the mobile. Setting the S flag to 1 in the P-P Registration Request message will cause the mobile's serving PDSN to tunnel copies of the same user PPP frames simultaneously to:

- The target PDSN over the P-P connection
- The source PCF/BSC over the mobile's old A10 connection between the serving PDSN and the source PCF

Bicasting user PPP frames to both the target PDSN and the source PCF/BSC allows the mobile to receive user PPP frames as soon as it connects to the target PDSN. The bicasting will change into unicasting when either the P-P connection or the mobile's A10 connection between the serving PDSN and the source PCF/BSC is closed.

The target PDSN will de-capsulate the PPP frames received from the serving PDSN over the P-P connection. If the mobile has established radio connections with the target PCF/BSC, the target PDSN will tunnel the packets received from the serving PDSN to the target PCF/BSC, which will in turn tunnel the PPP frames toward the mobile. If the mobile has not yet established radio connection with the target PCF/BSC, the target PDSN will discard the PPP frames received from the serving PDSN.

Upon receiving a P-P Registration Reply message from the serving PDSN indicating successful establishment of the P-P connection for the mobile, the target PDSN will create a binding record for the mobile by creating an association between the identity of the mobile, address of the mobile's serving PDSN, and the identifiers of the P-P connection for the mobile, and identifiers of the mobile's A10 connection between the target PDSN and the target PCF/BSC. Such a binding will enable the target PDSN to match the PPP frames received from the mobile's serving PDSN over the P-P connections to a particular mobile and then tunnel these PPP frames over the A10 connection to the target PCF.

The P-P connection for a mobile will be maintained and the mobile's serving PDSN can continue to remain unchanged as long as:

- The mobile's R-P (A10) between the target PDSN and the target PCF/BSC, referred to as the P-P connection's corresponding R-P (A10) connection, exists and
- The mobile's Packet Data Service State remains in ACTIVE state.

To maintain a P-P connection, the target PDSN refreshes the P-P connection by sending P-P Registration Requests periodically to the serving PDSN. The target PDSN or the serving PDSN can release a P-P connection when its corresponding A10 connection on the target PDSN is removed or when the mobile is changing into DORMANT state.

When the mobile plans to transition into DORMANT state, its serving PDSN will have to be changed to the target PDSN first. Recall that when a mobile is in DORMANT state, no traffic radio connection nor A8 connection will be maintained for the mobile. However, the mobile needs to maintain a PPP connection to its serving PDSN. Also, an A10 connection between a PCF and the mobile's serving PDSN needs to be maintained. As an A10 connection has already been established between the target PCF/BSC and the target PDSN during the fast inter-PDSN handoff process, the mobile will only need to establish a PPP connection to the target PDSN before the mobile changes into DORMANT state.

When the target BSC receives indication from a mobile that the mobile is about to enter DORMANT state, the target PCF/BSC will send a A10 Registration Request to the target PDSN indicating that the mobile is "Going DORMANT." The "Going DORMANT" indication is carried in a Vendor/Organization Specific Extension (Section 4.2.2.7) to the A10 Registration Request message. The target PDSN will in turn send a P-P Registration Request to the serving PDSN with an indication that the mobile is "Going DORMANT" and with the accounting-related information. Again, the "Going DORMANT" indication and the accounting-related information is carried in a Vendor/Organization Specific Extension to the P-P Registration Request message. The target PDSN will then initiate the establishment of a PPP connection with the mobile. The target PDSN becomes the serving PDSN for the mobile after a PPP connection is established between the mobile and the target PDSN. Simultaneously, the target PDSN will initiate the release of the P-P connection with the serving PDSN.

A target PDSN releases a P-P connection by sending a P-P Registration Request message with a zero Lifetime to the serving PDSN. Upon receiving such a P-P Registration Request message, the serving PDSN removes the binding record for the mobile and returns a P-P Registration Reply message to the target PDSN to trigger the target PDSN to remove its binding record for the mobile. If the target PDSN does not receive a P-P Registration Reply message after retransmitting a configurable number of P-P Registration Request messages, the target PDSN will assume that the P-P connection is no longer active and will remove its binding record for P-P connection.

The serving PDSN may initiate the release of a P-P connection for a number of reasons. For example, a serving PDSN can initiate the release of a P-P connection if the mobile returns to a radio access network that is served by the serving PDSN, if the existing PPP connection to the mobile expires, or when either the mobile or the serving PDSN chooses to close the PPP connection for any reason.

A serving PDSN initiates the release of a P-P connection by sending a P-P Registration Update message to the target PDSN. The target PDSN will remove its binding information for this P-P connection and reply with a P-P Registration Acknowledge message to the serving PDSN. The target PDSN will then send a P-P Registration Request with a zero Lifetime containing any accounting-related information to the serving PDSN. This will cause the serving PDSN to remove all its binding information for the P-P connection and reply with a P-P Registration Reply message to the target PDSN. If the serving PDSN does not receive a P-P Registration Acknowledge message after retransmitting a configurable number of P-P Registration Update messages, the serving PDSN will assume that the P-P connection is no longer active and proceeds to remove the binding information for this P-P connection.

The mobile's serving PDSN can continue to remain unchanged as long as the mobile's Packet Data Service State remains in ACTIVE state, even when the mobile moves away from its current target PDSN (let's call it target PDSN 1) to a new target PDSN (let's call it target PDSN 2). As illustrated in Figure 4.64, target PDSN 2 can use the same procedure described above to establish a P-P connection to the

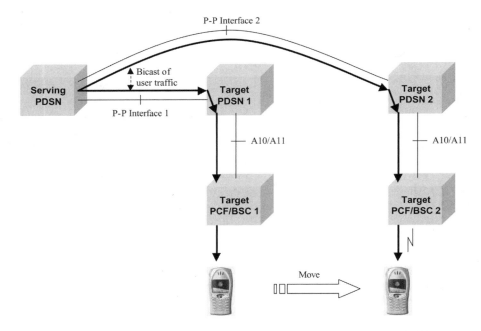

Fig. 4.64 *3GPP2 fast inter-PDSN handoff from target PDSN 1 to target PDSN 2*

mobile's serving PDSN. As shown in Figure 4.64, the mobile's serving PDSN can bicast user PPP frames to both target PDSN 1 and target PDSN 2.

Bicasting of user traffic changes into unicast when one of the P-P connections is released. For example, after the mobile has moved to PDSN 2 and is no longer able to receive user data from target PCF/BSC1, the mobile's A10 connection on target PDSN 1 will be released by target PDSN 1 after its Lifetime expires (Section 4.4.3.3). This will trigger target PDSN 1 to initiate the process to delete the mobile's P-P connection between target PDSN 1 and the mobile's serving PDSN. Removal of this P-P connection will also cause the serving PDSN to stop bicasting of user traffic and to begin to unicast user traffic only to target PDSN 2.

4.4.5   Paging and Sending User Data to a Dormant Mobile

The current 3GPP2 packet data network architecture does not have its own paging protocol. In fact, the packet data network is unaware of any paging process at all. Instead, paging is initiated and carried out inside the radio access network. Paging is carried out by circuit-switched network entities (i.e., the MSC and the BSC) using the existing paging protocol and procedures designed for circuit-switched services.

The PDSN is unaware of a mobile's Packet Data Service State (i.e., whether a mobile is DORMANT or ACTIVE) at all. The PDSN will always know the serving

PCF for every mobile regardless of whether the mobile is in DORMANT or ACTIVE state. Dormant mobiles ensure that the PDSN knows its source PCF by performing Packet Zone updates whenever it crosses a Packet Zone boundary (Section 4.4.3.4). As each Packet Zone is served by one PCF, Packet Zone update will occur whenever a mobile moves from one PCF to another. Packet Zone Update will also trigger a dormant mobile to perform dormant handoff from the old PCF to the new PCF, as illustrated in Figure 4.61. This handoff process will ensure that the PDSN maintains an A10 connection to the current source PCF for the mobile. Therefore, from the PDSN's perspective, no paging is needed as it always knows where to forward the packets destined to every mobile. In particular, a PDSN always forwards the IP packets destined to any dormant or active mobile along the existing PPP connection and the existing A10 connection for the mobile toward the PCF.

The PCF will try to further forward the user data toward the mobile. However, because the mobile is in DORMANT state, no A8 connection between the PCF and any BSC will exist for the mobile. Therefore, the PCF will issue an A9 Base Station (BS) Service Request to the last BSC (let's call it BSC 1) used by the PCF to exchange user data with the mobile to trigger BSC 1 to initiate the process to locate the mobile and to allocate all the resources needed for the mobile to receive user packets.

The BSC, which receives the A9 BS Service Request message, will initiate the BS initiated Mobile-terminated Call Setup Procedure used in the circuit-switched portion of the 3GPP2 network to locate the mobile and to set up the network resources for the mobile. The Mobile-terminated Call Setup Procedure is performed by the BSCs and the MSC using the A1 signaling interface between the BSCs and the MSC.

As illustrated in Figure 4.65 [3], [5], [6], the BSC initiates the Mobile-terminated Call Setup Procedure by sending a BS Service Request over the A1 signaling interface to the MSC to ask the MSC to help set up a data call to the dormant mobile. The MSC will acknowledge the receipt of the request by sending back a BS Service Response message to the BSC. At this point, the BSC will send an A9 BS Service Response message to the PCF to inform the PCF that the BSC is in the process of locating and connecting to the destination mobile.

In the mean time, the MSC will initiate the paging process to locate the dormant mobile. While the mobile is in DORMANT state, it may have moved away from BSC 1 and may be currently connected to a different BSC (let's call it BSC 2). As the MSC controls all handoffs from one BSC to another in the 3GPP2 network regardless of whether the mobile is in ACTIVE or DORMANT state, the MSC knows to which BSC the mobile is currently connected. Therefore, the MSC initiates the paging process by sending a Paging Request to the BSC to which the mobile is currently connected.

When a BSC receives a Paging Request message from the MSC, it will broadcast a Page Message over its paging channel to all the mobiles within its coverage area. The Page Message will carry an indication to inform the mobile that the mobile is being paged for packet data services.

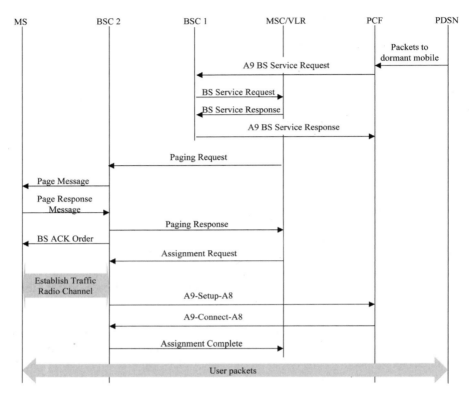

Fig. 4.65 *3GPP2 paging for packet data services*

When a dormant mobile receives a Page Message intended to it, it will respond by returning a Page Response Message to the BSC from which the Page Message was received. The BSC will acknowledge the receipt of this Page Response Message by returning a BS ACK Order message to the mobile. The BSC will also inform the MSC that it has found the mobile and ask the MSC to initiate the process to set up the traffic radio channel to the mobile by sending a Paging Response message to the MSC. The MSC sends an Assignment Request to the BSC to request the assignment of radio resources and the A8 connection for the mobile.

Upon receiving the Assignment Request, the BSC will initiate the procedures to set up the traffic radio channel and the A8 connection for the mobile. The radio resources may be set up first. Then the BSC will initiate the process to establish the A8 connection by sending an A9-Setup-A8 message to the PCF. Once the radio channel and the A8 connection are both established, the mobile and the network will be able to exchange user packets. Now, the BSC will inform the MSC of the completion of the resource assignment by sending an Assignment Complete message to the MSC.

4.5 MOBILITY MANAGEMENT IN MWIF NETWORKS

The MWIF architecture uses IP-based protocols defined or being developed by the IETF to support mobility. The main functional entities for mobility management in a MWIF network architecture are as follows:

- *Mobile Attendant (MA)*: The Mobile Attendant resides in the Access Gateway. A Mobile Attendant provides mobility support functions inside an access network. It acts as a Mobile IP Foreign Agent. It also acts as a proxy to relay mobility management messages between a mobile and its Home Mobility Manager.
- *Home Mobility Manager (HMM)*: The Home Mobility Manager supports the movement of a mobile terminal from one Access Gateway to another or from one administrative domain to another. It acts as the Mobile IP Home Agent.
- *Home IP Address Manager*: The Home IP Address Manager assigns *home* IP addresses to mobile terminals dynamically.
- *IP Address Manager*: The IP Address Manager resides in the Access Gateway and dynamically assigns local IP addresses to mobile terminals that a mobile can use to receive IP packets from the local IP network.
- *Location Server*: The Local Server maintains dynamic information, e.g., a mobile terminal's current location and geographical position, for supporting terminal and service mobility. It also provides location information to other authorized network entities upon request.
- *Geographical Location Manager (GLM)*: The GLM determines and supplies a mobile's geographical position.
- *Global Name Server (GNS)*: The GNS provides address mapping services. The IP Domain Name System (DNS) is considered to be part of the GNS. The GNS performs the following mapping services:
  - Between E.164 telephone numbers to IP addresses or URLs.
  - From URLs to Application Functional Entities.
  - For the same subscriber, maps between any two of its following addresses or identifiers: URL, E.164 telephone number, IP address, Subscriber Identity.
- *Service Discovery Server*: The Service Discovery Server enables a mobile terminal or a core network entity to discover network services, their attributes, and addresses.

Figure 4.66 illustrates the interactions among these functional mobility management entities. Each interface reference point is marked by their reference number defined by the MWIF.

MWIF recommended IETF protocols for the interface references point between the mobility management functional entities [34]. Many of these protocols are also

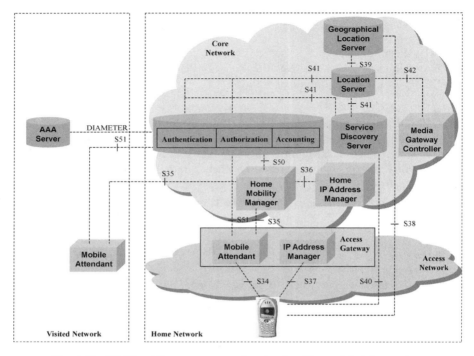

Fig. 4.66   *MWIF mobility management functional entities and their interactions*

used over other protocol reference points in the MWIF network architecture, as discussed in Chapter 2, "Wireless IP Network Architectures."

- *Reference points S34 and S35*: These reference points are used to exchange mobility management messages. The MWIF recommends Mobile IP (v4 or v6) as the protocol over these reference points.
- *Reference points S36 and S37*: These reference points are used for a mobile or a network node to access an IP Address Manager or a Home IP Address Manager. The MWIF recommends DHCP as the protocol for dynamic IP address assignment and therefore DHCP as the protocol over these reference points.
- *Reference points S38 and S39*: These reference points are used for a mobile or a network node to access the Geographical Location Server. The MWIF recommends that LDAP [28], DIAMETER [16], or SLP [25] be the protocol over these reference points. DIAMETER can be supported over TCP or SCTP [36] over IP.
- *Reference points S40 and S41*: These protocol reference points are used for a mobile or a network node to access a Location Server or a Service Discover

Server. The MWIF recommends that SLP [25], DHCP, or DNS be used as the protocol over these reference points.

- *Reference point S42*: This reference point is used for the Media Gateway Controller to access the Location Server. The MWIF recommends that LDAP or TRIP be used as the protocol over this protocol reference point.
- *Reference points S50 and S51*: These reference points are used for the network nodes to access the Authentication, Authorization and Accounting servers. The MWIF recommends that the DIAMETER protocol be the signaling protocol over these protocol reference points.

4.5.1 Handoffs

MWIF recommends that Mobile IP (v4 or v6) be used to support handoff from one Access Gateway to another in the same or different administrative domains. Mobility within the same area served by an Access Gateway has not been considered explicitly by WMIF. Instead, it is left for the Radio Access Network to implement any mobility management mechanism deemed appropriate in a specific Radio Access Network.

Figure 4.67 illustrates the inter-Access Gateway handoff process using Mobile IP [34]. When a mobile moves to a new RAN served by a new Access Gateway in a visited network, it first needs to gain access to the new RAN. Then, the mobile will

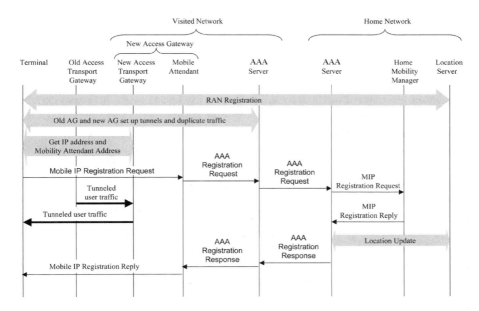

Fig. 4.67 *MWIF handoff procedure*

need to acquire a new care-of address from the visited network and register the new care-of address with its Home Mobility Manager (i.e., its Mobile IP Home Agent).

Unlike the basic Mobile IP, MWIF uses a two-level Mobile IP registration. In particular, a mobile sends its Mobile IP Registration Request (assuming Mobile IPv4 is used, for example) to the Mobile Attendant in the visited network. Instead of forwarding the mobile's Mobile IP Registration Request directly to the mobile's Home Mobility Manager (i.e., the mobile's Mobile IP Home Agent), the Mobile Attendant in the visited network sends the request to the local Authentication Server in the visited network. This will trigger the local Authentication Server to contact the Authorization Server in the mobile's home network to determine whether the mobile should be authorized to access the new serving access network.

Upon positive authentication in the mobile's home network, the mobile's home Authentication Server will forward the Mobile IP Registration Request received from the visited network to the Home Mobility Manager in the mobile's home network.

The Home Mobility Manager in the mobile's home network will perform standard Mobile IP processing to register the mobile's new care-of address upon positively authenticating the received Mobile IP Registration Request. This Home Mobility Manager will then respond by sending a Mobile IP Registration Reply to the Authentication Server in the mobile's home network. This Authentication Server will in turn relay the message to the Authentication Server in the mobile's visited network, which will further forward the message to the Mobile Attendant in the visited network. The Mobile Attendant will then forward the Mobile IP Registration Reply message to the mobile to complete the handoff process.

4.6 COMPARISON OF MOBILITY MANAGEMENT IN IP, 3GPP, AND 3GPP2 NETWORKS

This section discusses some fundamental similarities and differences among the mobility management methodologies for IP, and the packet data networks defined by 3GPP and 3GPP2. In particular, we consider Mobile IP, Mobile IP Regional Registration, SIP-based terminal mobility, mobility management in 3GPP and 3GPP2 packet networks, Cellular IP, and HAWAII.

We begin by comparing the basic mobility management architectures used in these mobility management approaches to deliver packets to mobiles and to manage the change of the packet delivery path to a mobile. For illustration purposes, we refer to the network entities that participate in the processing of the mobility management protocol messages as mobility protocol entities. Take Mobile IP, for example, the Home Agent (HA), Foreign Agent (FA), and the mobile are the mobility protocol entities. Other network nodes (e.g., intermediate IP routers) along the paths that interconnect these mobility protocol entities are not aware of Mobile IP and therefore are not considered mobility protocol entities.

The basic mobility management architectures used in Mobile IP, Mobile IP Regional Registration, and SIP mobility are illustrated in Figure 4.68. For SIP

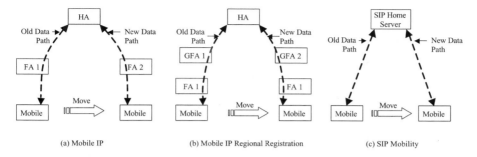

(a) Mobile IP   (b) Mobile IP Regional Registration   (c) SIP Mobility

Fig. 4.68 *Simplified mobility management models used by Mobile IP, Mobile IP Regional Registration, and SIP mobility*

mobility, Figure 4.68 shows only a home SIP server. It does not show the proxy SIP servers that may be used in visited networks. The basic mobility management architectures used in 3GPP packet network, 3GPP2 packet network, Cellular IP, and HAWAII are illustrated in Figure 4.69.

A key similarity shared by all mobility management methodologies shown in Figures 4.68 and 4.69 is that they all use the *Relayed Delivery* strategy discussed in Section 4.1.3 as the basic strategy for delivering signaling packets, user application packets, or both signaling and user application packets to mobiles. In particular, a *mobility anchor point* is used to track mobiles' locations and to relay packets to mobiles. For example:

- With Mobile IP and Mobile IP Regional Registration, packets destined to a mobile are first routed to the mobile's home agent, which then tunnels the packets to the mobile. Mobile IP with route optimization allows a correspondent host to learn a mobile's current location and then use Direct Delivery (Section 4.1.3) to send packets directly to the mobile.

- With SIP mobility, before a correspondent host knows a destination's current location, it always sends its initial SIP signaling messages to the destination's SIP home server. Depending on the type of SIP server used in the destination mobile's home network, the SIP home server may either relay the signaling messages to the mobile or return the mobile's current location to the correspondent host so that the correspondent host can contact the mobile directly. In either case, as soon as the correspondent host gets in touch with the mobile, the mobile can inform the correspondent host of its current location. The correspondent host can then send future signaling messages and user packets directly to the mobile without having to go through the destination mobile's SIP home server.

- With Cellular IP, packets to and from a mobile are routed first to a gateway router, which then relays the packets to their final destinations.

- With HAWAII, packets to and from a mobile are routed first to a Domain Root Router, which then relays the packets to their final destinations.

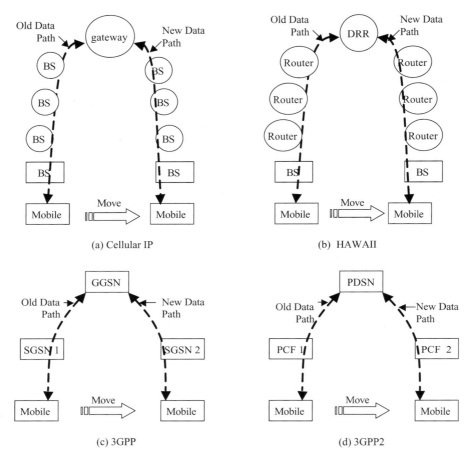

Fig. 4.69 *Simplified mobility management models used in 3GPP, 3GPP2, and IP micromobility management protocols*

- In 3GPP, all user packets are sent routed to a GGSN, which then relays them to their final destinations.
- In 3GPP2, all packets are routed first to a PDSN, which then forwards them to their final destinations.

A key difference among the mobility management methodologies shown in Figures 4.68 and 4.69 is in the ways packets are transported from one mobility protocol entity to another. For example:

- With Mobile IP and Mobile IP Regional Registration, home agents use IP-in-IP tunnels to tunnel packets to mobiles' current care-of addresses.

- With SIP mobility, a SIP server uses regular IP routing and forwarding to transport packets to another SIP server or a destination user application.
- 3GPP and 3GPP2 use a host-specific route to exchange user packets between a mobility anchor point (i.e., a GGSN in 3GPP and a PDSN in 3GPP2) and a mobile. However, they implement the host-specific routes in different ways:
  - 3GPP uses GPRS-specific protocols to implement the tunnels inside the packet core network and between the packet core network and the RAN.
  - 3GPP2 uses an IP tunneling protocol, GRE, defined by the IETF to implement the IP tunnels used to transport user packets over the core network. It then uses a PPP connection over these IP tunnels in the core network and radio bearers in the RAN to exchange packets between a mobile and its mobility anchor point, a PDSN.
- Cellular IP and HAWAII use IP-layer host-specific routes from a mobility anchor point (i.e., a Gateway in Cellular IP or a DRR in HAWAII) to a mobile for delivering packets from the anchor point to the mobile.

Another main difference is how location management is related to route management.

- With Mobile IP, Mobile IP Regional Registration, and SIP mobility, location management is separated from IP-layer routing. Centralized servers (e.g., Mobile IP home agent, SIP server) are used to maintain location information. Packets from one mobility protocol entity to another are either routed via regular IP routing (in the case of SIP mobility) or transported over IP tunnels (in the case of Mobile IP and Mobile IP Regional Registration).
- Cellular IP and HAWAII integrate location management with IP-layer routing. No separate location servers are used. Instead, the network maintains mobiles' locations implicitly by maintaining an up-to-date host-specific route to each active mobile.
- 3GPP and 3GPP2 packet networks also use host-specific routes. But the host-specific routes in the IP core networks are implemented as tunnels over an IP transport layer rather than IP-layer host-specific routes as in Cellular IP or HAWAII. Location management is then integrated with the management of these host-specific routes.

A third key difference among the mobility management methodologies shown in Figures 4.68 and 4.69 is on whether and how paging is supported.

- Mobile IP and Mobile IP Regional Registration do not support paging on their own, even though extensions to MIPv4 have been proposed recently to add paging functions to MIPv4.
- In a 3GPP packet core network, the edge routers (i.e., the SGSNs) in the packet core network are responsible for initiating paging operations. Location

management is integrated with host-specific routing in the core network in a way that eliminates the need for the GGSNs to be involved in the paging process.

- 3GPP2 integrates location management with host-specific routing in a way that the IP core network does not have to be concerned with paging. Instead, when mobile-bound packets enter a cdma2000 RAN, the RAN will carry out paging when necessary, using the paging protocol and procedures available inside the RAN.

- Cellular IP and HAWAII both defined their own paging procedures, which make use of their specific host-specific routing mechanisms.

REFERENCES

1. 3rd Generation Partnership Project 2 (3GPP2). Wireless IP network standard. 3GPP2 P.S0001, Version 1.0, December 1999.

2. 3rd Generation Partnership Project 2 (3GPP2). Wireless IP architecture based on IETF protocols. 3GPP2 P.R0001, Version 1.0.0, July 2000.

3. 3rd Generation Partnership Project 2 (3GPP2). 3GPP2 access network interfaces interoperability specification, revision A (3G-IOS v4.1.1). 3GPP2 A.S0001-A, Version 2.0, June 2001.

4. 3rd Generation Partnership Project 2 (3GPP2). cdma2000—layer 3 signaling, revision A, July 2001.

5. 3rd Generation Partnership Project 2 (3GPP2). 3GPP2 interoperability specifications (IOS) for CDMA 2000 access network interfaces—part 6 (A8 and A9 interfaces). 3GPP2 A.S0016-0, Version 2.0, May 2002.

6. 3rd Generation Partnership Project 2 (3GPP2). 3GPP2 interoperability specifications (IOS) for CDMA 2000 access network interfaces—part 7 (A10 and A11 interfaces). 3GPP2 A.S0017-0, Version 2.0, May 2002.

7. 3rd Generation Partnership Project (3GPP), Technical Specification Group, Core Networks. Numbering, addressing and identification, release 5. 3GPP TS 23.003, Version 5.3.0, June 2002.

8. 3rd Generation Partnership Project (3GPP), Technical Specification Group Radio Access Network. Radio resource control (RRC); protocol specification, release 5. 3GPP TS 25.331, Version 5.1.0, June 2002.

9. 3rd Generation Partnership Project (3GPP), Technical Specification Group, Services and System Aspects. Architecture requirements, release 5. 3GPP TS 23.221, Version 5.5.0, June 2002.

10. 3rd Generation Partnership Project (3GPP), Technical Specification Group, Services and System Aspects. General packet radio service (GPRS) service description, stage 2, release 5. 3GPP TS 23.060, Version 5.2.0, June 2002.

11. A. Bar-Noy, I. Kessler, and M. Sidi. Mobile users: to update or not to update? *ACM/Balzer Journal of Wireless Networks*, 1(2):175–185, July 1995.

12. B. Aboba and M. Beadles. The network access identifier. IETF RFC 2486, January 1999.

13. I.F. Akyildiz, J.S.M. Ho, and Y.-B. Lin. Movement-based location update and selective paging for PCS networks. *IEEE/ACM Transactions on Networking*, 4(4):629–638, August 1996.

14. C. Perkins. Minimal encapsulation within IP. IETF RFC 2004, October 1996.

15. P. Calhoun and C. Perkins. Mobile IP network access identifier extension for IPv4. IETF RFC 2794, March 2000.

16. P.R. Calhoun, J. Loughney, E. Guttman, G. Zorn, and J. Arkko. Diameter base protocol. IETF Internet Draft, <draft-ietf-aaa-diameter-17.txt> work in progress, December 2002.

17. A.T. Campbell, J. Gomez, S. Kim, A.G. Valko, C.-Y. Wan, and Z. Turanyi. Design, implementation and evaluation of cellular IP. *IEEE Personal Communications*, August 2000.

18. C. Castelluccia. Extending mobile IP with adaptive individual paging: a performance analysis. INRIA RT-0236, November 1999.

19. S. Deering. ICMP router discovery messages. IETF RFC 1256, September 1991.

20. S. Deering and R. Hinden. Internet protocol, version 6 (IPv6) specification. IETF RFC 2460, December 1998.

21. G. Dommety and K. Leung. Mobile IP vendor/organization-specific extensions. IETF RFC 3115, April 2001.

22. R. Droms. Dynamic host configuration protocol. IETF RFC 2131, March 1997.

23. A. Dutta, F. Vakil, J.-C. Chen, M. Tauil, S. Baba, N. Nakajima, and H. Schulzrinne. Application layer mobility management scheme for wireless Internet. IEEE 3G Wireless 2001, May 2001.

24. E. Gustafsson, A. Jonsson, and C.E. Perkins. Mobile IPv4 regional registration. IETF Internet Draft, <draft-ietf-mobileip-reg-tunnel-06.txt> work in progress, March 2002.

25. E. Guttman, C. Perkins, J. Veizades, and M. Day. Service location protocol, version 2. IETF RFC 2608, June 1999.

26. S. Hanks, T. Li, D. Farinacci, and P. Traina. Generic routing encapsulation (GRE). IETF RFC 1701, October 1994.

27. J.S.M. Ho and I.F. Akyildiz. A mobile user location update and paging mechanisms under delay constraints. *ACM/Baltzer Journal of Wireless Networks*, 1(4):413–425, December 1995.

28. J. Hodges and R. Morgan. Lightweight directory access protocol (v3): technical specification. IETF RFC 3377, September 2002.

29. I.F. Akyildiz and J.S.M. Ho. Dynamic mobile user location update for wireless PCS networks. *ACM/Baltzer Journal of Wireless Networks*, 1(2):187–196, July 1995.

30. T. Imielinski and J. Navas. GPS-based addressing and routing. IETF RFC 2009, November 1996.

31. J. Kempf, C. Castelluccia, P. Mutaf, N. Nakajima, Y. Ohba, R. Ramjee, Y. Saifullah, B. Sarikaya, and X. Xu. Requirements and functional architecture for an IP host alerting protocol. IETF RFC 3154, August 2001.

32. H. Krawczyk, M. Bellare, and R. Canetti. HMAC: keyed-hashing for message authentication. IETF RFC 2104, February 1997.

33. B. Liang and Z.J. Haas. Predictive distance-based mobility management for PCS networks. In *Proc. IEEE INFOCOM*, pp. 1377–1384, New York, 1999.

34. Mobile Wireless Internet Forum. Network reference architecture. Technical Report MTR-004 Release 2.0, May 2001.

35. G. Montenegro. Reverse tunneling for mobile IP (revised). IETF RFC 3024, January 2001.

36. L. Ong, I. Rytina, M. Garcia, H. Schwarzbauer, L. Coene, H. Lin, I. Juhasz, M. Holdrege, and C. Sharp. Framework architecture for signaling transport. IETF RFC 2719, October 1999.

37. C. Perkins. IP mobility support for IPv4. IETF RFC 3344, August 2002.

38. C.E. Perkins. Mobile IP. *IEEE Communications Magazine*, 35(5):84–99, May 1997.

39. D.C. Plummer. Ethernet address resolution protocol: or converting network protocol addresses to 48.bit Ethernet addresses for transmission on Ethernet hardware. IETF RFC 826, November 1982.

40. J. Postel. Multi-LAN address resolution. IETF RFC 925, October 1984.

41. R. Ramjee, L. Li, L. La Porta, and S. Kasera. IP paging service for mobile host. In *Proc. ACM/IEEE International Conference on Mobile Computing and Networking (MobiCom)*, pp. 332–345, July 2001.

42. R. Ramjee, T.L. Porta, L. Salgarelli, S. Thuel, K. Varadhan, and L. Li. IP-based access network infrastructure for next generation wireless data networks. *IEEE Personal Communications*, August 2000.

43. R. Ramjee, T.F. La Porta, S. Thuel, K. Varadhan, and S.Y. Wang. HAWAII: a domain-based approach for supporting mobility in wide area wireless networks. In *Proc. IEEE International Conference on Network Protocols (ICNP'99)*, pp. 283–292, Toronto, Canada, November 1999.

44. C. Rose and R. Yates. Minimizing the average cost of paging under delay constraints. *ACM/Baltzer Journal of Wireless Networks*, 1(2):211–219, 1995.

45. C. Rose and R. Yates. Ensemble polling strategies for increased paging capacity in mobile communication networks. *ACM Wireless Networks*, 3(2):159–67, May 1997.

46. J. Rosenberg, H. Schulzrinne, G. Camarillo, A. Johnston, J. Peterson, R. Sparks, M. Handley, and E. Schooler. SIP: session initiation protocol. IETF RFC 3261, June 2002.

47. S. Donovan. The SIP INFO method. IETF RFC 2976, October 2000.

48. S. Thomson and T. Narten. IPv6 stateless address autoconfiguration. IETF RFC 2462, December 1998.

49. H. Schulzrinne and E. Wedlund. Application layer mobility support using SIP. *ACM Mobile Computing and Communications Review*, 4(3):47–57, July 2000.

50. T. Narten, E. Nordmark, and W. Thomson. Neighbor discovery for IP version 6 (IPv6). IETF RFC 2461, December 1998.

51. G. Varsamopoulos and S.K.S. Gupta. On dynamically adapting registration areas to user mobility patterns in PCS networks, In *Proc. Int'l Workshop on Collaboration and Mobile Computing (IWCMC'99)*, Aizu, Japan, Aug. 1999.

52. W. Richard Stevens. TCP/IP Illustrated, volume 1: the protocols. Addison-Wesley, Reading, Massachusetts, 1994.

53. E. Wedlund and H. Schulzrinne. Mobility support using SIP. In *Proc. ACM/IEEE International Conference on Wireless and Multimedia (WoWMoM'99)*, August 1999.

54. V.W.-S. Wong and V.C.M. Leung. Location management for next generation personal communication networks. IEEE Network, September 2000.

55. T. Zhang, S. wei Li, Y. Ohba, and N. Nakajima. A flexible and scalable IP paging protocol. In *Proc. IEEE GLOBECOM*, pp. 630–635, Taipei, Taiwan, November 2002.

56. X. Zhang, J.G. Castellanos, and A.T. Campbell. P-MIP: paging extensions for mobile IP. *ACM Mobile Networks and Applications*, 7(2):127–141, April 2002.

5

Security

This chapter discusses how *network security* is supported over the Internet, 3GPP networks, and 3GPP2 networks. Here, the term network security is used in its broadest sense and refers to the protection of network from unauthorized access. We first discuss the different aspects of security management, forms of security attacks, and some of the fundamental technologies used in most network security mechanisms. Then, we describe the specific security support mechanisms designed for the Internet, in a 3GPP network, and in a 3GPP2 network.

5.1 INTRODUCTION

5.1.1 Different Facets of Security

Network security has many different facets:

- *Authentication:* Authentication is an ability for communicating parties, including network operators and users, to validate each other's authentic identity.
- *Authorization:* Authorization is the ability for a party (e.g., a network provider) to determine whether a user should be allowed to access particular networks, network services, or information. Authorization is also referred to as *access control.*
- *Integrity:* Integrity refers to the protection of information from unauthorized change.

IP-Based Next-Generation Wireless Networks: Systems, Architectures, and Protocols,
By Jyh-Cheng Chen and Tao Zhang. ISBN 0-471-23526-1   © 2004 John Wiley & Sons, Inc.

- *Confidentiality or Privacy:* Information confidentiality is to keep the information private such that only authorized users can understand it. Therefore, confidentiality is also referred to as *privacy*. Confidentiality is often achieved by encryption.

- *Availability:* The network operators should prevent outside malicious users from blocking legitimate access to a network or a network service. *Denial-of-service*, for example, will deter legitimate users from accessing the network information and resources.

- *Nonrepudiation:* Nonrepudiation refers to the ability for a network to supply undeniable evidence to prove the message transmission and network access performed by a user.

When the Internet was designed, security was not a major concern. As a result, the Internet and other IP networks traditionally lack security management capabilities. However, recently, security has been one of the main focuses of the IETF. Significant achievements have been made. Today, a reasonably well thought out security management framework and detailed security measures have emerged to support all of the security services described above. Security management over the Internet will be discussed further in Section 5.2.

In wireless networks, security management has traditionally focused on authentication and privacy. In 2G systems, for example, encryption only applies to wireless channels. Therefore, one can still listen to other people's conversations by connecting to the core network, where signaling and user messages are not protected. Security management in 3GPP and 3GPP2 networks will be discussed further in Sections 5.7 and 5.8.

5.1.2   Security Attacks

Security attacks can be *passive attacks* or *active attacks*. As the name implies, a passive attack does not attempt to damage the attacked system. It just *eavesdrops the transmission* or monitors and analyzes the network traffic. The passive nature of these attacks makes them difficult to detect. Active attacks, on the other hand, involve the modification of information, interruption of information transmission, and fabrication of messages. The following are some of the common active attacks:

- *Denial-of-service (DoS):* A DoS attack seeks to prevent a service from being provided to one or more users or to cause significant disruptions to the services. For example, an attacker may initiate a large number of connections to a target destination continuously to overload the target to make it impossible or difficult for the target to provide any service. Legitimate users, therefore, are deterred from network access.

- *Masquerade:* An attacker first acquires the identity of a legitimate user. It then pretends to be an authorized user to access the network information and resources.

- *Man-in-the-middle:* An attacker positions forces between communicating parties to intercept and manipulate the messages transmitted between the communicating parties. For example, the attacker may delay, modify, or counterfeit the messages. The attacker may also divert the messages to other locations before relaying them between the legitimate communicating parties. Before such attacks are detected, the legitimate communicating parties believe that they are still sending messages to each other directly.

- *Replay:* An attacker intercepts and records the legitimate transmission. The attacker then *replays* (i.e., resends) the messages later on. Using replay attacks, an attacker could pretend to be an authorized user to access a network or information even when the captured transmission was encrypted and even when the attacker does not know the security key needed to decrypt the captured transmission. For example, an attacker could replay a banking transaction to duplicate the previous transaction.

5.1.3 Cryptography

Cryptography is the *study of mathematical techniques related to aspects of information security such as confidentiality, data integrity, entity authentication, and data origin authentication* [64]. Cryptography techniques are the cornerstones of most network security mechanisms. Therefore, this section provides a high-level description of two specific cryptography techniques, *encryption* and *message authentication*, which are the necessary components of most network security measures. Readers are referred to [64], [76], and [81] for mathematical details.

5.1.3.1 Encryption Encryption is a methodology for transforming the representation or appearance of letters or characters without changing the information context carried by these letters or characters. The original message is called *plaintext* or *cleartext*. The transformed message is called *ciphertext*. The process of transforming plaintext is called *enciphering* or *encryption*, and the reversed process is called *deciphering* or *decryption*. Encryption can be used to achieve information confidentiality.

Encryption algorithms can be classified into two broad categories: *secret-key algorithms* and *public-key algorithms*. Conventional encryption techniques employ secret-key algorithms. Using a secret-key algorithm, the communicating parties share the same secret key. The basic idea is to use *transposition ciphering* and/or *substitution ciphering* such that only those who know the secret key will be able to decrypt the message.

Transposition ciphering rearranges the characters in the plaintext to produce the ciphertext. A simple form of transposition ciphering is permutation. For instance, the function f below permutes the sequence of $i = 1, 2, 3, 4, 5, 6, 7$ into

$$f(i) = 2, 4, 1, 6, 5, 3, 7 \qquad (5.1)$$

A message of

$$m = \text{IP-Based Next-Generation Wireless Networks} \quad (5.2)$$

is therefore encrypted into

$$f(m) = \text{-IsPaBeNdt xe-nGaeretniioW ree lssNwektros} \quad (5.3)$$

Substitution ciphering, on the other hand, substitutes characters in the plaintext with other characters to produce the ciphertext. For example, one can define a mapping g as

$$g = \begin{pmatrix} A\ B\ C\ D\ E\ F\ G\ H\ I\ J\ K\ L\ M\ N\ O\ P\ Q\ R\ S\ T\ U\ V\ W\ X\ Y\ Z \\ D\ E\ F\ G\ H\ I\ J\ K\ L\ M\ N\ O\ P\ Q\ R\ S\ T\ U\ V\ W\ X\ Y\ Z\ A\ B\ C \end{pmatrix} \quad (5.4)$$

Using the mapping g, a message

$$m = \text{THIS IS A PLAINTEXT} \quad (5.5)$$

is therefore encrypted into

$$g(m) = \text{WKLV LV D SODLQWHAW} \quad (5.6)$$

As shown in Figure 5.1, a sender can encrypt a plaintext with a secret key, using, for example, the transposition ciphering or substitution ciphering methods described above. The ciphertext is then transmitted over the network. Only those with the same secret key are able to decrypt the ciphertext.

Two of the most widely used secret-key encryption mechanisms are the Data Encryption Standard (DES) [35] and the Advanced Encryption Standard (AES) [3]. With DES, the plaintext is divided into blocks of 64 bits in length. Each block is then permuted and encrypted with 16 identical iterations that combine both substitution and transposition ciphers. Because the *brute-force* deciphering attack employs exhaustive analysis of the key space, the DES with the key length of 56 bits becomes vulnerable to

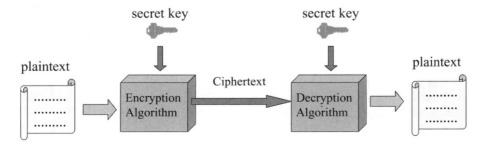

Fig. 5.1 Secret-key encryption and decryption

brute-force attack as the computing power in modern computers grows rapidly. Triple DES [35] thus applies the DES algorithms three times to make it more difficult to break. To overcome the weakness in DES, the National Institute of Standards and Technology (NIST) developed the AES in 1997 as a replacement of DES. The AES algorithm allows secret keys of different lengths to be used. The AES algorithms using 128-bit, 192-bit, and 256-bit secret keys are referred to as AES-128, AES-192, and AES-256.

Secret-key algorithms are *symmetric* in the sense that the same secret-key is used for both encryption and decryption.

A fundamental concern with secret-key algorithms is how to distribute the secret keys in a secure manner. If the secret keys can be transmitted over the network securely, it also implies that other messages should also be transmitted securely over the network, which suggests that there will be no need to utilize the secret-key algorithms to encrypt messages.

To overcome the limitations of secret-key algorithms, *public-key algorithms* are created. Using public-key encryption, each user has two keys: a *public key* and a *secret (private) key*. Only users themselves know their own *private keys*. The public key, however, is made publicly available to anyone and therefore can be distributed without having to worry about whether the public key will be revealed to attackers. As shown in Figure 5.2, when a sender wants to send a message to a receiver, the sender uses the receiver's public key to encrypt the message. Once encrypted, only those with the secret key are able to decrypt the ciphertext.

Compared with secret-key algorithms, public-key algorithms employ a more efficient way for key management. A well-known public-key encryption algorithm is the RSA[1] algorithm [75]. The RSA algorithm is based on the premise that it is extremely difficult to factor the product of two large prime numbers. Therefore, a secret key can be generated by two selected large prime numbers. The product of the two large prime numbers will be used as the public key. This way, knowing the public key does not allow one to easily derive the associated private key.

Public-key algorithms are *asymmetric* because different keys are used for encryption and decryption.

5.1.3.2 Message Authentication
Message authentication is a methodology to assure data integrity and to authenticate the data origin.

One way for message authentication is illustrated in Figure 5.3. It uses a secret key and the original message as inputs to generate a Message Authentication Code (MAC). An algorithm used to generate the MAC is called a MAC algorithm. The MAC is appended to the original message, which is then transmitted to the destination. Once received, the destination applies the same MAC algorithm with the same secret key to calculate a new MAC. If the message is altered during transmission, the MAC generated by the destination will be most likely different from the MAC carried in the message. If the MAC generated by the destination is the same as the MAC carried in the message, it indicates that the original data are not

---

[1]The algorithm is named after the inventors, Rivest, Shamir, and Adleman.

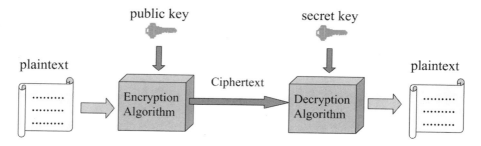

Fig. 5.2   *Public-key encryption and decryption*

likely modified by an unauthorized user in transmission. If the MAC generated by the destination matches with the MAC carried in the message, the destination also can authenticate the origin of the message because only the authentic source should have the same secret key.

A variation of message authentication code algorithm that has received much attention is *one-way hash function*. Essentially, a hash function generates a cryptographic checksum. A one-way hash function takes an arbitrarily long input message and produces a fixed-length, pseudorandom output called a hash. Knowing a hash, it is computationally difficult to find the message that produced the hash. In addition, it is almost impossible to find different messages that will generate the same hash.

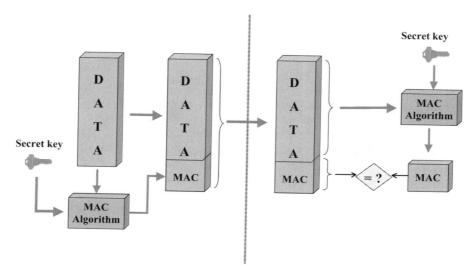

Fig. 5.3   *Message authentication code (MAC)*

A popular one-way hash function is *Message Digest 5 (MD5)* [74]. MD5 takes an input message of arbitrary length and produces an output of 128-bit *fingerprint* or *message digest*. The sender sends the original message and the message digest together to the destination. The destination then computes its own message digest from the received message. Any change to the original message during transmission will result in a message digest that is different from the one provided by the sender and sent together with the original message to the destination. The Secure Hash Algorithm (SHA-1)[2] [1] is similar to MD5 except that SHA-1 generates a 160-bit message digest.

Using a one-way hash function, a computer can authenticate a user without storing the user's password in cleartext. When an account is created and a user types in a password, the host stores the hash generated by a one-way hash function with the password as the input message. When the user logs into the host later, the host computes the hash with the password the user enters now as the input of the same one-way hash function. If the result is the same as the one stored in the host, the user is authenticated. Because the password is not stored in cleartext, user passwords will not be disclosed even when the password file is revealed.

Both MD5 and SHA-1 are *unkeyed* hash functions. That is, there is no secret key between the communicating parties. The algorithms do not include a secret key as an input. Combining with a shared secret key, the HMAC [56] is a mechanism for message authentication using cryptographic hash functions. As that listed in [56], the main goals of HMAC are as follows:

- To use, without modifications, available hash functions. In particular, hash functions that perform well in software, and for which code is freely and widely available.
- To preserve the original performance of the hash function without incurring a significant degradation.
- To use and handle keys in a simple way.
- To have a well-understood cryptographic analysis of the strength of the authentication mechanism based on reasonable assumptions on the underlying hash function.
- To allow for easy replaceability of the underlying hash function in case faster or more secure hash functions are found or required.

The NIST has also developed a HMAC specification [4] that is a generalization of the HMAC specified in IETF RFC 2104 [56].

In addition to hash functions, the public-key algorithms discussed in Section 5.1.3.1 can be used for integrity and message authentication as well. As illustrated in Figure 5.4, the sender could *sign* the plaintext with the sender's secret key. In other words, the sender uses a public-key encryption algorithm and takes its secret key and the original message as inputs to create a *digital signature*. The receiver then uses

---

[2]The one published by the NIST in 1993 is referred to as SHA. SHA-1 refers to the revision in 1995.

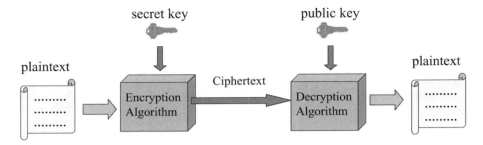

Fig. 5.4 *Integrity and authentication by public key*

the sender's public key to verify the signature. The receiver will not be able to decrypt the received message if the ciphertext is changed. In addition, only the sender owns the secret key. The data origin thus can be authenticated. The Digital Signature Standard (DSS) [2] specifies algorithms for computing digital signatures that can be used by a receiver to verify the identity of the signatory and the integrity of the data.

5.1.4 Public-Key Infrastructure (PKI)

The public-key encryption algorithms are now adopted as a key component in the security management infrastructure for the Internet. As public keys are distributed openly, a new issue arises: How can we assure that a public key is not forged? The Public-Key Infrastructure (PKI) can be used to establish digital identities that can be trusted.

With PKI, a trusted third party, which could be a government agency or financial institute, acts as a certification authority (CA). One can present a public key to the CA in a secure manner. The CA then issues a *digital certificate* or *public-key certificate (PKC)* that contains the user's public key to the user. The certificate is signed digitally by the CA. Therefore, it can be trusted. The user then can publish it for future security services. Anyone who needs the user's public key can obtain the certificate, which can be verified by the CA's digital signature.

One of the widely used standards for digital certification is the *X.509 Certificate* standardized by the ITU [50]. X.509 defines the information and its format that can go into a certificate. The structure of an X.509 certificate is illustrated in Figure 5.5, in which the *Signature* field contains the digital signature of the CA that issued the certificate.

5.2 INTERNET SECURITY

Today's Internet was designed initially to share computing resources so that expensive computing facilities can be better utilized. The primary objective was and still is to deliver packets from source to destination correctly. What is being delivered inside a packet was not a major concern. With this paradigm, packets are transmitted by intermediate nodes (e.g., routers) by simply inspecting the packet headers. A

Version
Certificate serial number
Signature algorithm identifier
Issuer name
Validity period
Subject name
Subject public-key information
Issuer unique identifier
Subject unique identifier
Extensions
Signature

Fig. 5.5 *Format of ITU X.509 certificate. Be produced with the kind permission of ITU.*

packet header contains necessary routing information so that each router can make a routing decision and find a way to forward this packet.

As one can see, today's Internet is not just a technology for connecting computers. It has been part of most people's daily life. As the World-wide Web (WWW) accelerates the Internet into commercial arena, the current network model, however, faces many new challenges. One of the essential concerns is security.

The Internet security comprises various aspects, including secure network infrastructure, secure transmission, and AAA (Authentication, Authorization, Accounting). When the packet transmission is secure, attackers may still be able to capture and interpret messages if they are stored in an unprotected system. In addition to securing the network, therefore, the operating systems, software, database, file systems, and other components should also be designed with security in mind to make the whole system secure.

The following sections discuss two main aspects of security over the Internet: secure transmission and AAA.

5.2.1 IP Security (IPsec)

To secure IP packet transmissions, the IETF has developed IP Security (IPsec) [53], [82]. IPsec provides data origin authentication, replay protection, data integrity, data confidentiality, and access control. IPsec is a suite of protocols, as illustrated in Figure 5.6, for protecting IP datagrams and higher-layer protocols. It consists of *security protocols, authentication and encryption algorithms, security associations*, and *key management*. IPsec is optional for IPv4 but mandatory in IPv6.

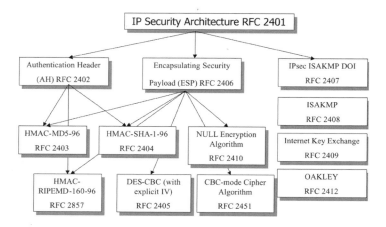

Fig. 5.6 *Family of IPsec protocols*

5.2.1.1 Security Protocols   As depicted in Figure 5.6, IPsec can support the following security protocols:

- *Authentication Header (AH):* AH [51] supports data integrity and authentication of the IP packets (Section 5.2.1.6).
- *Encapsulating Security Payload (ESP):* ESP [52] mainly provides confidentiality services, including confidentiality of message content and limited traffic flow confidentiality (Section 5.2.1.7).

Both AH and ESP could operate in *transport mode* or *tunnel mode*. In the transport mode, security is mainly applied to the IP payload to protect higher layer protocols. On the other hand, the tunnel mode applies to the encapsulated IP packet to protect the entire IP packet. Details of AH and ESP with different modes are discussed in Sections 5.2.1.6 and 5.2.1.7.

5.2.1.2 Authentication and Encryption Algorithms   IPsec specifies various options for cryptography algorithm. As illustrated in Figure 5.6, the bottom two rows indicate the authentication and encryption algorithms defined in IPsec. They include HMAC-MDS-96 [61], HMAC-SHA-1-96 [62], NULL encryption algorithm[3] [46], HMAC-RIPEMD-160-96 [54], DES-CBC [60], and CBC-Mode cipher algorithms [69]. Other algorithms may also be supported. For an implementation claimed to be compliant with ESP, DES-CBC, HMAC-MD5-96, HMAC-SHA-1-96, NULL encryption algorithm, and NULL authentication algorithm must be implemented to make an IPsec implementation compliant with ESP. When using ESP, however, encryption algorithm and authentication algorithm

---

[3]With NULL encryption, the encryption service is not offered.

cannot be both NULL. For a compliant AH implementation, HMAC-MD5-96 and HMAC-SHA-1-96 are mandatory.

5.2.1.3 Security Associations
Before two nodes can use IPsec to secure the traffic between them, a *Security Association (SA)* needs to be established between the two nodes. An SA is the critical building block of IPsec. It is a set of information maintained by the two nodes that defines which security services will be supported and how these security services will be provided. For example, it identifies which security mechanisms (e.g., cryptography algorithms, key management mechanisms) will be used to support the security services, the parameter values (e.g., security keys) needed by the security mechanisms, and how long these parameter values (e.g., the keys) will be valid. Security Associations are needed for both AH and ESP.

Because the information maintained in an SA generally is too much to fit into an IP header, the SAs are maintained in the *Security Association Database (SAD)*. Each SA is identified uniquely by a triplet that contains the following:

- A *security protocol identifier:* Identifies whether the security protocol is AH or ESP.
- *Destination IP address:* The IP address of the other node with which the SA is established.
- *Security Parameter Index (SPI):* An SPI is a 32-bit value that uniquely identifies one SA among different SAs terminating at the same destination.

Once an SA is created, the communicating parties utilize the triplet described above to identify the SA and to protect the transmission according to the parameters defined in the SA.

Essentially, a security association enforces *Security Policies* that define how communicating parties will communicate. Security policies are stored on a *Security Policy Database (SPD)*, which specifies the policies that determine the disposition of all inbound or outbound, IPsec or non-IPsec IP traffic. An SPD must discriminate among traffic that is protected by IPsec from the traffic that is not protected by IPsec. Each packet is either applied IPsec, allowed to bypass IPsec, or discarded. SPDs are identified by *Selectors.* Selectors are used to define the IP traffic for which an SPD should be applied to. The parameters of a Selector include the destination IP addresses, source IP addresses, name (e.g., user ID or system name), transport layer protocol, source and destination port numbers used by the transport layer protocol.

5.2.1.4 Key Management
Both manual and automated SA and key management are mandated in IPsec. A user may configure the security keys for IPsec manually. Manual keying, however, is only practical in a small and static network. Automated key management, therefore, is developed.

The Diffie-Hellman (DH) algorithm, a key-exchange algorithm, is adopted by most automated key management protocols. Using the DH algorithm, communicating parties can generate a security key by exchanging messages over an insecure network. Let us say that users Tao and Jyh-Cheng wish to generate a secret key. They first agree on a large prime number p and a number α such that α is primitive mod p. They then perform the following steps:

1. Tao chooses a random secret x and calculates A as follow:

$$A = \alpha^x \bmod p$$

   Tao then sends A to Jyh-Cheng.

2. Jyh-Cheng, similarly, chooses a random secret y and calculates B, as follows:

$$B = \alpha^y \bmod p$$

   Jyh-Cheng then sends B to Tao.

3. After Tao receives B, he calculates

$$K = B^x \bmod p$$

4. After Jyh-Cheng receives A, he calculates

$$K' = A^y \bmod p$$

Because both K and K' equal to $\alpha^{xy} \bmod p$, Tao and Jyh-Cheng, therefore, get the same secret key. Even though p, α, A, and B may be transmitted without protection, it is computationally difficult for attackers to calculate the discrete logarithm to get x and y.

OAKLEY [67] is a key exchange protocol based on the Diffie-Hellman algorithm. Using OAKLEY, two authenticated parties can agree on secure and secret keying material. OAKLEY describes a series of key exchanges that are called *modes*. Details of the services provided by each mode are specified in [67].

SKEME [55], a versatile key exchange technique, provides anonymity, repudiability, and quick key refreshment. Based on SKEME and OAKLEY, the *Internet Key Exchange (IKE)* [48] negotiates and provides keying material in a protected manner. IKE is also based on a Diffie-Hellman key exchange procedure. IKE is the default automated key management protocol selected for use with IPsec.

The *Internet Security Association and Key Management Protocol (ISAKMP)* [63] defines a framework for security association management and cryptographic key establishment. It defines the procedures and packet formats needed to establish, negotiate, modify, and delete SAs. These procedures and packet formats are independent of specific encryption and authentication mechanisms. ISAKMP does not define specific key exchange techniques. It is designed to support many different key exchanges.

The IPsec protocols AH and ESP are independent of the key management techniques described above.

5.2.1.5 Implementation

5.2.1.5 Implementation IPsec may be implemented on a host or a *security gateway* affording protection to IP traffic. A security gateway refers to an intermediate system such as a router or a firewall that implements IPsec protocols. There are various ways to implement IPsec. Examples are itemized as follows:

- *IP stack integration:* IPsec could be integrated into the network layer of the protocol to provide security services. This is more applicable for IPv6 because IPv6 is not fully deployed at this moment. In addition, this approach can be utilized for both hosts and security gateways.
- *Bump-in-the-stack (BITS):* By BITS, IPsec is inserted as a thin layer between IP layer and link layer. This implementation does not need to modify the IP layer, thus making it more appropriate for legacy systems. BITS, however, usually is adopted in hosts only.
- *Bump-in-the-wire (BITW):* This implementation assumes that IPsec is running on a separate device attached to the physical interface of a host or router. Also referred to as an *outboard crypto processor*, BITW is a common design in military systems. This implementation is applicable to both hosts and security gateways.

5.2.1.6 Authentication Header (AH)

5.2.1.6 Authentication Header (AH) Authentication Header (AH) [51] is used to provide data integrity and authentication of data origin. It also optionally provides replay detection. AH may be applied alone or in combination with the ESP to provide confidentiality service.

The *header format* of AH is shown in Figure 5.7. It contains the following fields:

- *Next Header:* The Next Header contains the value 51 in IPv4. For IPv6, the Next Header points to the next IPv6 extension header that follows the AH.
- *Payload Length:* As shown in Figure 5.7, the length of AH is variable. The Payload Length specifies the length of AH.
- *Security Parameter Index (SPI):* The SPI is as defined in Section 5.2.1.3.
- *Sequence Number:* The Sequence Number is a 32-bit value incremented by 1 for each packet sent. It is utilized to protect against replay attacks.
- *Authentication Data:* This field contains the output generated by the originator of the packet by applying the security algorithm on the original IP packet. The value carried in this field is also called the *Integrity Check Value (ICV)*. The Authentication Data field is a variable length field. However, its length must be an integral number of 32-bit words.

Next, we first examine how AH works in the transport mode in IPv4 networks. Figure 5.8 indicates the packet formats before and after AH is applied to the packet. In the example shown in Figure 5.8, TCP is used as an example for transport layer

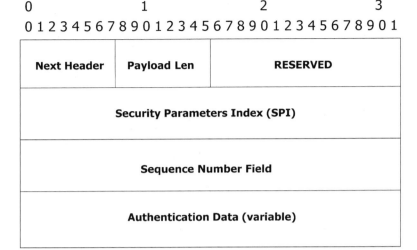

Fig. 5.7 *Header format of AH*

protocol. Any other transport layer protocol may also be used. As depicted in Figure 5.8, the AH header is inserted between IP header and TCP header. The authentication algorithm is applied to the entire IP packet as illustrated in Figure 5.8, excluding the mutable fields in the IP header. A mutable field is one that may need to be changed by the intermediate network nodes while the packet is in transit. Examples of mutable fields in IPv4 include the Type of Service (TOS) field, the Time to Live (TTL) field, the flags, the Fragment Offset, and the Header Checksum in the IP header.

The authentication algorithm is specified by the SA and may be based on MAC or HMAC as that discussed in Section 5.1.3.2. Upon receiving a packet, the destination recalculates the ICV by using the same algorithm specified by the SA, and it

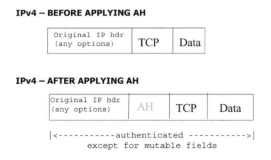

Fig. 5.8 *AH in IPv4 in transport mode*

compares the calculated ICV with the one carried in the AH in the received packet. The destination would know that the packet is modified by unauthorized user(s) if the ICVs are different.

For IPv6, the AH provides identical security protection as for IPv4. The fundamental difference between using AH in IPv4 and IPv6 is in how the AH is placed in a packet. For IPv6, the AH would be one of the IPv6 extension headers that is inserted between the IPv6 header and the header of the upper layer protocol (e.g., TCP or UDP). This is illustrated in Figure 5.9.

Figure 5.10 depicts the structure of an IPv4 or IPv6 packet protected by the AH in the *tunnel mode*. In the tunnel mode, each SA is applied to an IP tunnel. The AH is used to protect the IP packet that goes through this IP tunnel.

We call an IP packet an original IP packet (or an inner IP packet) before the packet enters the IP tunnel. The original IP header shown in Figure 5.10 is the IP header of the original IP packet. This IP header specifies the ultimate destination of the original IP packet. To send the original IP packet through the IP tunnel, the original IP packet is encapsulated by adding a new IP header to the original IP packet. The resulting IP packet is called an encapsulating packet or an outer packet. When the AH is applied to the IP tunnel, the AH is placed between the new IP header and the original IP header to protect the encapsulating IP packet, including the new IP header. In particular, the ICV is calculated by applying the security algorithm over the entire encapsulating packet except the mutable fields in the *new IP header*. All other aspects of the AH operations are identical to the transport mode.

5.2.1.7 Encapsulating Security Payload (ESP)

The Encapsulating Security Payload (ESP) [52] is designed mainly to provide confidentiality (encryption) and authentication. As in AH, the authentication service includes data integrity and data origin authentication. The major difference between ESP and AH is in the protection coverage of authentication data. ESP does not protect IP header unless it is encapsulated in the tunnel mode. ESP also supports replay detection. Although various security services are possible, each SA specifies which specific security

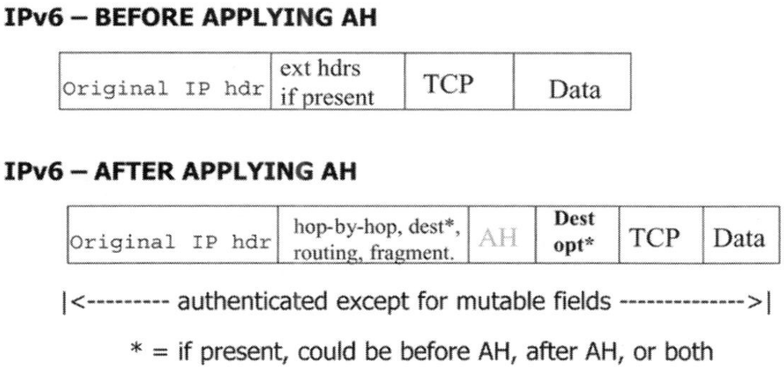

IPv6 – BEFORE APPLYING AH

Original IP hdr	ext hdrs if present	TCP	Data

IPv6 – AFTER APPLYING AH

Original IP hdr	hop-by-hop, dest*, routing, fragment.	AH	**Dest opt***	TCP	Data

|<--------- authenticated except for mutable fields -------------->|

* = if present, could be before AH, after AH, or both

Fig. 5.9 *AH in IPv6 in transport mode*

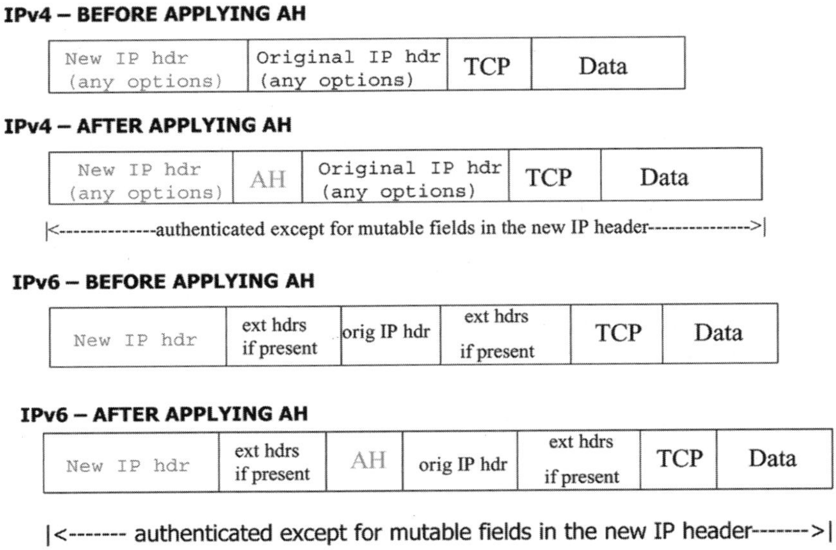

IPv4 – BEFORE APPLYING AH

New IP hdr (any options)	Original IP hdr (any options)	TCP	Data

IPv4 – AFTER APPLYING AH

New IP hdr (any options)	AH	Original IP hdr (any options)	TCP	Data

|<--------------authenticated except for mutable fields in the new IP header-------------->|

IPv6 – BEFORE APPLYING AH

New IP hdr	ext hdrs if present	orig IP hdr	ext hdrs if present	TCP	Data

IPv6 – AFTER APPLYING AH

New IP hdr	ext hdrs if present	AH	orig IP hdr	ext hdrs if present	TCP	Data

|<------- authenticated except for mutable fields in the new IP header------->|

Fig. 5.10 *AH in tunnel mode*

services will be actually provided. However, either confidentiality or authentication must be selected.

The *packet format* of ESP is shown in Figure 5.11. When ESP is applied to a packet, the resulting packet (Figure 5.11) contains the following four components:

- *ESP Header:* The ESP Header contains an SPI and a Sequence Number. The SPI is defined in Section 5.2.1.3. The Sequence Number is used in the same way as the Sequence Number in the AH to prevent replay attacks.

- *Payload Data:* In the transport mode, the Payload Data is the transport-layer data to be transported in this IP packet. In the tunnel mode, the Payload Data is the encapsulated IP packet.

- *ESP Trailer:* The trailer consists of the following fields: Padding, Pad Length, and Next Header. Because some encryption algorithms require that the plaintext be a multiple of some number of bytes, padding in the trailer adds extra bytes to the payload data to make the length of the payload appropriate for the encryption algorithm. Padding may be used to conceal the actual length of the payload as well. In addition, it also let the fields of *Pad Length* and *Next Header* right align within a 4-byte word. The *Pad Length* field indicates the number of padding bytes immediately preceding it. The *Next Header* contains a value 50 in IPv4. For IPv6, the Next Header points to the next IPv6 extension header that follows immediately the ESP Header.

- *Authentication Data:* The Authentication Data field contains the ICV produced by the security algorithm.

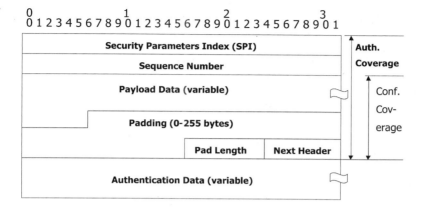

Fig. 5.11 *ESP packet format*

Next, we first examine how ESP in the transport mode is used in IPv4. Figure 5.12 indicates the packet formats before and after applying ESP. In Figure 5.12, the transport-layer protocol is assumed to be TCP for illustration purpose only. As depicted in Figure 5.12, ESP header is inserted between IP header and transport-layer protocol header (i.e., TCP header in this example). The authentication algorithm is applied from the ESP header to the ESP trailer as illustrated in Figure 5.12, excluding the mutable fields that are allowed to be changed while the packet is in transit. In other words, the portion of the packet from the ESP Header to the ESP Trailer is protected. Figure 5.12 also indicates that the portion of the packet after the ESP Header up to the ESP Trailer is encrypted.

Upon receiving the packet, the destination recomputes the ICV and compares it with the ICV carried in the Authentication Data field of the received packet. The destination also decrypts the packet to get the original data.

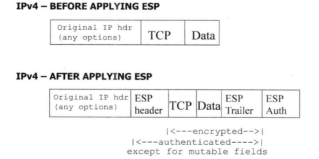

Fig. 5.12 *ESP in IPv4 in transport mode*

The ESP provides the same functionality in IPv6 as in IPv4. The fundamental difference is that ESP would be one of the IPv6 extension headers. The format of an IPv6 packet after ESP is applied to the packet is illustrated in Figure 5.13.

Figure 5.14 depicts the packet formats of ESP in *tunnel mode* for both IPv4 and IPv6. In the tunnel mode, an SA is applied to an IP tunnel. The operation of ESP in the tunnel mode is basically the same as that in transport mode. The difference is that, in the tunnel mode, the ESP Header is placed between the new (outer) IP header and the original (inner) IP header. The ICV is computed by applying the authentication algorithm over the portion of the packet from ESP Header to ESP Trailer excluding the mutable fields. The data after ESP Header up to the ESP Trailer are encrypted.

5.2.1.8 Traffic Processing

As stated in Section 5.2.1.3, the SPD must be consulted for all traffic, including inbound or outbound, IPsec or non-IPsec IP traffic. Based on the SPD, each packet is either applied IPsec or allowed to bypass IPsec. The packet is discarded if no policy is found matching the packet.

When a node wants to send a packet to another node, this outbound packet will be compared against the SPD to determine what processing is required for the packet. If IPsec processing is required, the packet is mapped into an existing SA or SA bundle in the SAD. If there is no SA that is currently available, a new SA or SA bundle is created for the packet. To create a new SA or SA bundle, the communicating nodes may need to use IKE to establish IKE SA (SA bundle) first. After the IKE SA is authenticated, IPsec SA is created under the protection of IKE SA. Once IPsec SA is established, either AH or ESP (or both) can be used in either transport mode or tunnel mode to protect the packets.

Upon receiving a packet, the receiver uses the SPI in the AH or ESP Header in the received packet to look up the SA in SAD for the packet. Based on the SA, it performs the IPsec processing that might decrypt the packet or check the ICV.

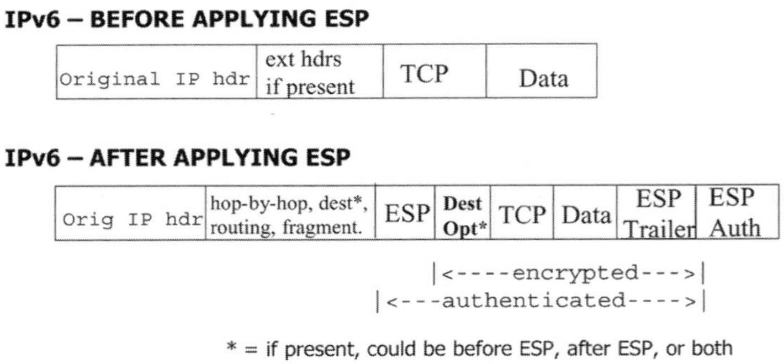

IPv6 – BEFORE APPLYING ESP

Original IP hdr	ext hdrs if present	TCP	Data

IPv6 – AFTER APPLYING ESP

Orig IP hdr	hop-by-hop, dest*, routing, fragment.	ESP	Dest Opt*	TCP	Data	ESP Trailer	ESP Auth

```
                                  |<----encrypted--->|
                                  |<---authenticated---->|
```

* = if present, could be before ESP, after ESP, or both

Fig. 5.13 *ESP in IPv6 in transport mode*

IPv4 – BEFORE APPLYING ESP

New IP hdr (any options)	Original IP hdr (any options)	TCP	Data

IPv4 – AFTER APPLYING ESP

New IP hdr (any options)	ESP header	Original IP hdr (any options)	TCP	Data	ESP Trailer	ESP Auth

```
                     |<------------------ encrypted ------------------>|
                     |<--------- authenticated except for mutable fields -------->|
```

IPv6 – BEFORE APPLYING ESP

New IP hdr	ext hdrs if present	Orig IP hdr	ext hdrs if present	TCP	Data

IPv6 – AFTER APPLYING ESP

New IP hdr	New ext hdrs	ESP header	Orig IP hdr	Orig ext hdrs	TCP	Data	ESP Trailer	ESP Auth

```
                     |<------------------ encrypted ------------------>|
                     |<----- authenticated except for mutable fields ------->|
```

Fig. 5.14 *ESP in tunnel mode*

5.2.1.9 IPsec Applications

This section illustrates four exemplary applications of IPsec. As specified in IPsec, these four scenarios *must* be supported by compliant IPsec hosts or security gateways.

Figure 5.15 depicts how end-to-end security could be supported using IPsec between two hosts. Once the SA between two hosts is established, they can utilize either AH or ESP in transport mode or tunnel mode to protect the data. Figure 5.15 also shows possible operations of IPsec in different modes and the associated packet formats. IPsec does not mandate the support of nesting, that is, combining both AH and ESP. However, AH must apply after ESP if AH and ESP are combined. It is shown in example 3 in the transport mode in Figure 5.15.

Figure 5.16 illustrates how IPsec may be used to establish Virtual Private Networks (VPN) between two network domains provided by different network operators. In particular, VPNs can be established by establishing IPsec Security Associations between security gateways in different network domains. For example, a security gateway may be a firewall that separates two networks (e.g., a private and a public network). The traffic between two different networks is tunneled and protected by the security gateways. As indicated in Figure 5.16, the original IP packet including the header of IP1 is encapsulated by AH or ESP, which further is encapsulated by the outer header of IP2 generated by the security gateway. The packet is transmitted in a secured manner to the destined security gateway, which then decapsulates the packet to restore the original packet and then routes it further toward to the destination host based on the original header of IP1.

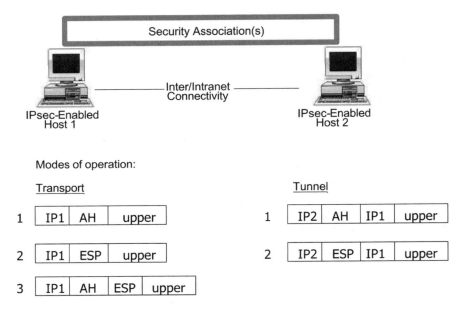

Fig. 5.15 End-to-end security

In the prior example, packets are not protected inside each private network. The premise is that the private network is trustable. Figure 5.17 shows an example in that IPsec is employed from host to host and from security gateway to security gateway. This scenario essentially couples the previous two examples to provide both end-to-end security and VPN protection.

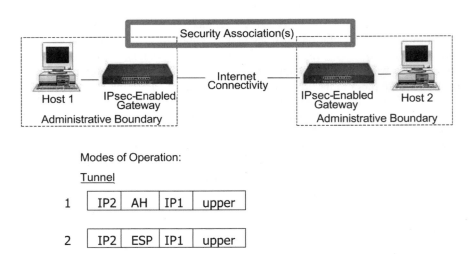

Fig. 5.16 VPN (virtual private network) with IPsec

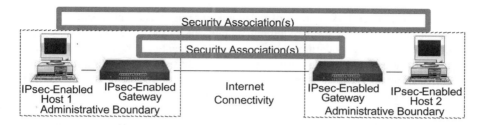

Fig. 5.17 End-to-end with VPN security

Figure 5.18 depicts an example in which a remote host connects to a private network behind a security gateway. This may be the case when a mobile host connects to its home enterprise network through a firewall. In this scenario, the sender must apply the transport mode before applying the tunnel mode. Figure 5.18 shows that the end-to-end security between two end hosts is protected by the transport mode. Tunnel mode is required between the remote host and the security gateway.

5.2.2 Authentication, Authorization, and Accounting (AAA)

In addition to IPsec, which secures the packet transmissions over the Internet, the importance of AAA cannot be underemphasized. Today, the widely used protocol for AAA over IP networks is the Remote Authentication Dial In User Service (RADIUS) [73]. RADIUS was originally developed for remote dial-up access. Its main purpose was to verify user name and password and to deliver configuration information detailing the type of service provided to the user. It, however, may suffer performance degradation and data loss in a large system. The AAA Working Group in the IETF thus is developing a new AAA protocol to address the deficiency in RADIUS.

In April 2000, the IETF AAA Working Group solicited submissions of candidate AAA protocols. Four proposals, including SNMP [57], RADIUS Enhancement (Radius++), Diameter [43], and COPS [45], were received. An evaluation team was organized to evaluate the proposals received. The evaluation results were published as IETF RFC 3127 [65] in June 2001. It suggests that SNMP and Radius++ are not

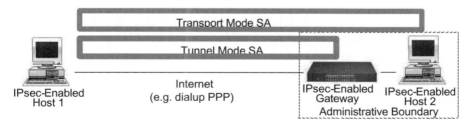

Fig. 5.18 Secured remote access

acceptable as a general AAA protocol and will require significant enhancements before they can meet all of the AAA requirements laid out by the IETF. SNMP has some utility to be used for accounting, but it requires significant modifications to become a viable authentication and authorization protocol. Radius++ also needs a fair amount of engineering to make it meet all of the AAA requirements. That engineering, however, would turn it into a protocol like Diameter. Both COPS and Diameter are believed by the IETF to be acceptable as an AAA protocol for the Internet. COPS still requires some minor to medium engineering to make it fully comply with the IETF AAA requirements. Diameter needs some minor engineering to bring it into complete compliance with the requirements. The evaluation team expressed a slight preference for Diameter. Since then, a lot of effort has been devoted to develop the Diameter protocol. Although Diameter has yet to become an Internet Standard, both 3GPP and 3GPP2 have chosen Diameter as their AAA protocol for supporting packet data services. We describe Diameter further in Section 5.2.2.1.

5.2.2.1 *Diameter*   Diameter [43] provides a base protocol to support AAA services. The Diameter base protocol is typically not used alone. Unless it is used only for accounting, the Diameter base protocol is extended for each particular application. For example, the Diameter base protocol has been extended with a Diameter Mobile IP Application in order to support AAA with the Mobile IP protocol (Section 5.2.2.2).

Unlike RADIUS, Diameter utilizes TCP and SCTP (Stream Control Transmission Protocol) [72] instead of UDP to transport messages so that Diameter messages can be transported reliably.

Figure 5.19 illustrates the main network elements used by Diameter and a typical message flow between these network elements. Diameter is a peer-to-peer protocol. A Diameter Peer could be a Diameter Client, Agent, or Server. A network entity that generates the Diameter request messages is regarded as a *Diameter Client*. A Diameter client is a device performing access control at the edge of a network. For example, it could be a Network Access Server (NAS) or a Mobile IP Foreign Agent (FA). The Diameter request message may be sent to a *Diameter Agent* that provides

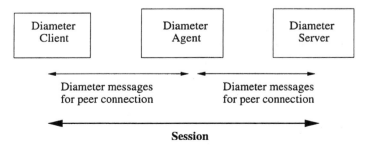

Fig. 5.19 *Typical flows in Diameter*

relay, *proxy*, *redirect*, or *translation* services. The Diameter request is processed by the *Diameter Server*. In addition to Diameter applications, a Diameter server must support the Diameter base protocol. As shown in Figure 5.19, there is a direct *connection* between each pair of Diameter peers to exchange Diameter messages. A *session* is a logical application-layer path between a client and a server.

The Diameter protocol uses the Network Access Identifier (NAI) [37] to identify a user and route Diameter messages. A *relay agent* simply routes the Diameter request messages to the destination Diameter server by looking up a routing table it maintains. This routing table is called a *Realm Routing Table*. The *realm* refers to the realm portion of an NAI. In general, a realm is an administrative domain that provides services to a user. A *proxy agent* relays the Diameter request messages in a way similar to that of a relay agent. It, however, can modify the messages to implement policy enforcement related to resource usage and resource provisioning. If policies are violated, a proxy agent can reject the request without further sending it to the Diameter server. When a Diameter request is received by a *redirect agent*, it replies with an answer to instruct the sending peer where to forward this request. The *translation agent* provides the translation between two protocols. When a *translation agent* receives a request message in RADIUS format, for instance, it translates the request to a Diameter message.

In Diameter, the message exchanged between peers is primarily in the form of *Attribute Value Pairs (AVPs)*. Each AVP contains a header and a protocol-specific data. It is used to transport specific AAA information. Some AVPs have been defined for the Diameter base protocol. New Diameter applications can be designed by reusing existing AVPs and/or creating new AVPs. This makes the base protocol extensible to incorporate with different applications.

The authentication process of a Diameter server depends on the particular Diameter application in use. Typically, the initial Diameter request message and all of the subsequent messages from a Diameter client to a Diameter server include a Session- Id that is used to identify the session. A Result-Code AVP is included in the answer message from a Diameter server to a Diameter client to indicate success or failure of a request.

The Diameter protocol provides end-to-end security for its own messages. To secure Diameter messages, Diameter servers must support Transport Layer Security (TLS) [44] and IPsec. For IPsec, the Diameter must support ESP in transport mode with certain encryption and authentication algorithms to provide per-packet authentication, integrity, and confidentiality. The Diameter implementation must also support IKE for peer authentication, key management, and negotiation of security association. For TLS, the server must request a certificate from the client to perform mutual authentication for the TLS session establishment.

5.2.2.2 Diameter with Mobile IPv4 Application

The IETF AAA working group has been working on the Diameter Mobile IPv4 application [42] that will allow Diameter to be used to authenticate, authorize, and collect accounting information for Mobile IPv4 services. Figure 5.20 depicts the interaction between Mobile IPv4 and AAA servers. Mobile IPv4 Foreign Agent (FA) and Home Agent (HA) are connected to the local AAA server (AAAL) and home AAA server (AAAH), respectively. Diameter is used as the protocol between the AAA servers, between FA and AAAL,

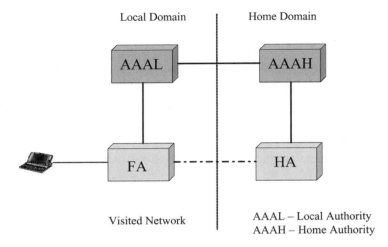

Local Domain　　　　　Home Domain

AAAL　　　　　AAAH

FA　　　　　HA

Visited Network

AAAL – Local Authority
AAAH – Home Authority

Fig. 5.20 *Mobile IP with AAA*

and between HA and AAAH. Each time a Mobile IP Registration Request is received by a Mobile IP mobility agent (FA or HA), the mobility agent will consult with its AAA server (AAAL or AAAH) before a transaction is granted.

Figure 5.21 illustrates how Diameter is used with Mobile IP to support interrealm (domain) mobility. After roaming to a foreign realm, a Mobile Node (MN) issues a

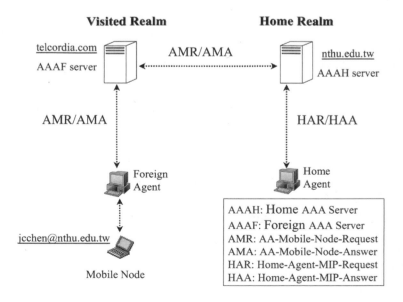

Visited Realm　　　　　**Home Realm**

telcordia.com　　　　　AMR/AMA　　　　　nthu.edu.tw
AAAF server　　　　　　　　　　　　　　AAAH server

AMR/AMA　　　　　　　　　　　　　HAR/HAA

Foreign
Agent

Home
Agent

jcchen@nthu.edu.tw

Mobile Node

AAAH: Home AAA Server
AAAF: Foreign AAA Server
AMR: AA-Mobile-Node-Request
AMA: AA-Mobile-Node-Answer
HAR: Home-Agent-MIP-Request
HAA: Home-Agent-MIP-Answer

Fig. 5.21 *Interrealm mobility with Diameter*

Mobile IP Registration Request to the FA to request services from foreign realm. The FA then sends a Diameter AMR (AA-Mobile-Node-Request) message to the local Diameter server, known as AAAF. Upon receiving the AMR, the AAAF determines whether the AMR should be processed locally or forwarded to the mobile's home Diameter server, i.e., the mobile's AAAH. Figure 5.21 shows that the AAAF passively awaits for a request from the FA. The FA may also challenge the MN by using the *Mobile IP Challenge/Response* defined in IETF RFC 3012 [40]. The MN should respond with *MN-AAA authentication extension*. In either case, if the authentication data supplied by the MN is invalid, the AAAH returns an AMA (AA-Mobile-Node-Answer) with the Result-Code AVP set to DIAMETER_AUTHENTICATION_RE-JECTED to deny the MN's access. On the other hand, if the MN has been successfully authenticated, the AAAH will send a HAR (Home-Agent-MIP-Request) to the HA (Home Agent). This HAR comprises the Mobile IP Registration Request encapsulated in the MIP-Reg-Request AVP. Upon receipt of the HAR, the HA processes the MIP-Reg-Request AVP and encapsulates the Registration Reply within the MIP-Reg-Reply AVP in an HAA (Home-Agent-MIP-Answer) message, which is sent to the AAAH. When receiving the HAA, the AAAH issues an AMA message to the FA through the AAAF. The FA then sends the Mobile IP Registration Reply to the MN to indicate that the MN is authenticated and authorized. Figure 5.22 depicts the message exchange for the architecture defined in Figure 5.21.

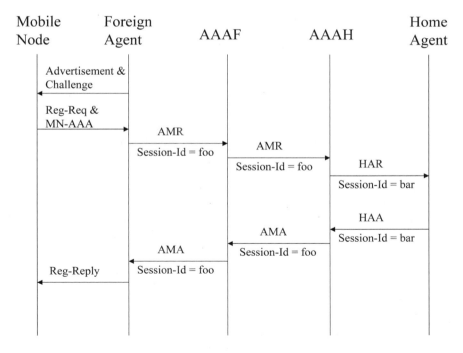

Fig. 5.22 Mobile IPv4/Diameter message exchange

To scale wireless data access across different administrative domains, the number of security associations should be kept small. For example, there should be no shared security association preconfigured between HA and FA, and between FA and MN. The Diameter Mobile IPv4 application utilizes a Key Distribution Center (KDC) to achieve this goal. After MN is successfully authenticated and authorized, the home Diameter server allocates the session keys. Three keys are generated: the MN-HA key (K1), the MN-FA key (K2), and the FA-HA key (K3).[4] K1 is used between MN and HA. K2 is used between MN and FA. Similarly, K3 is used between FA and HA. The keys destined for FA and HA are transmitted via the Diameter protocol and must be encrypted by IPsec or TLS in a network without Diameter agents. If Diameter agents exist, it is recommended that the Diameter CMS (Cryptographic Message Syntax) Security Application [41] be used. The keys for the MN (K1 and K2) must be propagated via the Mobile IP protocol. Instead of using them directly as the session keys, they are used as a random value, which is called *nonce* or *key material*, to derive the actual session keys [70]. The MN and the AAAH will use the nonce and the long-term shared secret key, which is preconfigured between the MN and the AAAH, to derive the MN-HA and MN-FA session keys. Once the session keys have been delivered and established, the mobile node can exchange Mobile IP registration information directly without the Diameter infrastructure. The session keys, however, have a limited lifetime. If the lifetime expires, the procedures described above must be invoked again to acquire the new session keys.

5.3 SECURITY IN WIRELESS NETWORKS

Many security issues in wireless networks are essentially the same as that in wired networks. However, the open nature of wireless channels makes a wireless system more vulnerable to threats such as unauthorized access to and manipulation of sensitive data and services. It is also possible for an attacker to deploy a fraud wireless base station to deceive wireless users to gather secret information. In addition, more elements in a wireless network are vulnerable to security attacks than in wired networks. These elements include, for example, the radio interface, the removable modules (e.g., SIM or USIM) on a mobile terminal that store confidential information, and the mobile terminal. Take radio interface, for example, an attacker could simply jam the radio so no communication is possible over the wireless channel.

Figure 5.23 [39] shows a generic security model used in today's 2G systems. This generic model is also the basis for security management in 3G systems. As indicated in the figure, there are three steps before user data can be transmitted [39]:

- Security provisioning
- Local registration
- Authentication and key agreement (AKA)

---

[4]The foreign Diameter server may generate K3 in some cases.

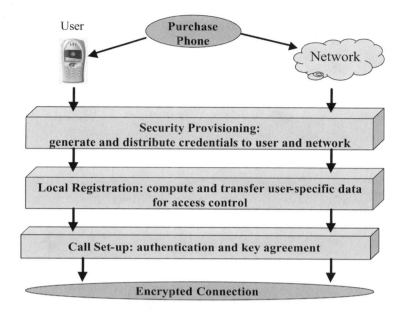

Fig. 5.23 *Generic security model in cellular systems. © 2003 IEEE.*

Security provisioning concerns the generation and distribution of credentials to both users and the network. Figure 5.24 [39] illustrates the security provisioning approaches in GSM and IS-41.

- *In GSM:* There is a secret key called K_i shared between the network operator and the user. A user's secret key along with the user's other identities such as IMSI and MSISDN are stored in a SIM card that is issued to the user by the user's service provider. The SIM card will be inserted to the mobile device the user wants to use. The secret key never leaves the SIM card. On the network side, the network provider is responsible for safeguarding the secret key to ensure that it will never be revealed to unauthorized parties. Once the secret key K_i is provisioned, further security operations are accomplished based on K_i.

- *In IS-41:* IS-41 also uses a secret key called *Authentication Key (A-key)* shared by the user and the network provider to support security operations. The difference between GSM and IS-41 in how the secret key is managed is that there is no smart card (SIM card) in IS-41-based 2G systems. The secret key is programmed into the handset manually either by the user or the network operator.

With the provisioned security information, a user can perform registration with the network in order to gain permission to use the network.

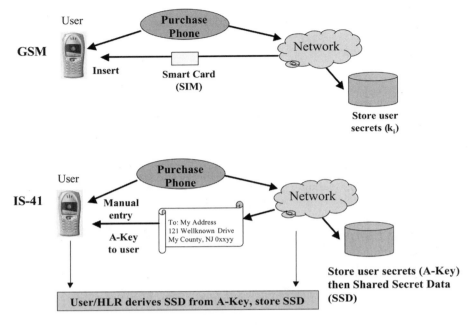

Fig. 5.24 *Key generation and distribution.* © *2003 IEEE.*

Once a user registers with the network for access control, the AKA protocol is executed to authenticate the user and determine if the user is authorized for the call the user is requesting. The AKA procedures for GSM and IS-41 are illustrated in Figure 5.25 [39].

- *For GSM:* Once the network receives a request for call setup, it challenges the user and expects a correct response from the user. The challenge essentially is a random number generated by the network. Based on this random number, the shared secret key K_i, and a same algorithm in both user and network, the response generated by the user should be the same as the one calculated in the network. If not, the user fails the authentication procedure and its call setup request will be denied. An attacker who intercepts the challenge message will not be able to generate the correct response without the shared secret key K_i. Based on the random number and the K_i, another cryptographic key will be derived in both user and network to encrypt user traffic.
- *For IS-41:* The AKA procedure in IS-41 is similar to that in GSM. The difference is that IS-41 uses a global challenge that is broadcast periodically by the network. A user picks up the challenge and sends the call setup request with the response embedded to the network.

Details of 2G security, including 2.5G of GPRS, are further discussed in Sections 5.4 to 5.6. The security management in 3GPP and 3GPP2 are then elaborated.

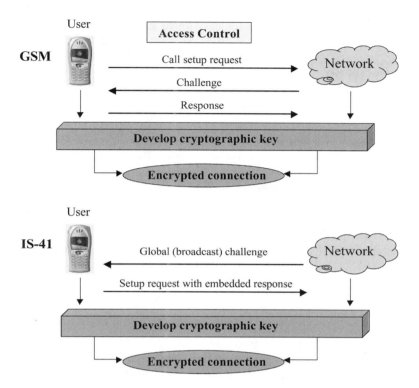

Fig. 5.25 *Authentication and key agreement (AKA).* ©2003 IEEE.

5.4 SECURITY IN IS-41

As discussed in Chapter 1, North America has two major 2G radio systems: IS-136 based on TDMA and IS-95 based on CDMA [47], [58]. The core networks of IS-136 and IS-95, however, are both based on IS-41 [80], [83]. This section reviews the authentication and privacy mechanisms in IS-41 [66], [80], [83].

Because IS-41 specifies the standard of a core network that could be deployed with different RANs (radio access networks), the authentication and privacy in IS-41 is independent of the air interfaces. Using the preprogrammed Authentication Key (A-key), subscribers do not need to be manually involved in the authentication process. That is, subscribers do not need to enter any username or password for authentication. On the network side, the A-key of each user is stored in an Authentication Center (AC). The AC is the primary functional entity for authentication and privacy in IS-41. Authentication and privacy are provided using the Cellular Authentication and Voice Encryption (CAVE) algorithm, which will be examined later. In IS-41, the authentication process might be executed in various events, including user registration, call origination, and call termination.

5.4.1 Secret Keys

The A-key is a 64-bit permanent secret number shared by Mobile Station (MS) and AC. The installation of A-key in MS is not standardized. As mentioned earlier, it could be programmed manually. The process of manual programming is specified in TIA/EIA TSB50 [87]. The IS-725 [49] defines the *over-the-air service provisioning* (*OTASP*) method for A-key programming using the Diffie-Hellman (DH) key agreement procedure.

In addition to the A-key, there is another secret number, which is called *Shared Secret Data* (*SSD*). SSD is a 128-bit temporary secret key calculated in both MS and AC. It can be modified by the network at any time. It can also be shared with a foreign (visited) network, such as by a VLR in a foreign network. The SSD comprises two parts, and each has 64 bits. The first part is used for authentication and is named as *SSD-A*. The second part, called *SSD-B*, is used to support privacy.

Figure 5.26 illustrates how an SSD is generated. The network generates the SSD using the CAVE algorithm with the following inputs:

- *A-key*
- *The mobile station's Electronic Serial Number (ESN):* An ESN is a 32-bit number permanently stored in a terminal by the manufactures to uniquely identify the terminal. The highest 8 bits of ESN identifies the manufacturer, and the remaining bits are assigned by the manufacturer to uniquely identify each

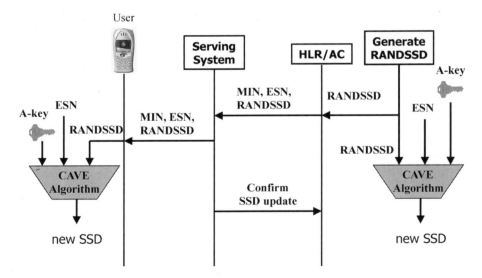

Fig. 5.26 *Generation of shared secret data (SSD). Reproduced with permission of the McGraw-Hill Companies.*

terminal produced by the manufacturer. It can be viewed as the hardware number of the terminal.

- *A random number (RANDSSD):* A random number *RANDSSD* is used as one of the inputs to ensure that the SSD generated each time for the subscriber is different from the one generated for the same subscriber before.

The *RANDSSD* is also propagated to HLR/AC, which retrieves the mobile's ESN and Mobile ID Number (MIN). MIN is a 10-digit North American Numbering Plan (NANP) number that represents a mobile terminal's identification and directory number. It is assigned by a network operator and programmed into a mobile terminal when the mobile terminal is purchased by the user.

The triplet of *RANDSSD*, MIN, and ESN are further propagated to the serving system (MSC, BS, etc.), and then to the MS. The serving system could be either the mobile's home network or the visited network. The MS uses the same algorithm and the same inputs as those used by the network to generate the SSD. As the A-key is never transmitted over the air, attackers would not generate the same SSD unless the A-key is stolen.

The SSD for a mobile can be updated by the network. Figure 5.27 [80] shows the message flow for SSD update. The network first produces a new SSD (*SSD_new*) using the procedure described above. The AC then sends the random number *RANDSSD* used to generate the new SSD to the mobile to order the update of its SSD. Using the *RANDSSD* with the A-key and ESN stored locally on the mobile as inputs to the CAVE algorithm, the mobile can generate the same *SSD_new* as the one generated by the AC. The mobile, however, will not adopt this new SSD until it verifies the SSD

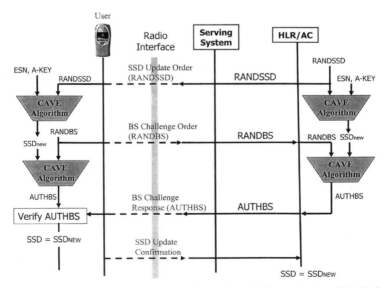

Fig. 5.27 *Update of shared secret data (SSD). Reproduced with permission of the McGraw-Hill Companies.*

Update Order received from the network. This is because although attackers may not know the mobile's A-key, they could simply send a deceitful SSD update order to the mobile. Such a deceitful SSD Update Order could cause the MS to generate a new SSD that is different with the one in the network. To prevent this type of attack, the MS issues a *BS Challenge Order* to the network, which contains another random number called *RANDBS*. The network (i.e., the AC in the network) uses SSD_{new} and *RANDBS* as inputs to its CAVE algorithm to generate the authentication result *AUTHBS* to be sent back to the mobile. The MS then verifies the response *AUTHBS* from the network. If the network's response is the same as the *AUTHBS* calculated by the mobile using SSD_{new} and *RANDBS* as inputs, the mobile will update its SSD with SSD_{new}. Because the A-key is a secret between the user and the network, the attacker would not generate same SSD_{new}. It, therefore, would not produce the same *AUTHBS*. As a result, an attacker will not be able to cause a mobile to update its SSD.

5.4.2   Authentication

Section 5.4.1 described how an SSD is generated and updated. User authentication in IS-41 is based on the SSD and will be explained in this section.

Before IS-41 was introduced, user authentication in 1G systems has a significant weakness. For example, to authenticate a user in AMPS, the user's MIN is used like a username and the user's ESN is used as a password. However, the user's ESN and MIN are sent over the air to the network in order to authenticate the user. This means that an attacker could easily steal a user's MIN and ESN by, for example, scanning the radio, and then clone them to other terminals.

To overcome the deficiency described above, IS-41 uses a new challenge-response technique for user authentication.

To authenticate a user in IS-41, the network issues a *challenge* message to the user. The challenge message contains a random number as that discussed in Figure 5.27. The user should be able to generate a correct *response* based on the shared secret data, which is never transmitted over the air. If the response is incorrect, the user fails the authentication and is denied for network access.

There are two types of challenges in IS-41: *global challenge* and *unique challenge:*

- *Global challenge:* Figure 5.28 illustrates the process of global challenge. A challenge (random number) is generated by the serving system. The challenge is broadcast and updated periodically to all mobile stations using a particular radio control channel. The MS takes the random number along with SSDA, ESN, and MIN as the inputs for the CAVE algorithm. As mentioned in Section 5.4.1, SSD-A is the 64 most-significant-bits in SSD. The authentication result is sent back to the serving system, which relays the authentication result and the random number to the AC. The AC then performs the same calculation by using the CAVE algorithm. It further compares its calculation with the one sent by the MS to either accept or reject the MS.

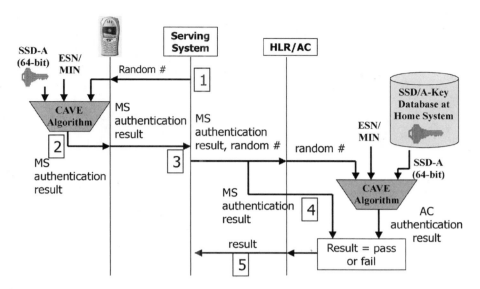

Fig. 5.28 *Global challenge in CAVE algorithm. Reproduced with permission of the McGraw-Hill Companies.*

- *Unique challenge*: The unique challenge is depicted in Figure 5.29. Unlike global challenge, the process is initiated by the home network. The AC directs the serving system to issue a challenge to a *particular* MS, which either is requesting service or is already engaged in a call. Both MS and AC calculate the authentication result using the CAVE algorithm. The authentication results derived by the AC and the MS are sent to the serving system. By verifying the results, the serving system either accepts or rejects the MS.

5.4.3 Privacy

Recall that privacy refers to confidentiality service to prevent eavesdropping. The same CAVE algorithm used for authentication is also utilized for Voice Privacy (VP) and Signaling Message Encryption (SME).

To encrypt voice conversation, a mask referred to as the *Voice Privacy Mask* (*VPMASK*) is generated using the CAVE algorithm with SSD-B to encrypt voice traffic. SSD-B is the 64 least-significant-bits in SSD.

Unlike voice traffic, only certain fields of signaling messages are encrypted. Privacy of signaling messages are protected by a *Signaling Message Encryption Key* (*SMEKEY*). SMEKEY is also generated using the CAVE algorithm with SSD-B. The Cellular Message Encryption Algorithm (CMEA) [36], [84] then adopts SMEKEY to encrypt the signaling messages to be protected.

Unlike the Internet, the core network of IS-41 is accessible only by a limited number of people. In IS-41, therefore, voice privacy and signaling message encryption are employed only between MS and the serving BS.

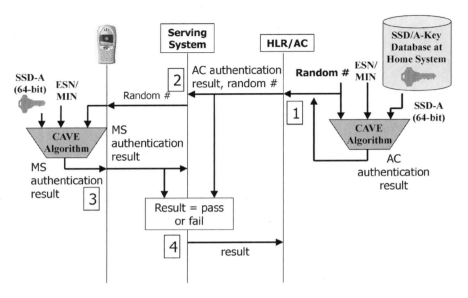

Fig. 5.29 *Unique challenge in CAVE algorithm. Reproduced with permission of the McGraw-Hill Companies.*

To close this section, we point out weaknesses in the security management of IS-41 [68]. First, the distribution of A-keys to mobiles is a critical process. Disclosure of an A-key would make the security techniques worthless. Second, IS-41 uses the same algorithm for both authentication and privacy. Breaking this algorithm means that both authentication and privacy are broken. By decoupling them, the authentication and privacy algorithms can also evolve independently. Third, the authentication process based on the generation and periodic update of SSD incurs additional complexity. Fourth, the 64-bit of SSD-A/SSD-B might not be long enough. Such a short key is vulnerable to brute-force attacks that carry out exhaustive analysis of the key space. Fifth, the security management architecture in IS-41 does not support *mutual authentication*. It allows a network to authenticate a user, but it does not provide sufficient mechanisms for a user to authenticate messages from a network. As mentioned earlier, it is possible for an attacker to deploy a fraud base station to deceive wireless users for secret information. Finally, the current security mechanisms in IS-41 assume that the signaling inside an administrative domain and between two administrative domains is secured. It is also assumed that a mobile's home network can trust each visited network to get and use cryptographic keys to authenticate users.

5.5 SECURITY IN GSM

Security management in GSM emphasizes on authentication and key agreement (AKA) and privacy [39]. As introduced in Section 5.3, AKA provides a methodology to authenticate a user and to generate a new key for the encryption of the user's traffic. As in IS-41, encryption in GSM is only employed over the wireless channels to prevent eavesdropping from the open air space.

Unlike IS-41, GSM uses three algorithms for security management:

- *A3 Algorithm:* A3 is used for authentication.
- *A5 Algorithm:* A5 is a stream cipher algorithm used to encrypt the user traffic.
- *A8 Algorithm:* A8 is used to generate a cipher key.

As shown in Figure 5.30, a user and the network share a 128-bit long secret key called K_i. To authenticate a user, the network sends a challenge, which comprises a 128-bit random number to the user. The user uses its K_i and the random number received from the network as the inputs to the A3 algorithm. The user then sends the output of its A3 algorithm to the network as the user's response to the challenge from the network. The network also uses its K_i and the same random number it sent to the user in the challenge message to its A3 algorithm. The network will then compare the response from the user with the output of its own A3 algorithm to decide whether to accept the connection request or not.

The A8 algorithm in the network takes the same inputs as the A3 algorithm and generates a 64-bit cipher key called K_c. The plaintext, i.e., user traffic, will be encrypted using the A5 algorithm. In addition to the plaintext, the A5 algorithm also takes K_c and a 22-bit counter value as its input. The changes in counter value can be used to prevent replay attacks. Similarly, the user also uses its own A8 algorithm to generate a 64-bit cipher key K_c. The user's A8 algorithm takes the same secret key

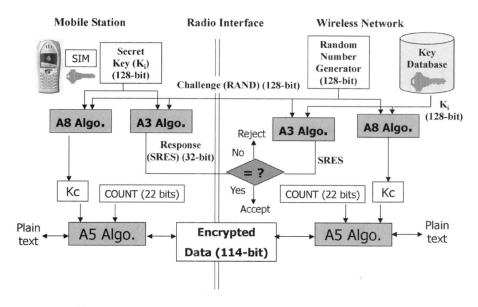

Fig. 5.30 *GSM security algorithms*

K_i and the same random number as used by the network to generate its K_c. Therefore, the K_c generated by the user should be identical to the K_c generated by the network. Therefore, the network is capable of decrypting the encrypted text.

K_c is recomputed whenever a new challenge (random number) is generated and sent to a user by the network. This makes it more difficult for an attacker to crack (determine) the value of K_c. As one can see, the shared secret key of K_i is never transmitted over the wireless channel.

A major weakness of GSM security management is the lack of *mutual authentication* as in IS-41. The GSM security management architecture also assumes that signaling messages inside an administrative domain and between two different administrative domains are secured. It is also assumed that a mobile's home network trusts each visited network to get and use cryptographic keys to authenticate users. Besides, the 64-bit in K_c might not be long enough.

5.6   SECURITY IN GPRS

GPRS is an extension of GSM. The authentication mechanisms used in GPRS are essentially the same as that in GSM. A major difference is in the encryption algorithm used for wireless channel access. GPRS uses the GPRS Encryption Algorithm (GEA) [13] to support confidentiality service. In addition to GSM authentication, GPRS can also utilize UMTS authentication, which will be examined in Section 5.7.3.1.

Figure 5.31 illustrates the GPRS Encryption Algorithm. GEA is a *symmetric stream cipher* algorithm. As depicted in Figure 5.31, it does not depend on plaintext as one of the inputs. Instead, GEA takes the K_c generated by the A8 algorithm along with other

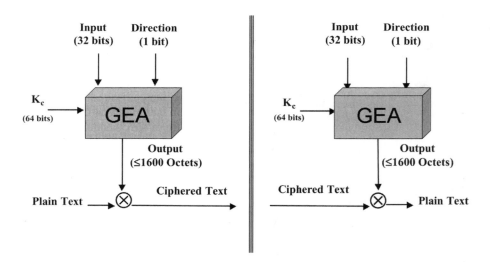

Fig. 5.31   *GPRS encryption algorithm (GEA)*

parameters to generate an output stream, which essentially is a *mask* to conceal the actual content of the plaintext by bit-wise exclusive OR (XOR). The receiver uses the same GEA algorithm and the same inputs to generate the same mask of output stream, which is applied with XOR operation against the ciphered text. Applying the same operation of XOR twice restores the plaintext from the ciphertext.

The major advantage of GEA over the A5 algorithm in GSM is that the output of GEA could be generated before the actual plaintext is known. The encryption is simply a fast bit-operation over the plaintext.

As shown in Figure 5.31, the ciphering for uplink and downlink are independent, which are differentiated by the 1-bit *direction*. The 32-bit *input* depends on the type of Logical Link Control (LLC) frame, which might be *I-frame* for user data or *UI-frame* for signaling and user data. The maximum length of the GEA *output* stream is 1600 bytes, which is the same as the maximum length of LLC payload. The GEA is installed in MS and SGSN. Similar to GSM, encryption is restricted to the communications between MS and SGSN only. Traffic is not protected inside the core network.

5.7 SECURITY IN 3GPP

The section illustrates the security management in 3GPP.

5.7.1 Security Principles

As discussed in Section 5.5, there are some weaknesses in GSM security. The lack of mutual authentication makes it possible to deploy a false base station to conceive users. There is no security in the core network, so cipher keys, authentication data, and signaling messages are transmitted without security protection within networks. Although data are encrypted, integrity check is not provided. The key length might also not be long enough. In addition, GSM security is not flexible enough to be upgraded and improved over time.

As specified in 3GPP TS 33.120 [19], 3GPP security will build on the security management mechanisms used in the second-generation systems, with improvements to address their limitations. 3GPP will not develop brand new security techniques. Instead, security elements within GSM and other second-generation systems that have proved to be robust and necessary will be adopted. It addresses and corrects perceived weaknesses in 2G systems and will offer new security features that will secure new services offered by 3G. Details of 3GPP security requirements and possible threats are discussed in 3GPP TS 21.133 [20].

Before examining the 3GPP security architecture, definitions specified in 3GPP TS 33.102 [25] are itemized. Although some terms have been defined earlier, we provide a complete list of definitions as that in 3GPP TS 33.102 without rewording to prevent loss of original means.

- *Confidentiality*: The property that information is not made available or disclosed to unauthorized individuals, entities, or processes.

- *Data integrity:* The property that data have not been altered in an unauthorized manner.

- *Data origin authentication:* The corroboration that the source of data received is as claimed.

- *Entity authentication:* The provision of assurance of the claimed identity of an entity.

- *Key freshness:* A key is fresh if it can be guaranteed to be new, as opposed to an old key being reused through actions of either an adversary or authorized party.

- *UMTS entity authentication and key agreement:* Entity authentication according to TS 33.102.

- *GSM entity authentication and key agreement:* The entity Authentication and Key Agreement procedure to provide authentication of a SIM to a serving network domain and to generate the key K_c in accordance to the mechanisms specified in GSM 03.20 [14].

- *User:* Within the context of TS 33.102, a user is either a UMTS subscriber or a GSM subscriber or a physical person as defined in TR 21.905 [17].

- *UMTS subscriber:* A Mobile Equipment (ME) with a UMTS IC Card (UICC) inserted and activated USIM-application.

- *GSM subscriber:* A Mobile Equipment with a SIM inserted or a Mobile Equipment with a UICC inserted and activated SIM-application.

- *UMTS security context:* A state that is established between a user and a serving network domain as a result of the execution of UMTS AKA. At both ends, "UMTS security context data" is stored, which consists at least of the UMTS cipher/integrity keys CK and IK and the Key Set Identifier (KSI). One is still in a UMTS security context, if the keys CK/IK are converted into K_c to work with a GSM BSS.

- *GSM security context:* A state that is established between a user and a serving network domain usually as a result of the execution of GSM AKA. At both ends, "GSM security context data" is stored, which consists at least of the GSM cipher key K_c and the Cipher Key Sequence Number (CKSN).

- *Quintet, UMTS authentication vector:* Temporary authentication and key agreement data that enable a VLR/SGSN to engage in UMTS AKA with a particular user. A quintet consists of five elements: (a) a network challenge RAND, (b) an expected user response XRES, (c) a cipher key CK, (d) an integrity key IK, and (e) a network authentication token AUTN.

- *Triplet, GSM authentication vector:* Temporary authentication and key agreement data that enables a VLR/SGSN to engage in GSM AKA with a particular user. A triplet consists of three elements: (a) a network challenge RAND, (b) an expected user response SRES, and (c) a cipher key K_c.

- *Authentication vector:* Either a quintet or a triplet.

- *Temporary authentication data:* Either UMTS or GSM security context data or UMTS or GSM authentication vectors.

- *R98 − :* Refers to a network node or ME that conforms to R97 or R98 specifications.
- *R99 + :* Refers to a network node or ME that conforms to R99 or later specifications.
- *R99 + ME capable of UMTS AKA*: Either a R99 + UMTS only ME, a R99 + GSM/UMTS ME, or a R99 + GSM only ME that does support USIM-ME interface.
- *R99 + ME not capable of UMTS AKA*: A R99 + GSM only ME that does not support USIM-ME interface.

5.7.2 Security Architecture

Figure 5.32 depicts the 3GPP security architecture, which has five security features [25]:

(I) Network access security: The network access security primarily specifies radio-related access link security. It provides users with secure access to 3G services and protects against attacks on wireless links.

(II) Network domain security: The network domain security protects against attacks on wireline networks between different domains. As shown in Figure 5.32, the network domain security provides security services between *Home Environment (HE)* domain and *Serving Network (SN)* domain.

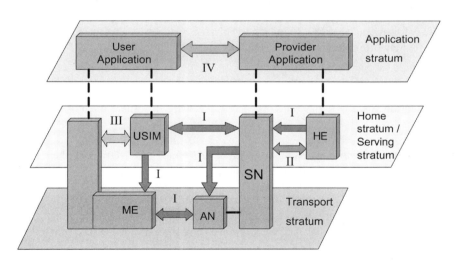

HE: Home environment   SN: Serving Network

AN: Access Network   ME: Mobile Equipment

USIM: Universal Subscriber Identification Module

Fig. 5.32 *3GPP security architecture*

(III)   User domain security: The user domain security is to protect the access to mobile stations.

(IV)   Application domain security: The application domain security provides a set of security features that ensure secure transmissions of user application traffic between user and provider domains.

(V)   Visibility and configurability of security: The visibility and configurability of security (not shown in Figure 5.32) refer to the features that enable a user to be informed whether a security feature is in operation or not, and allow a user to configure the security features.

- *Visibility of security* refers to the ability for the network to indicate to users of the availability of access network encryption and the level of security provided by the network. Upon a call setup, for instance, the user should be informed whether the user data will be encrypted and the level of security provided. This is particularly important when a mobile roams into a network with lower security level than the network it was using, e.g., when roaming from a 3G network to a 2G network.

- *Configurability of security* refers to the capabilities that allow a user to configure the security features so that the user can decide whether to accept or reject nonciphered incoming calls. In addition, the user should also be able to determine whether to set up a call if ciphering is not enabled by the network, and control which ciphering algorithms are acceptable for use.

The following sections focus on (I) network access security (Section 5.7.3) and (II) network domain security (Section 5.7.4).

5.7.3   Network Access Security

This section discusses the network access security specified in 3GPP. We will describe:

- *Authentication and Key Agreement (AKA):* The AKA provides *mutual authentication* for both user and network to authenticate each other. It also generates security keys for other security services. (Section 5.7.3.1)
- *UMTS Encryption Algorithm (UEA):* The UEA provides access link confidentiality (privacy) service. (Section 5.7.3.2)
- *UMTS Integrity Algorithm (UIA):* The UIA provide access link integrity service. (Section 5.7.3.3)

5.7.3.1   Authentication and Key Agreement (AKA)   One of the key objectives of 3GPP AKA [25] is to achieve maximum compatibility with current GSM security mechanism. The main purpose of 3GPP AKA is similar to that in 2G systems. 3GPP AKA, however, provides mutual authentication to allow user and network to authenticate each other. In addition, two keys are generated in 3GPP AKA: CK for encryption and IK for integrity. There is also a shared secret key called K, which functions like the A-key in IS-41 (Section 5.4) and K_i in GSM (Section 5.5). The secret key K is shared by the user and the network and available only to the Authentication Center (AuC) in the user's Home Environment (HE) and the USIM on mobile terminal.

Figure 5.33 depicts a simplified message flow showing how AKA is used by a VLR or SGSN in a visited network, which could collaborate with the HE/HLR in a mobile's home network to authenticate the user. As indicated in the figure, there are two phases: *distribution of authentication vectors* and *authentication and key establishment*. The 3GPP AKA could be used for both the circuit mode and the packet mode in the 3GPP network architecture. Therefore, Figure 5.33 shows that the authentication vectors (AVs) could be sent from HE/HLR to either a SGSN (for packet-switched services) or a VLR (for circuit-switched services) in the visited network. As specified in Section 5.7.1, a quintet of authentication vector (AV) consists of the following components:

1. RAND: a random number used to authenticate user
2. XRES: an expected response from the user

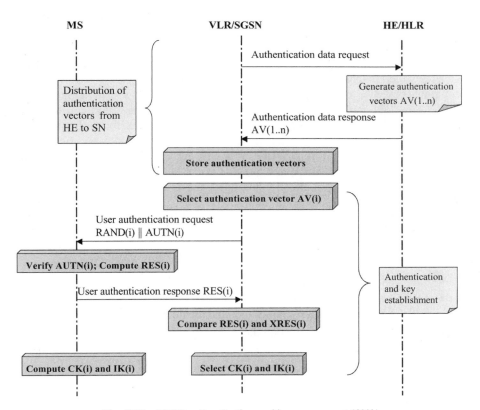

Fig. 5.33 *3GPP authentication and key agreement (AKA)*

3. CK: a cipher key
4. IK: an integrity key
5. AUTN: an authentication token utilized to authenticate the network

Upon receiving an authentication data request from VLR/SGSN, the HE/HLR distributes a set of authentication vectors to the SGSN/VLR. The request from VLR/SGSN includes the mobile's IMSI (International Mobile Subscriber Identity) to specify the AVs for a particular user. The request also indicates whether the mobile is operating in circuit-switching mode or packet-switching mode. The AuC may precompute the AVs or perform the computation on demand. An ordered array of n number of authentication vectors $AV(1, \ldots, n)$ are sent to the VLR/SGSN. The VLR/SGSN will store these authentication vectors such that it does not need to ask the mobile's HE/HLR for authentication data each time when it needs to authenticate the user. The authentication vector is ordered based on sequence number (SQN). Each AV is good for one AKA between the VLR/SGSN and the USIM.

To authenticate a user, the VLR/SGSN retrieves the next available authentication vector $AV(i)$ in the ordered array of $AV(1, \ldots, n)$, where i is the sequence number. The parameters of RAND and AUTN are transmitted to the mobile station. Based on the RAND, the shared secret key K, and other parameters, the mobile station verifies the correctness of AUTN. Because a fraud system would not be able to generate the correct AUTN without the secret key K, the mobile station thus could be able to authenticate the network. If the AUTN can be accepted, the mobile station generates the response (RES), cipher key (CK), and integrity key (IK), which are computed based on the RAND and the shared secret key K. As shown in Figure 5.33, the VLR/SGSN compares the RES with the XRES to authenticate the user. If the user is authenticated, the VLR/SGSN retrieves the CK and IK for further usage.

3GPP allows the VLR/SGSN to offer secured services even though the connection to AuC is unavailable. The CK and IK derived previously can be used by a VLR or SGSN for encryption and integrity check. Because the VLR/SGSN will not issue a new challenge to the mobile station, the CK and IK stored in the mobile station should be same as those in the VLR/SGSN.

Details of AV generation in AuC are illustrated in Figure 5.34. First, both the sequence number (SQN) and the random number (RAND) are generated. Then, five algorithms f1, f2, f3, f4, f5 are used to carry out the following operations:

- *Algorithm f1:* Used to generate the message authentication code (MAC).
- *Algorithm f2:* Used to generate the expected authentication response (XRES) that can be used by VLR/SGSN to authenticate a user.
- *Algorithm f3:* Used to generate the cipher key (CK).
- *Algorithm f4:* Used to generate the integrity key (IK).
- *Algorithm f5:* Used to generate the anonymity key (AK).

The inputs to algorithms f2, f3, f4, and f5 are a 128-bit random number (RAND) and the 128-bit shared secret key of K. Algorithm f2 is similar to the A3 algorithm in

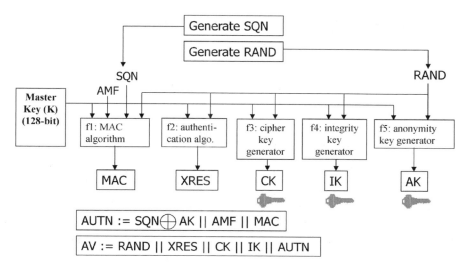

Fig. 5.34 *Generation of authentication vectors*

GSM (see Section 5.5), which calculates the expected response from the user for authentication. Algorithms f3 and f4 generate new keys to be used to encrypt user traffic and ensure data integrity. In particular, algorithm f3 generates a key called *CK* for ciphering and algorithm f4 generates a key called *IK* for ensuring integrity. This differs from GSM where a single key, K_c, generated by the A8 algorithm is used for ciphering only.

The AK generated by algorithm f5 is used to conceal the sequence number SQN because the sequence number may expose the identity and location of the user. If no concealment is needed, f5 could be null and $AK = 0$.

The inputs to algorithm f1 include RAND, K, SQN, and AMF (Authentication Management Field). The AMF can be used to support multiple authentication algorithms and keys, changing sequence number verification parameters, and setting threshold values to restrict the lifetime of cipher and integrity keys. Please refer to [25] for detailed usages of AMF. Algorithm f1 outputs a MAC, which is concatenated with AMF and SQN to form the AUTN. As depicted in Figure 5.34, the SQN is concealed by bit-wise XOR with AK. The AV is then generated based on RAND, XRES, CK, IK, and AUTN. The length of authentication parameters is listed in Table 5.1.

The authentication process in the USIM is depicted in Figure 5.35. As shown in Figure 5.33, both RAND and AUTN are delivered to the mobile station. To verify the AUTN, the mobile first generates the AK by taking the RAND and K as inputs to algorithm f5. The mobile can compute the SQN by employing XOR again with the AK. The SQN_{HE} (or simply S_{HE}) refers to the SQN embedded in the AUTN, which essentially is an individual counter for each user kept in AuC. The SQN_{MS} (or S_{MS}) is the highest sequence number the mobile station has accepted. To ensure the *freshness* of authentication keys, the mobile station compares the SQN_{HE} with

TABLE 5.1 Length of authentication parameters

Parameter	Length
K (authentication key)	128 bits
RAND (random challenge)	128 bits
SQN (sequence number)	48 bits
AK (anonymity key)	48 bits
AMF (authentication management field)	16 bits
MAC (message authentication code)	64 bits
CK (cipher key)	128 bits
IK (integrity key)	128 bits
RES (authentication response)	32−128 bits

SQN_{MS}. If the SQN_{HE} is not greater than the SQN_{MS}, the same AV has been used before. The mobile station thus aborts the authentication process and indicates synchronization failure to the network. The SQN_{HE} along with RAND, K, and AMF are inputs of the f1 algorithm, which produces the expected MAC (XMAC). If the XMAC is different with the MAC embedded in the AUTN, the network fails the authentication and no further process will be performed in the mobile station. Otherwise, the RES generated by the f2 algorithm is sent back to the network. CK and IK are also generated by the f3 algorithm and the f4 algorithm, respectively. The following sections discuss the UMTS Encryption Algorithm (UEA) and the UMTS Integrity Algorithm (UIA), which use CK and IK as one of the inputs.

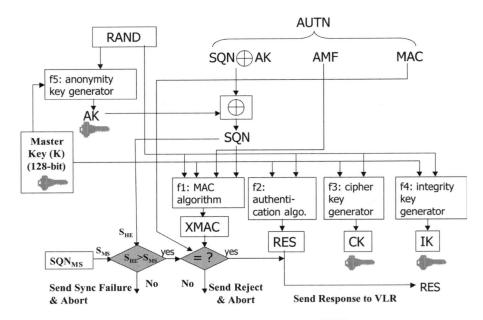

Fig. 5.35 *Authentication process in USIM*

5.7.3.2 UMTS Encryption Algorithm (UEA) The encryption algorithm in 3GPP is referred to as the UMTS Encryption Algorithm (UEA), which provides confidentiality (privacy) service for user traffic and certain signaling messages in dedicated wireless channels [25]. The operation of UEA is similar to the GPRS Encryption Algorithm (GEA) discussed in Section 5.6. Both UEA and GEA do not depend on plaintext as one of the inputs. Instead, a ciphering key and some other parameters are used to derive a keystream. CK is the ciphering key in UEA, and K_c is the ciphering key in GEA. The derived keystream essentially is a mask, which is applied to the plaintext using bit-wise XOR to conceal the actual content. By applying the same operation of XOR again, the ciphertext is decrypted.

Figure 5.36 illustrates the operation of UEA, which is referred to as algorithm f8. The algorithm f8 is used to protect transmissions between mobile station and Radio Network Controller (RNC). As discussed in Section 5.7.3.1, the ciphering key CK is generated and transferred to the VLR/SGSN by AuC as one of the AV quintet. It is further transferred to the RNC by the VLR/SGSN using a secure mode Radio Access Network Application Part (RANAP) message [16]. The encryption is then employed on dedicated channels between the mobile and the RNC.

In addition to the ciphering key CK, Figure 5.36 indicates that the algorithm f8 also takes the following inputs:

- *COUNT-C:* To prevent using the same keystream (mask) for all blocks of the plaintext, COUNT-C, a sequence number, changes with each PDU. The COUNT-C consists of *short sequence number* and *long sequence number*, which represent the least significant bits and the most significant bits of COUNT-C, respectively. Because the algorithm f8 could be executed in either MAC layer or RLC layer, the short sequence number is either the Connection

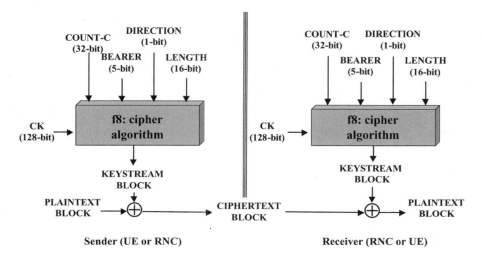

Fig. 5.36 *UMTS encryption algorithm (UEA)*

Frame Number (CFN) in MAC layer or the RLC Sequence Number (RLC SN) in RLC layer. The long sequence number is also called Hyper Frame Number (HFN). It is incremented when the short sequence number wraps around. The HFN is reset to zero when a new CK is generated.

- *BEARER:* Because radio bearers are multiplexed on a single physical layer frame, the parameter of BEARER is used to identify each radio bearer associated with the same user.

- *DIRECTION:* The 1-bit DIRECTION is an input to discriminate uplink and downlink transmission. Its value is 0 for transmissions from UE to RNC and 1 for transmissions from RNC to UE.

  By using BEARER and DIRECTION, different uplink and downlink logical channels will have different keystreams for encryption.

- *LENGTH:* The LENGTH determines the length of the output of the algorithm f8, i.e., the length of the keystream. It determines the length only and will not affect the actual bits in the keystream.

Currently, 3GPP only specifies one f8 algorithm, which is based on the KASUMI algorithm [26], [27]. As described in Section 5.7.2, the mobile should be able to indicate whether encryption is used.

5.7.3.3 UMTS Integrity Algorithm (UIA)

The UMTS Integrity Algorithm (UIA) provides integrity service [25]. Like UEA, UIA is implemented in mobile and RNC. The integrity key IK is transferred to the RNC from VLR/SGSN. The integrity protection is applied at the Radio Resource Control (RRC) layer to protect most signaling messages.

As indicated in Figure 5.37, the message to be protected is calculated with IK and other parameters using the integrity algorithm f9. Similar to that shown in Figure 5.3 in Section 5.1.3.2, an Integrity Message Authentication Code (MAC-I) generated by the algorithm f9 is appended to the original message. The original message along with the MAC-I are transmitted to the destination. With the same integrity key IK and the same input parameters, the destination calculates the expected MAC-I (XMAC-I) and compares the XMAC-I with the MAC-I carried in the received message. Any unauthorized modification thus can be detected if MAC-I and XMAC-I are different. The UIA can also be used to authenticate the origin of the source because only the alleged user should have the same integrity key kept in the network.

In addition to IK and the message to be protected, Figure 5.37 shows that the algorithm f9 also has the following input parameters:

- *COUNT-I:* The purpose of COUNT-I in algorithm f9 is similar to the COUNT-C in algorithm f8. Same as COUNT-C, COUNT-I is also composed of a *short sequence number* and a *long sequence number.* The short sequence number in COUNT-I is the 4 least significant bits that contain the RRC Sequence Number (RRC SN). The long sequence number is the 28 most significant bits that

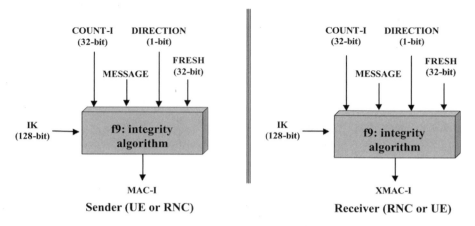

Fig. 5.37 *UMTS integrity algorithm (UIA)*

enclose the RRC Hyper Frame Number (RRC HFN). The long sequence number is incremented when RRC SN wraps around. Because COUNT-I changes for each RRC PDU, it could be used to protect against the replaying of old PDUs.

- *FRESH:* When a connection is set up, RNC generates a random value called FRESH and sends it to the mobile in RRC security mode. There is one FRESH for each user. It is used by both the RNC and the mobile for the duration of a single connection. Therefore the user or an attacker who masquerades as the user would not be able to reuse the old MAC-Is. Thus, they cannot replay the previous sent signaling messages.

- *DIRECTION:* Same as algorithm f8, the DIRECTION is used to distinguish uplink and downlink connections.

The input parameters of algorithm f9 do not include the BEARER parameter. This is because the bearer identity is already embedded in the message to be protected, which is also one of the inputs of the algorithm. Currently, 3GPP only defines one f9 algorithm, which is based on the same KASUMI algorithm adopted in algorithm f8 [26], [27].

5.7.4 Network Domain Security

As mentioned earlier, there is no end-to-end security in 2G systems. Because the core networks in most 2G systems are accessible only by a limited number of people, the *network domain security* is not a major concern. Security protection is employed only in wireless channels. Messages including secret keys are transmitted in cleartext inside the core network.

As the core networks evolve to be IP-based, which is open and easily accessible, protection for core network traffic, especially signaling messages, becomes important. The 3GPP specification TS 33.210 [33] defines the architecture for the UMTS network domain security. More specifically, IPsec is adopted for the network domain security to protect IP-based protocols. In addition to IPsec, MAP Security (MAPsec) [34] has been developed to protect the Mobile Application Part (MAP) messages in Signaling System No. 7 (SS7) networks. IPsec is elaborated in Section 5.2.1. The next section examines the MAPsec.

5.7.4.1 MAP Security (MAPsec)

MAP Security (MAPsec) [34] specifies the protection of the MAP protocol [15] and the security management procedures in SS7 networks. Because MAPsec protects the MAP protocol at the application layer, it is independent of the network- and transport-layer protocols. The operation of MAPsec is similar to IPsec. There are Security Policy Database (SPD) and Security Association Database (SAD) as that in IPsec. The Security Associations (SAs), which specify keys, algorithms, protection profiles, and other parameters, should be established between communicating parties before MAPsec can be employed. The Security Parameter Index (SPI) in the message header is then utilized to identify the established SA stored in the SAD to protect the MAP signaling messages. Like IPsec, the security services provided in MAPsec include confidentiality, integrity, data origin authentication, and anti-replay protection.

Figure 5.38 illustrates the MAPsec functional architecture. The figure depicts two exemplary Public Land Mobile Networks (PLMNs). Each PLMN consists of one Key Administration Center (KAC) and several MAP Network Entities (MAP NEs). For inter-PLMN communications, the KAC negotiates the MAPsec SA on behalf of the MAP NEs in the PLMN. The KAC then distributes the negotiated MAPsec SA along with policy information to all MAP NEs within the same PLMN. The same MAPsec SA should be used by all MAP NEs in the same PLMN for communications with the MAP NEs in other PLMN. When a MAP NE wants to establish a secure connection to another MAP NE located in a different PLMN, it will request a MAPsec SA from the KAC if there is no appropriate MAPsec SA in the NE's local SAD. The KAC responsible for this MAP NE will either provide an existing MAPsec SA or negotiate a new MAPsec SA for the MAP NE. The procedure for intra-PLMN communications is similar to inter-PLMN communications. The MAPsec SA for the communicating MAP NEs is also created by the KAC. To balance traffic load and to improve system reliability, there can be more than one KAC in a PLMN. Each KAC is responsible for a well-defined set of MAP NEs.

There are three interfaces defined in MAPsec as depicted in Figure 5.38. The Z_d-interface is the interface between KACs. It is used to negotiate MAPsec SAs. The Internet Key Exchange (IKE) [48] should be used for Z_d-interface. The Z_e-interface is used to transfer MAPsec SAs and security policy from KAC to MAP NEs in a secure manner. The Z_f-interface is located between MAP NEs. By utilizing the received MAPsec SAs, the MAP NEs protect the MAP operations by the Z_f-interface. As shown in Figure 5.38, Z_d-interface is managed by KAC. Z_f-interface is managed by MAP NE. Both KAC and MAP NE manage Z_e-interface.

MAPsec can operate in three different modes. In the *Protection Mode 0*, there is no security service. The *Protection Mode 1* provides integrity and authentication

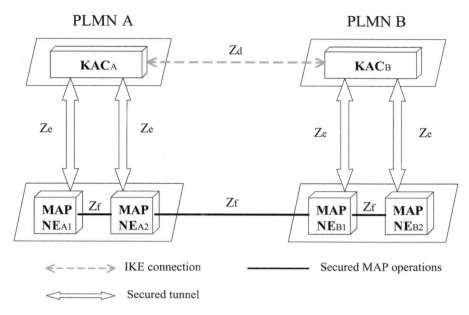

PLMN A   PLMN B

<------ -----> IKE connection   ———— Secured MAP operations

<-------------> Secured tunnel

Fig. 5.38 *MAPsec architecture*

services. In addition to integrity and authentication services, the *Protection Mode 2* supports confidentiality service as well. The encryption algorithm and integrity algorithm in MAPsec are referred to as algorithm f6 and algorithm f7, respectively. 3GPP TS 33.200 [34] specifies that the AES in counter mode with 128-bit key length must be implemented for algorithm f6. For algorithm f7, the AES in CBC MAC mode with a 128-bit key is mandatory.

Figure 5.39 shows the message format of the Protection Mode 1, in which the security header and payload are protected by the message authentication code MAC-M. The MAC-M is 32-bit long. It is generated by the f7 algorithm with the MAP Integrity Key (MIK) specified in the MAPsec SA. The MAPsec SA also specifies which f7 algorithm should be used. Figure 5.40 depicts the message format of the Protection Mode 2 in MAPsec. The payload is first encrypted by employing the f6 algorithm with the MAP Encryption Key (MEK) defined in the MAPsec SA. The ciphertext along with the security header is then protected by the f7 algorithm as that in Protection Mode 1.

5.7.5 Summary

The 3GPP security overcomes many weaknesses in 2G systems. Mutual authentication is supported in 3GPP. The ciphering key and integrity key are extended to 128 bits as well. Both encryption and integrity are employed for link access security. Each AV is utilized only for one AKA. Thus, security keys are always refreshed. In addition, IPsec and MAPsec are used to secure traffic in network domain.

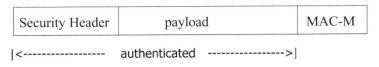

Security Header	payload	MAC-M

|<----------------- authenticated ----------------->|

Fig. 5.39 *Protection mode 1 in MAPsec*

Figure 5.41 summarizes the AKA and the network access security in 3GPP. AVs are transmitted from HLR/AuC to VLR/SGSN. The 3GPP AKA is executed when a user is attached to a UTRAN. The UMTS security context[5] between the user and the serving network domain is then established. Traffic between UTRAN and ME is protected by UEA and UIA. IPsec and MAPsec are adopted to protect traffic inside the core network. Figure 5.42 illustrates the network access security for a UMTS user in a network comprising both UMTS and GSM. The UE may execute either UMTS security or GSM security mechanisms.

Although we have described the major techniques developed for 3GPP security, there are still aspects not discussed in this section. Security in IMS [21], for instance, specifies the security features and mechanisms for security access to the IMS (IP Multimedia Subsystem). Because SIP is chosen as the signaling protocol for creating and terminating multimedia sessions, the IMS security deals with the protection of SIP signaling messages and the authentication between subscribers and the IMS. Although techniques have been developed to encrypt and protect communications, the lawful interception [24], [23], [22] defines architecture and functions to legally intercept telecommunications traffic and other related information. The lawful interception provides means for an authorized person to access sensitive information and monitor other users. It, however, should be compliant with the national or regional laws and technical regulations. The cryptographic algorithms specified by 3GPP are also not detailed. Related information can be found in [18] and [26]–[30].

5.8 SECURITY IN 3GPP2

In 1991, TIA established the Ad Hoc Authentication Group (AHAG) to work on encryption-related specifications in accordance with U.S. and Canadian law. Before the Security Working Group (WG4) in the TSG-S (Technical Specification Group – Service and System Aspects) was organized in August 2001, 3GPP2 relied on the TIA AHAG for security-related standards [71]. The TSG-S WG4 has assumed most of the work previously done by the AHAG to standardize security specifications for 3GPP2. The AHAG continues as a TIA support group. The TSG-S WG4 has specified cryptographic algorithms [7], [11], [9], [8]. In addition, the TSG-S WG4 is now also defining security requirements for the all-IP networks. Specifications are

[5]Please refer to Section 5.7.1 for the definition of *UMTS security context*.

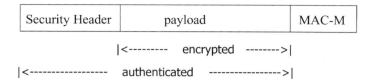

Security Header	payload	MAC-M

```
              |<---------  encrypted  -------->|
|<------------------  authenticated  ----------------->|
```

Fig. 5.40 *Protection mode 2 in MAPsec*

expected to be released in the near future. Next, we will discuss the network access security and network domain security in Sections 5.8.1 and 5.8.2, respectively.

5.8.1   Network Access Security

This section examines the network access security specified in 3GPP2. As in 3GPP, we will describe:

- Authentication and key agreement in Section 5.8.1.1
- Privacy service in Section 5.8.1.2
- Integrity service in Section 5.8.1.3

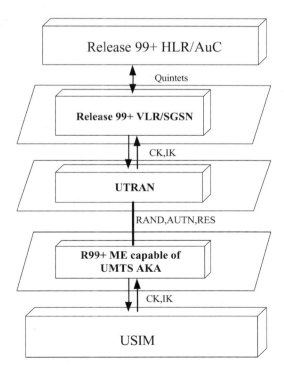

Fig. 5.41 *3GPP network access security*

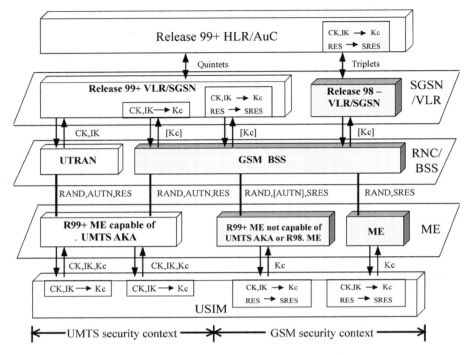

Fig. 5.42   *3GPP/GSM network access security*

5.8.1.1   *Authentication and Key Agreement (AKA)*   Recall that in Chapter 1, the core network and RAN of 3GPP2 evolve from IS-41 and IS-95, respectively. As shown in Figure 1.11 in Chapter 1, the specifications of 3GPP2 C.S001-0005 (also known as IS-2000 [86]) Revisions 0, A, and B specify the operations of cdma2000 1x and 3x. There are two evolutions of cdma2000 1x. The cdma2000 1x EV-DO, specified by IS-856 [85]/3GPP2 C.S0024 [6], defined the enhancement of cdma2000 1x for data only (DO). It is based on the HDR developed by QUALCOMM for direct Internet access. The IS-2000/3GPP2 C.S001-0005 Revision C specifies cdma2000 1x EV-DV, the evolution of cdma2000 1x for both data and voice (DV) enhancement. In addition to the conventional circuit-switching network, the packet-switching network based on IP is also incorporated in cdma2000 1x EV-DV.

Next, we describe the AKA procedures for various revisions of cdma2000.

- *cdma2000 1x and 3x*

  The process for authentication and key agreement in cdma2000 1x and 3x, i.e., 3GPP2 Revision B and earlier, is based on the legacy IS-41 CAVE algorithm. It is explained in Section 5.4.

- *Enhanced Subscriber Authentication (ESA)*

  In S.R0032 [5], 3GPP2 defines the requirements for Enhanced Subscriber Authentication (ESA) to replace the CAVE algorithm. The major requirements include:

  - The *root authentication key*[6] should be known only to the MS and the Authentication Center (AC) in the home network.
  - The ESA should employ 128-bit authentication key and support mutual authentication.
  - The ESA should be backward compatible with CAVE.
  - The ESA should be able to negotiate the cryptographic algorithms.
  - The ESA algorithm should be published openly and commercially available, and it should have been scrutinized thoroughly.

Because the 3GPP AKA discussed in Section 5.7.3.1 has met most of the requirements identified in the ESA, the 3GPP2 Revision C (cdma2000 1x EV-DV) adopts the 3GPP AKA as the basis for 3GPP2 AKA. Algorithms used for 3GPP2 AKA are specified in 3GPP2 S.S0055 [9]. In addition to 3GPP AKA, 3GPP2 Revision C also optionally supports the User Identity Module (UIM) authentication to verify the presence of UIM. In addition to the cryptographic keys of CK, IK, and AK, the 3GPP2 AKA can optionally generate an UIM Authentication Key (UAK), which is a 128-bit key utilized to authenticate the UIM. UAK is generated by the algorithm referred to as algorithm f11 [38]. Algorithm f11 takes the same inputs for generating AK, CK, and IK to generate the UAK. UMAC, a MAC generated by the SHA-1 hash algorithm with the UAK, is then transmitted to the network for authentication. In mobile station, UMAC can only be computed by the UIM. By using the same algorithm and the same inputs, the network calculates the UMAC and compares the result with the one received from the mobile. The presence of UIM thus can be verified.

Same as that in IS-41, the provisioning of the root authentication key (A-key) in 3GPP2 is a major operational concern. Different options are available depending on business models. It might be installed by the manufacturer. It could be programmed by a keypad or a provisioning device by the service representative at the point of sale. In addition, Over-The-Air Service Provisioning (OTASP) [12] utilizes a 512-bit Diffie-Hellman key agreement algorithm to install keys. As that in GSM, a removable UIM that stores subscriber's credentials could also be issued as a token. Please refer to Section 5.2.1.4 for details of the Diffie-Hellman algorithm. The OTASP can also be utilized by the network operator to disable services to a cloned mobile by changing the authentication key of the legitimate subscriber.

---

[6]The root authentication key specified in the ESA is equivalent to the A-key defined in IS-41.

- *cdma2000 1x EV-DO*

  Because the cdma2000 1xEV-DO specified in IS-856/3GPP2 C.S0024 provides IP-based data services, the authentication essentially is based on the Internet model. The mobile first initiates the radio connection without authentication or encryption. Once the initial radio session is established, PPP/LCP [78], [77], [88] session in RAN is set up. The session is authenticated optionally by the PPP Challenge Handshake Authentication Protocol (CHAP) [79] by the visited AAA server. After setting up the PPP/LCP session in the RAN, the PPP/LCP session to the PDSN is established. The PPP/LCP session to the PDSN is authenticated by the home AAA server using either CHAP or Mobile IP authentication. The key exchange in cdma2000 1x EV-DO is also based on the Diffie-Hellman algorithm.

5.8.1.2 Privacy   This section discusses the encryption techniques in cdma2000 to provide privacy service.

- *cdma2000 1x and 3x*

  Privacy in cdma2000 1x and 3x is based on the legacy CAVE algorithm specified in IS-41.

  In addition to the legacy CAVE encryption, the spreading nature of CDMA makes the wireless channels more resistant to eavesdropping. For instance, cdma2000 1x employs *Long Code*, which is a 42-bit Pseudo-Random Noise (PN) sequence, to scramble voice and data traffic. Traffic is scrambled at a rate of 19.2 Kilo-symbols per second (Ksps) for forward link. For reverse link, traffic is scrambled at 1.2288 Mega-chips per second (Mcps).

  To prevent interception of voice traffic, more specifically, a *Private Long Code Mask (PLCM)* derived from SSD-B and the CAVE algorithm is used to change the characteristics of a Long Code in both the mobile and the network. The PLCM replaces the conventional CDMA Long Code to scramble voice traffic. Because the PLCM is known only to the mobile and the network, it provides an extra level of privacy protection to voice traffic in cdma2000 1x systems.

  To protect signaling traffic, the Enhanced Cellular Message Encryption Algorithm (ECMEA) [7] is adopted. The ECMEA is an enhancement to the CMEA used in IS-41 (Section 5.4.3). The CMEA Key (CMEAKEY), generated by the CAVE algorithm, is used as the secret key for ECMEA.

  Data Privacy (DP) is based on the ORYX algorithm [84] and the Data Key (DKEY) to generate the encryption mask to encrypt and decrypt data traffic. Figure 5.43 illustrates the encryption of voice, data, and signaling messages in 3GPP2.

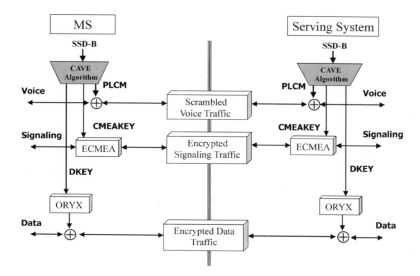

Fig. 5.43 *Privacy in 3GPP2*

- *Enhanced Subscriber Privacy (ESP)*

  In addition to the legacy IS-41 privacy techniques, 3GPP2 S.R0032 [5] specifies the requirements for Enhanced Subscriber Privacy (ESP) to encrypt user and signaling traffic. The primary features of ESP are as follows:
    - Keys for ESP may be based on the root authentication key but should be cryptographically decoupled from the keys used for authentication.
    - The privacy key in MS can be modified under control of the home system.
    - Keys for ESP are changed with each new security association.
    - Privacy keys for each call are established when a mobile is authenticated.
    - Privacy keys for control channel are established after a mobile is authenticated successfully.

  The ESP is applicable to all services, including voice, data, and signaling. The algorithm adopted for enhanced privacy in 3GPP2 Revision B and later is the 128-bit Rijndael AES algorithm [38]. If the legacy CAVE is used for authentication, the encryption key should be formed by repeating the 64-bit CMEAKEY twice. Otherwise the 128-bit key CK generated by 3GPP2 AKA should be used as the encryption key for the Rijndael algorithm.
- cdma2000 1x EV-DO

  As mentioned earlier, in cdma2000 1x EV-DO, the Diffie-Hellman algorithm is used for key exchange when establishing a session. Subsequent RAN-domain messages can be encrypted using the negotiated keys. The encryption protocol based on AES is specified in 3GPP2 C.S0039 [10].

5.8.1.3 Integrity Because security in 3GPP2 Revision B and earlier essentially is based on the legacy CAVE algorithm, they do not provide integrity service. For cdma2000 1x EV-DO, 3GPP2 C.S0024 [6] does not specify the integrity service either. The integrity technique specified in 3GPP2 Revision C is similar to 3GPP UIA discussed in Section 5.7.3.3. Instead of using the KASUMI algorithm, 3GPP2 adopts the SHA-1 to provide integrity service [38].

5.8.2 Network Domain Security

Like 3GPP, techniques described in Sections 5.8.1.2 and 5.8.1.3 apply to air interface only. The network domain security, however, is primarily based on the Internet models. IPsec, for instance, would be needed to provide end-to-end security for packet domain. This section reviews network security–related aspects specified by 3GPP2.

Figure 5.44 illustrates the security architecture for circuit-switching domain. It is essentially based on IS-41. The security architecture for packet-switching domain is depicted in Figure 5.45. It is based on the AAA model developed for IP networks.

Recall that 3GPP2 allows a user to use either Simple IP or Mobile IP to access a 3GPP2 packet data network. To authenticate a mobile station that uses either Simple IP or Mobile IP, a static security association between the mobile station and the mobile's home IP network should be established. For Mobile IP access, the visited network should use Mobile IP *Foreign Agent Challenge* to authenticate and authorize the mobile station. For Simple IP access, the visited network may use the PPP Challenge Handshake Authentication Protocol (CHAP) [79] or the Password Authentication Protocol (PAP) [59] defined for PPP to authenticate and authorize

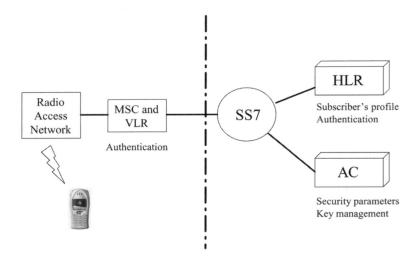

Fig. 5.44 Security architecture for circuit-switching domain. Reproduced with the kind permission of ITU.

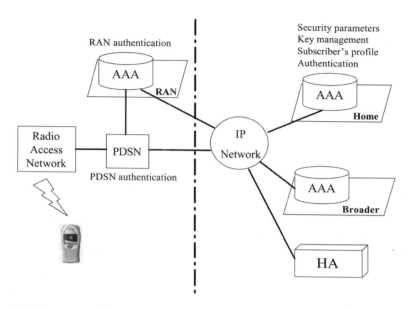

Fig. 5.45 *Security architecture for packet-switching domain. Reproduced with the kind permission of ITU.*

the mobile station. However, if the mobile station does not support CHAP or PAP, there is no IP network authentication.

As discussed in Chapter 2, a user maintains a PPP session with a PDSN in the visited 3GPP2 network in order to send and receive user packets over the visited network. The PDSN is also responsible to perform authentication, authorization, and accounting for the user. For this purpose, the PDSN is connected to an AAA server. Before a PPP session can be established between a PDSN and a user, CHAP and PAP can be used to authenticate the user in the same manner as they are used for conventional PPP session authentication. As a mobile station may not support CHAP and PAP, the PDSN should support a configuration option to allow an MS to receive Simple IP service without CHAP or PAP. The PDSN should also support IPsec and IKE. A security association between the PDSN in a visited network and the mobile's Mobile IP HA may be established using X.509-based certificates. Alternatively, a shared secret for IKE may be statically configured or dynamically provisioned by the mobile's Home AAA server. IPsec Encapsulating Security Payload (ESP) is preferred over IPsec Authentication Header (AH) as the security algorithm. To ensure backward compatibility, IPsec AH should also be implemented. The PDSN should act as a AAA client for the AAA server.

Figure 5.46 further illustrates the relationship between AAA server and other network entities. The AAA server may support both the IP domain and the legacy circuit-switching domain. In addition to providing IP-based authentication, authorization, and accounting, the AAA server also maintains security associations with peer AAA entities to support intra-administrative domain and inter-administrative

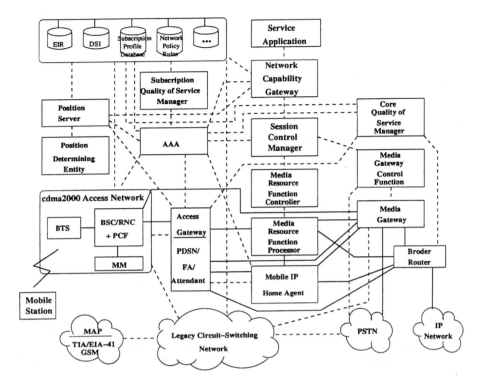

Fig. 5.46 *AAA in 3GPP2 architecture*

domain AAA functions. In some cases, an AAA broker shown in Figure 5.45 should be used to process and forward security information between the service provider network and the home IP network. The AAA messages transmitted between AAA servers may be encrypted. Because AAA brokers may not be reliable, the AAA messages to and from AAA brokers may be further protected with digital signatures and additional encryption using, for example, the public key mechanisms.

The TSG-S WG4 of 3GPP2 is still working on several aspects related to network security such as IP Multimedia Domain (MMD) Security Framework, Packet Data Security Framework, and Broadcast/Multicast Service Security Framework. It is expected that techniques developed in 3GPP will be adopted or modified for 3GPP2.

REFERENCES

1. FIPS 180-1. Secure hash standard (SHS). National Institute of Standards and Technology (NIST), April 1995.
2. FIPS 186-2. Digital signature standard (DSS). National Institute of Standards and Technology (NIST), January 2000.

3. FIPS 197. Advanced encryption standard (AES). National Institute of Standards and Technology (NIST), November 2001.

4. FIPS 198. The keyed-hash message authentication code (HMAC). National Institute of Standards and Technology (NIST), March 2002.

5. 3rd Generation Partnership Project 2 (3GPP2). Enhanced subscriber authentication (ESA) and enhanced subscriber privacy (ESP). 3GPP2 S.R0032, Version 1.0, December 2000.

6. 3rd Generation Partnership Project 2 (3GPP2). cdma2000 high rate packet data air interface specification. 3GPP2 C.S0024-0, Version 4.0, October 2002.

7. 3rd Generation Partnership Project 2 (3GPP2). Common cryptographic algorithms. 3GPP2 S.S0053, Version 1.0, January 2002.

8. 3rd Generation Partnership Project 2 (3GPP2). Common security algorithms. 3GPP2 S.S0078-0, Version 1.0, December 2002.

9. 3rd Generation Partnership Project 2 (3GPP2). Enhanced cryptographic algorithms. 3GPP2 S.S0055, Version 1.0, January 2002.

10. 3rd Generation Partnership Project 2 (3GPP2). Enhanced subscriber privacy for cdma2000 high rate packet data. 3GPP2 C.S0039, Version 1.0, September 2002.

11. 3rd Generation Partnership Project 2 (3GPP2). Interface specification for common cryptographic algorithms. 3GPP2 S.S0054, Version 1.0, January 2002.

12. 3rd Generation Partnership Project 2 (3GPP2). Over-the-air service provisioning of mobile stations in spread spectrum standards. 3GPP2 C.S0016-B, Version 1.0, October 2002.

13. 3rd Generation Partnership Project (3GPP), Digital Cellular Telecommunications system (phase 2+). GPRS ciphering algorithm requirements, release 1999. 3GPP TS 01.61, Version 8.0.0, April 2000.

14. 3rd Generation Partnership Project (3GPP), Digital Cellular Telecommunications system (phase 2+). Security related network functions, release 1999. 3GPP TS 03.20, Version 8.1.0, October 2000.

15. 3rd Generation Partnership Project (3GPP), Technical Specification Group Core Network. Mobile application part (MAP) specification, release 5. 3GPP TS 29.002, Version 5.2.0, June 2002.

16. 3rd Generation Partnership Project (3GPP), Technical Specification Group, Radio Access Network. UTRAN Iu interface RANAP signaling, release 5. 3GPP TS 25.413, Version 5.1.0, June 2002.

17. 3rd Generation Partnership Project (3GPP), Technical Specification Group Services and System Aspect. Vocabulary for 3GPP specifications, release 5. 3GPP TR 21.905, Version 5.5.0, September 2002.

18. 3rd Generation Partnership Project (3GPP), Technical Specification Group Services and System Aspects, 3G security. Cryptographic algorithm requirements, release 4. 3GPP TS 33.105, Version 4.1.0, June 2001.

19. 3rd Generation Partnership Project (3GPP), Technical Specification Group Services and System Aspects, 3G security. Security principles and objectives, release 4. 3GPP TS 33.120, Version 4.0.0, March 2001.

20. 3rd Generation Partnership Project (3GPP), Technical Specification Group Services and System Aspects, 3G Security. Security threats and requirements, release 4. 3GPP TS 21.133, Version 4.1.0, December 2001.

21. 3rd Generation Partnership Project (3GPP), Technical Specification Group Services and System Aspects, 3G security. Access security for IP-based services, release 5. 3GPP TS 33.203, Version 5.3.0, September 2002.

22. 3rd Generation Partnership Project (3GPP), Technical Specification Group Services and System Aspects, 3G security. Handover interface for lawful interception, release 5. 3GPP TS 33.108, Version 5.1.0, September 2002.

23. 3rd Generation Partnership Project (3GPP), Technical Specification Group Services and System Aspects, 3G security. Lawful interception architecture and functions, release 5. 3GPP TS 33.107, Version 5.4.0, September 2002.

24. 3rd Generation Partnership Project (3GPP), Technical Specification Group Services and System Aspects, 3G security. Lawful interception requirements, release 5. 3GPP TS 33.106, Version 5.1.0, September 2002.

25. 3rd Generation Partnership Project (3GPP), Technical Specification Group Services and System Aspects, 3G security. Security architecture, release 5. 3GPP TS 33.102, Version 5.0.0, June 2002.

26. 3rd Generation Partnership Project (3GPP), Technical Specification Group Services and System Aspects, 3G security. Specification of the 3GPP confidentiality and integrity algorithms, document 1: f8 and f9 Specification, release 5. 3GPP TS 35.201, Version 5.0.0, March 2002.

27. 3rd Generation Partnership Project (3GPP), Technical Specification Group Services and System Aspects, 3G security. Specification of the 3GPP confidentiality and integrity algorithms, document 2: KASUMI Specification, release 5. 3GPP TS 35.202, Version 5.0.0, June 2002.

28. 3rd Generation Partnership Project (3GPP), Technical Specification Group Services and System Aspects, 3G security. Specification of the 3GPP confidentiality and integrity algorithms, document 3: implementors' test data, release 5. 3GPP TS 35.203, Version 5.0.0, June 2002.

29. 3rd Generation Partnership Project (3GPP), Technical Specification Group Services and System Aspects, 3G security. Specification of the 3GPP confidentiality and integrity algorithms, document 4: design conformance test data, release 5. 3GPP TS 35.204, Version 5.0.0, June 2002.

30. 3rd Generation Partnership Project (3GPP), Technical Specification Group Services and System Aspects, 3G security, Network Domain Security. Criteria for cryptographic algorithm design process, release 4. 3GPP TR 33.901, Version 4.0.0, September 2001.

31. 3rd Generation Partnership Project (3GPP), Technical Specification Group Services and System Aspects, 3G security, Network Domain Security. General report on the design, specification and evaluation of 3GPP standard confidentiality and integrity algorithms, release 4. 3GPP TR 33.908, Version 4.0.0, September 2001.

32. 3rd Generation Partnership Project (3GPP), Technical Specification Group Services and System Aspects, 3G security, Network Domain Security. Report on the design and evaluation of the MILENAGE algorithm set, deliverable 5: an example algorithm for the 3GPP authentication and key generation functions, release 4. 3GPP TR 33.909, Version 4.0.1, June 2001.

33. 3rd Generation Partnership Project (3GPP), Technical Specification Group Services and System Aspects, 3G security, Network Domain Security. IP network layer security, release 5. 3GPP TS 33.210, Version 5.0.0, June 2002.

34. 3rd Generation Partnership Project (3GPP), Technical Specification Group Services and System Aspects, 3G security, Network Domain Security. MAP application layer security, release 5. 3GPP TS 33.200, Version 5.0.0, March 2002.

35. FIPS 46-3. Data encryption standard (DES). National Institute of Standards and Technology (NIST), October 1999.

36. TIA/EIA IS-54 Appendix A. Dual-mode cellular system: authentication, message encryption, voice privacy mask generation, shared secret data generation, A-key verification, and test data, February 1992.

37. B. Aboba and M. Beadles. The network access identifier. IETF RFC 2486, January 1999.

38. TIA TR45 AHAG. Enhanced cryptographic algorithms, revision B, 2001.

39. D. Brown. Techniques for privacy and authentication in personal communication systems. *IEEE Personal Communications*, pp. 6–10, August 1995.

40. C. Perkins and P. Calhoun. Mobile IP challenge/response extensions. IETF RFC 3012, November 2000.

41. P.R. Calhoun, S. Farrell, and W. Bulley. Diameter CMS security application. IETF Internet Draft, <draft-ietf-aaa-diameter-cms-sec-04.txt>, work in progress, March 2002.

42. P.R. Calhoun, T. Johansson, and C.E. Perkins. Diameter Mobile IP application. IETF Internet Draft, <draft-ietf-aaa-diameter-mobileip-14.txt>, work in progress, April 2003.

43. P.R. Calhoun, J. Loughney, E. Guttman, G. Zorn, and J. Arkko. Diameter base protocol. IETF Internet Draft, <draft-ietf-aaa-diameter-17.txt>, work in progress, December 2002.

44. T. Dierks and C. Allen. The TLS protocol. IETF RFC 2246, January 1999.

45. D. Durham, J. Boyle, R. Cohen, S. Herzog, R. Rajan, and A. Sastry. The COPS (common open policy service) protocol. IETF RFC 2748, January 2000.

46. R. Glenn and S. Kent. The NULL encryption algorithm and its use with IPsec. IETF RFC 2410, November 1998.

47. D. Goodman. *Wireless Personal Communications Systems.* Addison-Wesley Publishing Company, Reading, MA, 1997.

48. D. Harkins and D. Carrel. The Internet key exchange (IKE). IETF RFC 2409, November 1998.

49. TIA/EIA IS-725-A. Cellular radio telecommunications intersystem operations - over-the-air service provisioning (OTASP)&parameter administration (OTAPA), July 1999.

50. ITU-T Rec. X.509. The directory: public-key and attribute certificate frameworks, March 2000.

51. S. Kent and R. Atkinson. IP authentication header. IETF RFC 2402, November 1998.

52. S. Kent and R. Atkinson. IP encapsulating security payload (ESP). IETF RFC 2406, November 1998.

53. S. Kent and R. Atkinson. Security architecture for the Internet protocol. IETF RFC 2401, November 1998.

54. A. Keromytis and N. Provos. The use of HMAC-RIPEMD-160-96 within ESP and AH. IETF RFC 2857, June 2000.

55. H. Krawczyk. SKEME: a versatile secure key exchange mechanism for Internet. In *Proc. of IEEE Symposium on Network and Distributed Systems Security*, February 1996.

56. H. Krawczyk, M. Bellare, and R. Canetti. HMAC: keyed-hashing for message authentication. IETF RFC 2104, February 1997.

57. D. Levi, P. Meyer, and B. Stewart. Simple network management protocol (SNMP) applications. IETF RFC 3413, December 2002.

58. Y.-B. Lin and I. Chlamtac. *Wireless and Mobile Network Architectures.* John Wiley & Sons, Inc., New York, 2001.

59. B. Lloyd and W. Simpson. PPP authentication protocols. IETF RFC 1334, October 1992.

60. C. Madson and N. Doraswamy. The ESPDES-CBC cipher algorithm with explicit IV. IETF RFC 2405, November 1998.

61. C. Madson and R. Glenn. The use of HMAC-MD5-96 within ESP and AH. IETF RFC 2403, November 1998.

62. C. Madson and R. Glenn. The use of HMAC-SHA-1-96 within ESP and AH. IETF RFC 2404, November 1998.

63. D. Maughan, M. Schertler, M. Schneider, and J. Turner. Internet security association and key management protocol (ISAKMP). IETF RFC 2408, November 1998.

64. A.J. Menezes, P.C. van Oorschot, and S.A. Vanstone. *Handbook of applied cryptography.* CRC Press, Boca Raton, FL, 1996.

65. D. Mitton, M. St. Johns, S. Barkley, D. Nelson, B. Patil, M. Stevens, and B. Wolff. Authentication, authorization, and accounting: protocol evaluation. IETF RFC 3127, June 2001.

66. S. Mohan. Privacy and authentication protocols for PCS. *IEEE Personal Communications*, pp. 34–38, October 1996.

67. H. Orman. The OAKLEY key determination protocol. IETF RFC 2412, November 1998.

68. S. Patel. Weakness of North American wireless authentication protocol. *IEEE Personal Communications*, pp. 40–44, June 1997.

69. R. Pereira and R. Adams. The ESP CBC-Mode cipher algorithms. IETF RFC 2451, November 1998.

70. C.E. Perkins and P.R. Calhoun. AAA registration keys for Mobile IP. IETF Internet Draft, <draft-ietf-mobileip-aaa-key-11.txt>, work in progress, March 2003.

71. F. Quick. Security in cdma2000. In *ITU-T Workshop on Security*, Seoul, Korea, May 2002.

72. R. Stewart, Q. Xie, K. Morneault, C. Sharp, H. Schwarzbauer, T. Taylor, I. Rytina, M. Kalla, L. Zhang, and V. Paxson. Stream control transmission protocol. IETF RFC 2960, October 2000.

73. C. Rigney, S. Willens, A. Rubens, and W. Simpson. Remote authentication dial in user service (RADIUS). IETF RFC 2865, June 2000.

74. R. Rivest. The MD5 message-digest algorithm. IETF RFC 1321, April 1992.

75. R.L. Rivest, A. Shamir, and L. Adleman. A method for obtaining digital signatures and public-key cryptosystems. *Communications of the ACM*, 21(2):120–126, February 1978.

76. B. Schneier. *Applied cryptography.* John Wiley & Sons, Inc., New York, 1996.

77. W. Simpson. PPP LCP extensions. IETF RFC 1570, January 1994.

78. W. Simpson. The point-to-point protocol (PPP). IETF RFC 1661, July 1994.

79. W. Simpson. PPP challenge handshake authentication protocol (CHAP). IETF RFC 1994, August 1996.

80. R.A. Snyder and M.D. Gallagher. *Wireless telecommunications networking with ANSI-41.* McGraw-Hill, New York, 2001.

81. W. Stallings. *Cryptography and network security.* Prentice Hall, Englewood Cliggs, NJ, 1999.

82. R. Thayer, N. Doraswamy, and R. Glenn. IP security document roadmap. IETF RFC 2411, November 1998.

83. TIA/EIA-41-D. Cellular radiotelecommunications intersystem operations, December 1997.

84. TIA/EIA TR45.0.A. Common cryptographic algorithms, revision B, June 1995.

85. TIA/EIA TR45.4. cdma2000 high rate packet data air interface specification, November 2000.

86. TIA/EIA TR45.5. CDMA 2000 series, revision A, March 2000.

87. TIA/EIA TR45.3 TSB50. User interface for authentication key entry (R2002), March 1993.

88. G. Zorn. PPP LCP internationalization configuration option. IETF RFC 2484, January 1999.

6

Quality of Service

This chapter first examines techniques to support Internet Quality of Service (QoS), which are adopted as the basis for 3GPP and 3GPP2 QoS management. Architectures and mechanisms specified by 3GPP and 3GPP2 for QoS management are then discussed. The MWIF QoS is not included in this chapter because the MWIF specified only QoS requirements without getting into technical solutions.

6.1 INTERNET QoS

The Internet was originally designed to provide best-effort data delivery without QoS considerations. As the Internet evolves, many applications, especially real-time voice and multimedia applications, need some levels of control over end-to-end packet delays. Network and service providers are also looking for ways to distinguish the levels of services provided to different classes of subscribers (e.g., high paying vs. low paying customers).

Today, architectures for supporting QoS on the Internet can be broadly classified into two groups, which differ in techniques for resource provisioning and the granularity of service differentiation. The Integrated Services (Int-Serv) defined by the IETF seeks to guarantee fine levels of QoS using dynamic resource reservation. In particular, an application must set up an end-to-end path and reserve resources along the path before data transmission. The Resource Reservation Protocol (RSVP) is the signaling protocol used in Int-Serv by the end-user equipment and the network nodes to reserve resources to provide *fine-grain* per-flow QoS guarantees. The IETF

IP-Based Next-Generation Wireless Networks: Systems, Architectures, and Protocols,
By Jyh-Cheng Chen and Tao Zhang. ISBN 0-471-23526-1 © 2004 John Wiley & Sons, Inc.

Int-Serv working group was formed in early 1995. It was a very active group with significant developments in 1997, but it became quiescent since late 1997. Attentions have since been shifted to Differentiated Services (Diff-Serv), which seek to provide a *coarse-grain* QoS support. With Diff-Serv, packets are marked differently to create several packet classes for different treatment in the network. Packets of the same class are aggregated and forwarded together inside a Diff-Serv Domain.

In addition to Int-Serv and Diff-Serv, the IETF has also developed policy-based management, which leverages policy rules to automate and program network reactions. The policy-based QoS management could be utilized to monitor, control, allocate, and enforce use of network resources.

In the rest of this section, we will discuss Int-Serv, Diff-Serv, and policy-based QoS management.

6.1.1 Integrated Services (Int-Serv)

A main objective of Int-Serv is to support end-to-end QoS for a wide variety of IP applications, allowing the applications to specify traffic descriptors and service requirements of arbitrary granularity. Int-Serv seeks to provide such fine-grain QoS support to both unicast and multicast sessions and to accommodate heterogeneous receivers. The term of Integrated Services suggests that the QoS framework would support integrated services, including real-time as well as non-real-time services [19]. It was argued that guarantees cannot be achieved without resource reservations [19]. Therefore, essential to Int-Serv is the Resource Reservation Protocol (RSVP) [20], [45]. RSVP allows a user terminal to explicitly signal the network about the user's QoS requirements and allows the network nodes to signal each other to allocate resources along the traffic path. To achieve guaranteed levels of QoS, user application uses RSVP to set up an end-to-end path and to reserve resources along the path before it starts data transmission. RSVP can be operated in both IPv4 and IPv6.

Figure 6.1 depicts the Int-Serv architecture [20]. To establish a connection, the source application initiates the RSVP process to reserve resources. The RSVP QoS request is passed by the RSVP process to two local decision modules: *Admission Control* and *Policy Control*. The admission control determines whether the node, which may be a host or a router, has sufficient resources to support the requested QoS. Before reserving resources, the RSVP process also needs to consult with the policy control to ensure the user has administrative permission to make the reservation. If both admission control and policy control agree with the reservation setup, the RSVP process sets up parameters in the *Packet Classifier* and *Packet Scheduler* to obtain the desired QoS. As indicated in Figure 6.1, in a router, the RSVP process also needs to consult with the routing process to determine the route(s). The packet classifier sorts the data packet into different QoS classes. The packet scheduler (or other link-layer QoS mechanism) then schedules packets to output links according to the QoS parameters set by the RSVP process. The process

Host   Router

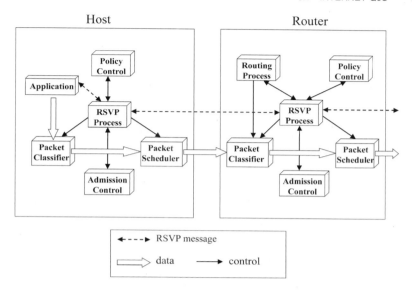

Fig. 6.1 *Int-Serv architecture*

essentially is similar to circuit-switched telephone networks, in which explicit signaling is executed to setup path and resources before user traffic is transmitted.

As shown in Figure 6.1, RSVP plays a central role in the Int-Serv QoS model. Figure 6.2 illustrates the RSVP operation. Before session setup, the sender first sends a *Path* message with QoS requirements to receiver. The Path messages carries the *Traffic Specification (Tspec)*, which specifies QoS parameters such as the average data rate, maximum packet size, and the peak data rate [45]. The Path message traverses intermediate routers and then reaches the receiver, which sends back a *Resv* message along the same path traversed by the Path message. The *Resv* message contains the *Reservation specification (Rspec)* computed based on the Tspec to reserve the resource in intermediate routers. As mentioned earlier, the choice of path is decided by the routing process depicted in Figure 6.1. RSVP simply signals for resource reservation along the path indicated by the routing process.

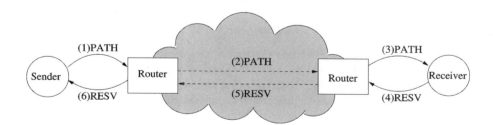

Fig. 6.2 *RSVP operation*

Based on the Rspec, routers on the path from receiver to sender allocate resources for the session. Once resources are reserved, QoS guarantee for a specific session thus is achieved. Each RSVP-enabled router maintains a *soft state* on the resources allocated to each RSVP session. In other words, the resources reserved on a router for a session will expire automatically if the router does not receive any traffic for this session within a pre-determined timeout period. The timer value could be negotiated or preconfigured. With RSVP, desired QoS are requested by applications in real-time. If a router fails to reserve the resources, it sends an error message back to the receiver.

In addition to the two fundamental messages of Path and Resv, there are other messages for maintaining the path and the resource reservation along the path. *ResvErr* and *PathErr*, for instance, are utilized to indicate errors and failures. *ResvTear* and *PathTear* are used to tear down reservation. The purpose of *ResvConf* is to confirm or refresh reservation. All messages are transported over UDP.

RSVP was designed specifically with multicast in mind as that multicast is one of the key requirements in Int-Serv [19]. As illustrated earlier, RSVP is receiver-oriented, in which the receiver computes the desired QoS and signals the intermediate routers to reserve the resources over the path. Therefore, there could be heterogeneous receivers in a multicast group. Different receivers can receive traffic from the same sender with different QoS requirements. Multiple reservations from heterogeneous receivers are merged at common ancestor nodes, which only need to reserve resources according to the most demanding receiver. This could reduce the number of states maintained at the common ancestor nodes.

There are two standardized service definitions based on Int-Serv/RSVP. The *Guaranteed Service* provides strict bounds on end-to-end queuing delays, which makes it possible to provide a service that guarantees both delay and bandwidth [40]. There should also be no queuing loss of packets as long as the traffic is conformed to the Tspec. The guaranteed service is suitable for real-time applications such as audio and video conferences. The *Controlled-load Service* does not explicitly specify target values for QoS parameters such as delay or loss [44]. It only assures minimal queuing loss. Although there is no strict delay guarantees, the delay experienced by most packets would not greatly exceed the minimum end-to-end delay. The controlled load service also employs capacity/admission control to assure QoS when the network is overloaded. Applications that are highly sensitive to overloaded conditions could utilize the controlled load service.

6.1.2 Differentiated Services (Diff-Serv)

Although Int-Serv/RSVP could provide strict and tight quantitative end-to-end QoS guarantees, maintaining states for each flow makes it difficult to scale to a large network. In addition, the RSVP-based resource reservation is complex and needs to be implemented on all immediate routers of each potential traffic path in order to provide Int-Serv. As a result, there has been little incentive to most commercial network providers to support RSVP although it is implemented by major network equipment vendors.

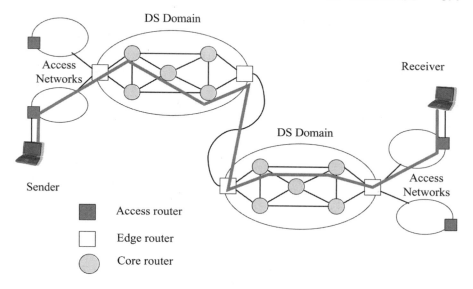

Fig. 6.3 *Diff-Serv architecture*

The Differentiated Services (DS) or Diff-Serv [18], on the other hand, uses a much coarser differentiation model where packets are classified into a relatively small number of classes at the network edge. Traffic states are maintained at the network edge only. It does not maintain any reservation state in intermediate routers.[1] The core routers in the network only need to deal with *aggregation* of traffic to perform mechanisms such as scheduling and buffer management. Diff-Serv therefore is more scalable than Int-Serv/RSVP.

Unlike Int-Serv, Diff-Serv does not mandate the use of any end-to-end signaling protocol and there is no need for cooperative signaling among all nodes for session setup. Figure 6.3 shows the functional architecture of Diff-Serv, which essentially resembles today's Internet service model. In the figure, traffic from a sender to a receiver passes through the access routers in the access networks, edge routers[2] in the edge/border of DS domains, and core routers inside each DS domain. A *DS domain* is a network domain that contains a contiguous set of node complaints with Diff-Serv mechanisms. Generally speaking, one DS domain is administrated by one ISP (Internet Service Provider).

Diff-Serv QoS specifications are carried in a *DS field* of the IP header. For IPv4 networks, the DS field is defined by the 8-bit *TOS (type of service)* in the IP header. For IPv6, the *Traffic Class* octet in the IP header of a packet is utilized as the DS field [38]. In Figure 6.3, the DS field is read by the access router connected to the sender to classify traffic flows into different service classes. The access router may also

---

[1] We simply use routers to represent all networking devices.

[2] Edge router and border router are used interchangeably in this book.

condition traffic according to the Service Level Specification (SLS) [30]. The DS field contains a DS Code Point (DSCP), a tag that specifies the forwarding *Per-Hop Behavior (PHB)* for that packet. A PHB might simply specify dropping precedence. It might also include other performance characteristics. When packets enter the DS domain, they pass through a Diff-Serv edge router. They are then forwarded by Diff-Serv core routers. If a packet is not already classified, the edge router performs classification. The edge router may also reclassify the packet if the classification by access router does not conform to the SLS. The edge router assigns packets to a *Behavior Aggregate (BA)*. A behavior aggregate is a collection of packets with the same DSCP and is associated with a specific PHB. Packets then are subject to the parameters described in a *Traffic Conditioning Specification (TCS)* between their Diff-Serv domain and the customer's access network. The edge router also performs important conditioning functions to keep PHBs *in profile* with the TCS. Conditioning functions include metering, marking, shaping, and dropping/policing. Ingress edge routers, hence, can assure traffic is conformed with the TCS. Inside the DS domain, packets are forwarded aggregately rather than individually according to the PHB. Traffic may pass through multiple DS domains maintained by different service providers as shown in Figure 6.3. The egress edge router of a DS domain may perform traffic conditioning according to a TCS with the next DS domain.

6.1.2.1 *Packet Classification and Conditioning*   Based on the functional model illustrated in Figure 6.3 and the discussion above, traffic *classification* and *conditioning* are the foremost processes in Diff-Serv. The details of traffic classification and conditioning are furthered depicted in Figure 6.4.

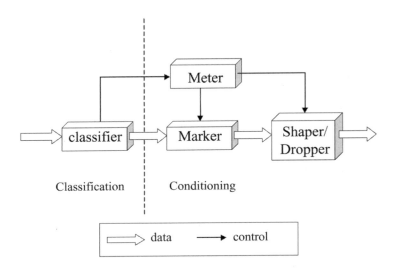

Fig. 6.4 *Packet classification and conditioning*

Packet classifier is *an entity that selects packets based on the content of packet headers according to defined rules* [18]. A packet classifier could be located in access router or ingress edge router. It classifies packets into different service classes based on the contents of the DS field and other fields in the IP headers of the packets, and then forwards them to a traffic conditioner for further processing. Two types of classifiers have been defined: *BA (Behavior Aggregate) Classifier* and *MF (Multi-Field) Classifier* [18]. The BA classifier sorts packets based on the DSCP only. The MF classifier, however, categorizes packets based on DS field and other IP header fields, such as source address, destination address, protocol ID, source port, and destination port.

The traffic conditioner executes control functions to assure that packets are compliant with contracted traffic profile. It measures the traffic load and marks/ remarks packets to be in-profile or out-of-profile. It may also delay or drop packets to enforce traffic characteristics to conform with the contracted profile. A traffic conditioner comprises meter, marker, dropper, and shaper. To truly reflect the operations of meter, marker, dropper, and shaper, the following definitions are excerpted from IETF RFC 2475 [18].

- *Metering* is the process of measuring the temporal properties (e.g., rate) of a traffic stream selected by a classifier. The instantaneous state of this process may be used to affect the operation of a marker, shaper, or dropper, or may be used for accounting and measurement purposes.
- *Meter* is a device that performs metering.
  Traffic meters measure the temporal properties of the stream of packets selected by a classifier against a traffic profile specified in a TCA.[3] A meter passes state information to other conditioning functions to trigger a particular action for each packet, which is either in- or out-of-profile (to some extent).
- *Marking/Premarking/Remarking* is the process of setting the DS codepoint in a packet based on defined rules.
- *Marker* is a device that performs marking.
  Packet markers set the DS field of a packet to a particular codepoint, adding the marked packet to a particular DS behavior aggregate. The marker may be configured to mark all packets that are steered to it to a single codepoint, or may be configured to mark a packet to one of a set of codepoints used to select a PHB in a PHB group, according to the state of a meter. When the marker changes the codepoint in a packet, it is said to have *remarked* the packet.
- *Shaping* is the process of delaying packets within a traffic stream to cause it to conform to some defined traffic profile.
- *Shaper* is a device that performs shaping.

---

[3]TCA stands for Traffic Conditioning Agreement, which has been replaced by TCS (Traffic Conditioning Specification) [30].

Shapers delay some or all of the packets in a traffic stream in order to bring the stream into compliance with a traffic profile. A shaper usually has a finite-size buffer, and packets may be discarded if there is not sufficient buffer space to hold the delayed packets.

- *Dropping* is the process of discarding packets based on specified rules.
- *Dropper* is a device that performs dropping.
  Droppers discard some or all of the packets in a traffic stream in order to bring the stream into compliance with a traffic profile. This process is known as *policing* the stream. Note that a dropper can be implemented as a special case of a shaper by setting the shaper buffer size to zero (or a few) packets.
- *Policing* is the process of discarding packets (by a dropper) within a traffic stream in accordance with the state of a corresponding meter enforcing a traffic profile.

Figure 6.4 illustrates that packets are classified first. Based on the measurement by the meter, packets are marked, shaped, and dropped before entering a DS domain. The implementation of traffic classifier and conditioner is not standardized in Diff-Serv. A pure shaper or other techniques such as *Leaky Bucket* [24], [16] could be adopted.

6.1.2.2 Per-Hop Behavior (PHB)

As stated earlier, within a DS domain, packets are forwarded based on Per-Hop Behavior (PHB). PHBs are the essential building blocks of Diff-Serv. Currently, there are two PHBs standardized by the IETF: Expedited Forwarding (EF) [25], [21] and Assured Forwarding (AF) [31]. EF is a PHB that can provide *premium* or *virtual leased line* service with low loss, low latency, and low jitter services. AF, on the other hand, specifies different levels of forwarding priority among classes.

The EF PHB is defined in [25]. In EF, packets should be served at least by the configured rate over a defined time interval regardless of the traffic load of other service types. To achieve this goal, a proper scheduling algorithm and buffer management mechanism should be designed to keep queue empty or minimize the queue length to within the available buffer space. If queue length is minimized, queuing delay is minimized. Once queuing delay is minimized, packet delay and jitter are minimized. The specification of EF provides a formal definition of the EF PHB. It defines the ideal time a packet should depart from the queue such that packet service rate on a given output interface is greater than the packet arrival rate at that interface. The ideal departure time of a packet simply is the arrival time plus the packet transmission time if the queue is empty. The packet should be scheduled to be transmitted as soon as it arrives at the queue. Otherwise the ideal departure time depends on the ideal departure time of a previous packet if the queue is not empty. The ideal departure time thus is computed iteratively. Please refer to the EF specification [25] for a rigorous mathematical definition of packet departure time. Although the behavior of EF is defined, the implementation, however, is not standardized. Practically, a PHB could be implemented based on particular packet scheduling and buffer management mechanisms. Some exemplary implementations of EF, such as strict priority FIFO queue, WF^{2}Q [17], Deficit Round Robin (DRR) [41], Start-Time Fair Queuing (SFQ) [29], and Self-Clocked Fair Queuing (SCFQ) [28], are discussed in [21]. Although the EF PHB is not mandatory for Diff-Serv architecture, when a DS-compliant component is claimed to implement the EF PHB, it must conform to the specification defined in RFC 3246 [25].

The AF PHB [31] defines four AF classes for packet delivery. Each class must be forwarded independently and different AF classes cannot be aggregated. Each class is configured with a certain amount of forwarding resources such as buffer space and bandwidth. There is no priority implied among classes although the forwarding resources allocated to each AF class might be different. There is also no reordering of packets belonging to the same AF class. Packets within each AF class are marked with three different levels of drop precedence. The recommended values of the AF DSCP are indicated in Table 6.1 [31]. When network is congested, packets are dropped according to the drop precedence. Packets with higher drop precedence are discarded earlier than packets with lower drop precedence. Thus, packet forwarding assurance in AF PHB depends on the forwarding resources allocated to the AF class, traffic load of the AF class, and the drop precedence of each packet. Active queue management such as Random Early Detection (RED) [27] is recommended to avoid abrupt change in dropping behavior. It should respond to long-term congestion by dropping packet, while react to short-term congestion by queuing packets. Like EF, the implementation of AF is not standardized. In addition, the AF PHB is not a requirement for Diff-Serv architecture. However, when a DS-compliant component is claimed to implement the AF PHB, it must conform to the specification defined in RFC 2597 [31].

To conclude Section 6.1.2, we point out that the charter of the IETF Diff-Serv working group is to standardize the semantics of packet marking and the forwarding behavior in DS-compliant nodes. By standardization, all routers would interpret the mapping between packet header and forwarding behavior uniformly such that multivendor equipments would employ same service definition. The standardization of PHB implementation is beyond the scope of the Diff-Serv working group. It is the vendor's choice of which specific algorithms to use to realize the PHBs. There is also no standard for end-to-end services defined by the IETF Diff-Serv standards. Along with traffic provisioning, PHBs are building blocks that vendors could utilize to provide various end-to-end services. Finally, the Diff-Serv working group does

TABLE 6.1 DSCP in AF PHB

	Class 1	Class 2	Class 3	Class 4
Low drop precedence	001010	010010	011010	100010
Medium drop precedence	001100	010100	011100	100100
High drop precedence	001110	010110	011110	100110

not intend to standardize QoS signaling mechanisms although such a signaling protocol has been proposed to extend RSVP for aggregate reservations.

6.1.3 Comparison of Int-Serv and Diff-Serv

The Int-Serv approach utilized RSVP to explicitly signal and dynamically allocate resources at each intermediate router along a traffic path. Int-Serv/RSVP enables each application to specify service requirements and traffic descriptors in arbitrary granularity. It supports deterministic end-to-end performance bounds for individualized services and maintains a separate reservation state at each intermediate router for each session. A serious shortcoming of this approach, however, is that it does not scale well in a large network. Implementations of RSVP have demonstrated that maintaining individual reservation state reduces the processing capability and places a heavy load for a router in a large network.

Int-Serv intends to provide explicit and tight end-to-end quantitative QoS guarantees. It requires all routers along a traffic path to be RSVP capable, i.e., to support RSVP to reserve resources and to guarantee the desired QoS. This makes Int-Serv/RSVP not an easy solution for incremental deployment.

The soft states cause all routers to monitor and update states constantly. It significantly increases the complexity of routers. In addition, to assure strict quantitative QoS bounds for specific applications is difficult. Theoretically, the computational complexity required to maintain bounded delay increases sharply as the number of sessions increases.

RSVP is driven by receivers. This feature makes it properly fit for receiver-based multicast. Heterogeneous receivers could specify different QoS requirements for a dynamic multicast group. However, the delay for setting up a reservation is arbitrary. Both Path and Resv messages must be sent before actual data traffic can be transmitted. The delays of Path and Resv messages, however, are unpredictable due to the nature of the Internet. RSVP also employs a very detailed service definition. The overhead to set up a session thus is also a concern.

Compared to Int-Serv, Diff-Serv is a much simpler and coarser method for providing service differentiation. With Diff-Serv, traffic is classified and conditioned at the network edge, and there is no per-flow state maintained in core routers within DS domains. Traffic is forwarded according to the PHB aggregately inside each DS domain. This leads to a more scalable solution. Diff-Serv envisions fewer traffic classes with coarse differentiation. The performance guarantees are also less stringent and more statistical in nature. In addition, services provided in Diff-Serv are based on long-term and static provisioning. Unlike Int-Serv/RSVP, the contracted QoS specifications in Diff-Serv are more static and could not be negotiated dynamically in real time or on a per-flow basis. With RSVP, applications, however, could initiate QoS reservation dynamically for each individual session. Diff-Serv might be more feasible to deploy than Int-Serv for today's Internet due to its simplicity. A comparison of Int-Serv and Diff-Serv is listed in Table 6.2.

TABLE 6.2 Int-Serv vs. Diff-Serv

Int-Serv	Diff-Serv
Per-flow QoS	Aggregated flows or *classes*
Fine-grain classification	Coarse-grain classification
Tight quantitative end-to-end QoS	Loose qualitative QoS
Short-term reservation	Long-term provisioning
Dynamic	Static

6.1.4 Policy-Based QoS Management

Policy-based management has been investigated extensively. A policy is *a definite goal, course, or method of action to guide and determine present and future decisions* [43]. Policies are typically expressed as a set of rules to administrate, manage, and control access to network resources [43], [36]. Policy rules could be utilized to automate network reaction to protocol events or user requests. Policy-based management is particularly useful for QoS management, e.g., for network managers and service providers to monitor, control, allocate, and enforce use of network resources.

The IETF has developed *Common Open Policy Service (COPS)*-based management to manage network resources [26]. COPS is a query and response protocol that is designed to exchange policy information between a Policy Decision Point (PDP) and Policy Enforcement Points (PEPs). A PDP is *a logical entity that makes policy decisions for itself or for other network elements that request such decisions* [46]. A PEP is *a logical entity that enforces policy decisions* [46]. COPS is based on a client-server model in which PDP is a policy server and PEPs are policy clients. Figure 6.5 illustrates the policy framework defined by the IETF. As depicted in the figure, PDP is a decision-making entity that retrieves and interprets policies stored in a policy repository. The Lightweight Directory Access Protocol (LDAP) [33] is a protocol that PDP could use to access the policy directory. Once policies are retrieved, the PDP detects policy conflicts and determines which policy is relevant. It then answers requests from PEPs using COPS protocol. The PEP then applies actions according to the decision made by the PDP. PEP also measures and audits the policy compliance and reports to the PDP as that PDP may need dynamic data to determine future decision.

COPS is transported over TCP because COPS messages need to be delivered reliably. COPS is a stateful protocol in which policies are enforced at PEP until explicitly revoked by PDP. COPS has built-in message-level security for authentication, replay protection, and message integrity. It can also utilize IPsec infrastructure to secure the message transmission.

COPS is a common protocol to communicate policies from a server to its clients. It is applicable to any generic mechanism rather than just QoS-related policies. The policy representation used in COPS is generic in nature such that the definition of

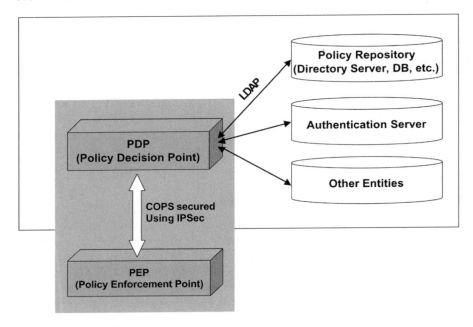

Fig. 6.5 IETF policy framework

policy is applicable to QoS, non-QoS, and even non-networking applications such as backup and auditing access

There are two major operation models in COPS: *outsourcing* model and *provisioning* model. The outsourcing model is triggered by an external event to address instantaneous network resource demands. For instance, when an RSVP message arrives at PEP, the PEP issues a COPS-RSVP request [32] to PDP to request the PDP to make a policy decision. The PDP would need to respond quickly so the end-to-end RSVP signaling would not be delayed excessively. Essentially, the outsourcing mode is *an execution model where a policy enforcement device issues a query to delegate a decision for a specific event to another component, external to it* [43]. The request for policy decision is issued dynamically in real time. This model could be utilized to provide a centralized real-time decision, such as admission control, to network elements.

The provisioning model is *an execution model where network elements are preconfigured, based on policy, prior to processing events* [43]. PEPs could request for policy decision from PDP earlier than the policy event is generated. That is, the PEPs would be preconfigured. A provisioning model could be utilized for near-real-time or non-real-time services to address mid-term or long-term network resource demands. For example, in Diff-Serv, policies that the edge routers need to enforce could be distributed prior to packets arriving. According to the policies, edge routers then classify and condition packets when they arrive. Compared to the outsourcing model, the policy enforcement is more static in the provisioning model.

6.2 QoS CHALLENGES IN WIRELESS IP NETWORKS

A major challenge in providing QoS in wireless IP networks is how to allocate network resources efficiently when users are moving. Take Int-Serv/RSVP, for instance, every change in a mobile's point of attachment requires the generation of new RSVP messages to reserve resources on a new path. Even though some modifications proposed in the literature can restrict such signaling only to the modified segment of the path [42], [23], [34], the approaches generally incur latency in end-to-end resource reallocation. RSVP Path message must be reinitiated at least locally. It could significantly increase the signaling load over the wireless interface. In addition, resources need to be reserved for the new path. The QoS cannot be guaranteed if the same level of resources cannot be allocated.

It is well understood that scalable resource management methods are needed to statistically assure consistent QoS for differentiated classes of mobile users. However, in wireless IP networks, multimedia traffic and a large number of small cells could make many commonly used assumptions such as Poisson handoff, Poisson call arrivals, stationary handoff traffic, and exponential channel and call holding time distributions no longer valid. In addition, large number of users, high degree of user mobility, and widely varying mobile velocities make it difficult to model mobility patterns in real time and to exchange mobility information among neighboring cells. Heterogeneous radio technologies in the same network (e.g., public WLANs inside cellular areas, Bluetooth and IEEE 802.11 in the same WLAN environment) make it even more difficult for different cells to exchange information needed by most existing solutions. New mechanisms such as *Localized Predictive Resource Reservation* [47] have been proposed as a simple yet scalable and flexible resource management technique for wireless IP networks.

Both 3GPP and 3GPP2 have adopted Diff-Serv to provide a QoS guarantee for their IP packet domains, although Int-Serv could be optionally supported as well. Compared with Int-Serv/RSVP, Diff-Serv is more suitable for a mobile environment. Mobile users negotiate with the network for SLS/SLA (Service Level Agreement), which is utilized to condition user traffic. A mobile does not need to signal solely for resource allocation along the new path once it moves. It, however, needs methods for dynamic QoS negotiation to dynamically adjust their QoS requirements because the contracted QoS sometimes may not be honored. If the originally requested QoS level cannot be authorized, it could negotiate a lower QoS level rather then relinquish the session. Dynamic QoS negotiation could also allow networks to utilize their wireless resources more efficiently. 3GPP TR 25.946 [11] and TR 25.851 [12] specify QoS negotiation and renegotiation at the radio link layer. Protocols for dynamic QoS negotiation, such as the Dynamic SLS Negotiation Protocol (DSNP), have also been proposed in the literature for dynamic QoS negotiation in IP layer [22].

The following sections introduce the QoS architectures defined by 3GPP and 3GPP2, respectively. QoS management, QoS classes, QoS attributes, and end-to-end IP QoS support pertinent to the architectures are discussed as well.

6.3 QoS IN 3GPP

This section examines the QoS architecture, QoS management, QoS classes, QoS attributes, and management of end-to-end IP QoS specified by 3GPP.

6.3.1 UMTS QoS Architecture

The QoS architecture of 3GPP is defined in 3GPP TS 23.107 [15]. As the core network evolves to be IP-based, more techniques have been developed to fulfill the QoS requirements. Users, however, are only typically concerned with what levels of QoS they experience. This places an important requirement for end-to-end QoS support. It is difficult to guarantee QoS if some portions of the network do not provide mechanisms to support QoS. For the IP network, users may also require different speeds for uplink and downlink. Therefore, it is also important to support asymmetric applications. Major QoS principles identified in [15] are itemized as follows:

- QoS has to be provided end-to-end.
- The QoS attributes are needed to support asymmetric bearers.
- The number of user-defined and controlled attributes should be as small as possible.
- The derivation and definition of QoS attributes from the application requirements have to be simple.
- It should be able to provide different levels of QoS using UMTS-specific control mechanisms that are not related to QoS mechanisms in the external networks.
- The QoS mechanisms have to allow efficient use of radio capacity and efficient resource utilization.
- It should allow independent evolution of core and access networks.
- The UMTS network should be evolved with minimized impact on the evolution of transport technologies in the wireline networks.
- The UMTS QoS control mechanisms shall be able to efficiently interwork with current QoS schemes.
- The overhead and additional complexity caused by the QoS scheme should be kept reasonably low, so as the amount of state information transmitted and stored in the network.
- The QoS behavior should be dynamic. That is, it should be possible to modify QoS attributes during an active session.

The QoS architecture is depicted in Figure 6.6, in which QoS functions are divided into different layers. Each bearer service provides its QoS services by utilizing the services furnished by lower layer(s). Figure 6.6 indicates that the end-to-end QoS consists of *Terminal Equipment (TE) to Mobile Terminal (MT) Local*

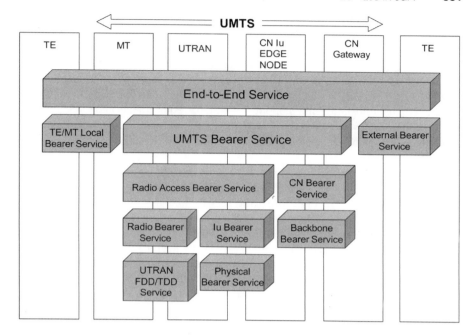

Fig. 6.6 *UMTS QoS architecture*

Bearer Service, *UMTS Bearer Service*, and *External Bearer Service*. Although QoS should be considered end-to-end, the *TE/MT Local Bearer Service* and *External Bearer Service* are outside the control of the UMTS network. They, therefore, are not further elaborated by 3GPP. From the UMTS perspective, the focus of end-to-end QoS is on the *UMTS Bearer Service* only. The design of UMTS bearer service should be independent with external QoS mechanisms. This suggests that the interworking with other bearer services is an important issue to achieve end-to-end QoS from source to destination. Studies have been carried out to investigate the interworking of UMTS QoS with other networks [35], [37], [39].

Figure 6.6 indicates that the UMTS Bearer Service comprises *Radio Access Bearer Service* and *Core Network (CN) Bearer Service*. The Radio Access Bearer Service provides confidential transport of signaling and user data between MT and CN I_u Edge Node.[4] The service is based on the characteristics of the radio interface. The CN Bearer Service connects the UMTS CN I_u Edge Node with the CN Gateway to the external network. It should efficiently control and utilize the backbone network in order to provide the contracted UMTS Bearer Service. The packet core network should support different backbone bearer services for a variety of QoS.

---

[4]I_u is the reference point between radio access network and core network.

6.3.2   UMTS QoS Management

The management of the UMTS Bearer Service includes management functions in *control plane* and *user plane*. The management functions seek to ensure the negotiated QoS between the UMTS Bearer Service and external services, including TE/MT Local Bearer Service and External Bearer Service. The end-to-end QoS is achieved by the translation and mapping of the QoS requirements and QoS attributes between the UMTS Bearer Service and external services. The control-plane and user-plane QoS management functions are illustrated in Figure 6.7 and Figure 6.8, respectively.

There are four major functional blocks in control plane, namely, *Service Manager, Translation Function, Admission/Capability Control*, and *Subscription Control*. As illustrated in Figure 6.7, the service manager is further classified as *UMTS Bearer Service (BS) Manager, Radio Access Bearer (RAB) Manager, Local Bearer Service (BS) Manager, Radio Bearer Service (BS) Manager*, etc. according to its functionality. To establish or modify a UMTS bearer service, the Translation Functions in the MT and the Gateway signal or negotiate with external bearer services. The service primitives and QoS attributes are converted between the UMTS Bearer Service and the external bearer services. The Translation Functions further signals/negotiates with the UMTS BS Managers in MT, CN Edge, and Gateway as depicted in Figure 6.7. Each UMTS BS Manager consults with its associated Admission/Capability Control to decide whether the requested services and desired resources are available and can be granted. The UMTS BS Manager in CN Edge also consults with the Subscription Control to check the administrative privileges for the requested services. Once all checks are positive, a UMTS bearer service could be established/modified. As shown in Figure 6.7, each UMTS BS Manager requests services from lower layers and translates its service attributes to

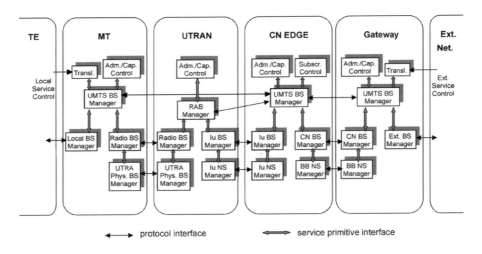

Fig. 6.7   *UMTS QoS management in control plane*

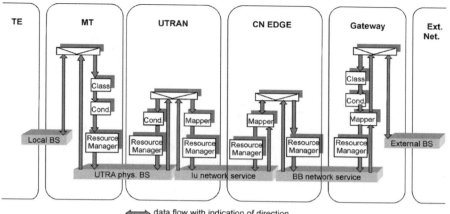

Fig. 6.8 *UMTS QoS management in user plane*

lower layers. For instance, the UMTS BS Manager in MT requests services from the Local BS Manager and the Radio BS Manager. The one in Gateway asks services from the CN BS Manager and the External BS Manager. In addition to the I_u BS manager and the CN Manager in the CN Edge, the UMTS BS Manager in CN Edge translates QoS attributes and requests services from the RAB Manager in UTRAN as well. The RAB Manager in UTRAN verifies with its associated Admission/ Capability Control to determine whether the requested services are supported and the desired resources are available.

The user plane ensures that the user data transmitted in UMTS Bearer Service conforms to the traffic characteristics and service attributes defined by the control plane. As illustrated in Figure 6.8, there are four major components, including *mapper, classifier, conditioner*, and *resource manager*, in user plane. Before entering the domain of UMTS Bearer Service, traffic is classified and conditioned in the MT and the Gateway. The functionalities of classification and conditioning essentially are similar to that described in Section 6.1.2.1 for Diff-Serv. Based on packet header or traffic characteristics, data are classified into different UMTS bearer services. They are then conditioned to ensure conformance with the negotiated QoS. For downlink traffic to MT, there is also a traffic conditioner in UTRAN. This is because the conditioner in the Gateway is for conditioning traffic that enters the core network from external networks. The output traffic from this conditioner in the gateway may not conform with the QoS attributes specified for downlink traffic in the UTRAN. As in Section 6.1.2.1, packets may be shaped or dropped. The algorithm for traffic conditioning is not specified by the standard. The mapper marks data in order to receive the intended QoS. The resource manager is responsible for managing and distributing resources according to the QoS requirements. Techniques performed by resource manager include scheduling, bandwidth management, and power control for the radio bearer.

6.3.3 UMTS QoS Classes

Data services over 2G cellular systems are typically transported over circuit-switched networks. Because of the nature of fixed data rate and continuous transmission in circuit-switching, traffic is easy to characterize. For future packet data and multimedia services, it is a challenge to classify traffic. 3GPP aims to define a simple yet effective way to categorize QoS classes.[5] 3GPP uses two major classes: real time and non-real time. According to delay sensitivity, real-time traffic is further classified as *Conversational Class* and *Streaming Class*. Also based on delay sensitivity, non-real-time traffic is categorized as *Interactive Class* and *Background Class*. It is clear that real-time traffic is more delay sensitive than is non-real-time traffic. Among these four classes, therefore, conversational class is most sensitive to delay, followed by streaming class, interactive class, and then background class.

For the conversational class, the time relation[6] between each information entity (packet, for example) of a packet stream needs to be preserved. The conversational class also requires low delay. The conversational class could be applied between peers or groups of end users. Examples of conversational-class applications include telephony speech, voice over IP, and video conferencing.

The streaming class is for one-way transport from a server to one or multiple destinations. Watching real-time video and listening to real-time audio are classified as this class. Because traffic could be synchronized by buffer management at the destination, the delay requirement is not as strict as that for the conversational class. It, however, still needs to preserve time relation between information entities of the stream.

The interactive class generally maintains a request-response pattern. For instance, an end-user, which is either a machine or a human, is online requesting data transfer from a remote server. A human may perform web browsing or database retrieval from a server. A machine, on the other hand, may poll for measurement records or perform automatic database inquiries. The fundamental characteristic of this class is to preserve payload content to ensure data correctness.

Same as the interactive class, the fundamental characteristic of background class is also to preserve payload content to ensure data correctness. Unlike the interactive class, the source does not expect the response within a tight time bound. A classic example of this traffic type is that an end-user, typically a computer, sends and receives data files in background. Typical examples include sending and receiving e-mail, sending and receiving SMS (short message service), downloading databases, and receiving measurement records. Table 6.3 summarizes the UMTS QoS classes.

6.3.4 QoS Attributes (QoS Profile)

This section briefly introduces the QoS attributes associated with the UMTS Bearer Service, Radio Access Bearer (RAB) Service, and Core Network (CN) Bearer

---

[5]In 3GPP specifications, QoS class is also referred to as traffic class.

[6]Time relation is also referred to as time variation or jitter.

TABLE 6.3   UMTS QoS classes

Traffic Class	Characteristics	Examples
Conversational	Preserve time relation between each information entity of the stream; Stringent and low delay	Voice; Video conferencing
Streaming	Preserve time relation between each information entity of the stream	Streaming video; Streaming audio
Interactive	Request response pattern; Preserve payload content	Web browsing; Database retrieval
Background	Not expecting response within a certain time; Preserve payload content	E-mail; SMS (Short Message Service)

Service. The bearer services discussed in this section are indicated in shaded areas in Figure 6.9.

Table 6.4 summarizes the QoS attributes defined for UMTS Bearer Service [15]. With these attributes, services provided by UMTS Bearer Service are explicitly specified. In Table 6.4, the *maximum bit rate* and *guaranteed bit rate* define the maximum and guaranteed bit rates as the names imply. The *delivery order* specifies whether the SDU (Service Data Unit) should be delivered in order. The *maximum SDU size* is the maximum allowable size of SDUs. The *SDU format information*

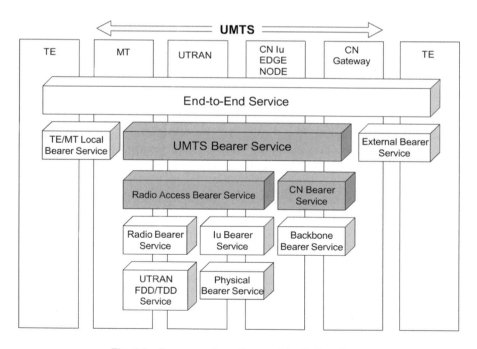

Fig. 6.9   *Bearer services discussed for QoS attributes*

TABLE 6.4   QoS attributes in UMTS bearer service

Traffic class	Conversational class	Streaming class	Interactive class	Background class
Maximum bit rate	x	x	x	x
Delivery order	x	x	x	x
Maximum SDU size	x	x	x	x
SDU format information	x	x		
SDU error ratio	x	x	x	x
Residual bit error ratio	x	x	x	x
Delivery of erroneous SDUs	x	x	x	x
Transfer delay	x	x		
Guaranteed bit rate	x	x		
Traffic handing priority			x	
Allocation/Retention priority	x	x	x	x
Source statistics descriptor	x	x		

indicates the possible actual sizes of SDUs, which might be useful for RLC operation in UTRAN. The *SDU error ratio* is the fraction of lost or detected erroneous SDUs. The *residual bit error ratio (BER)* expresses the undetected bit error ratio of a delivered SDU. The *delivery of erroneous SDUs* points out whether the detected erroneous SDU should be transmitted. The *transfer delay* is the maximum delay of 95th percentile of the delay distribution of all delivered SDUs. The *traffic handling priority* specifies the priority for SDUs, whereas the *allocation/retention priority* indicates the priority for allocation and retention of the UMTS bearer. The *source statistics descriptor* shows the traffic characteristics of SDUs. Studies have shown that speech holds a discontinuous behavior, in which there are talking and silent periods. By specifying the source characteristics, it helps the system in making a decision for admission control to achieve statistical multiplex gain. Please note the discussion of UMTS Bearer Service attributes is still going on and not finalized yet. The possible values of the attributes in UMTS bearer service are listed in Table 6.5.

The QoS attributes for Radio Access Bearer Service are summarized in Table 6.6, and the value ranges are listed in Table 6.7. They are similar to attributes in the UMTS Bearer Service. To map QoS attributes from UMTS Bearer Service to Radio Access Bearer Service, most of the attributes essentially are the same. The differences are in SDU error ratio, residual BER, and transfer delay. As illustrated in Figure 6.9, the UMTS Bearer Service comprises Radio Access Bearer Service and CN Bearer Service. Therefore, the transfer delay in the Radio Access Bearer Service should be smaller than the transfer delay in UMTS Bearer Service because extra delay will incur in core network for the UMTS Bearer Service. Similarly, the UMTS Bearer Service should accommodate larger error rates for SDU error ratio and residual BER than should the Radio Access Bearer Service.

TABLE 6.5 Values of UMTS bearer service attributes

Traffic class	Conversational class	Streaming class	Interactive class	Background class
Maximum bit rate (Kbps)	<2048	<2048	<2048-overhead	<2048-overhead
Delivery order	Yes/No	Yes/No	Yes/No	Yes/No
Maximum SDU size (octets)	≤1500 *or* 1502	≤1500 *or* 1502	≤1500 *or* 1502	≤1500 *or* 1502
Delivery of erroneous SDUs	Yes/No/−	Yes/No/−	Yes/No/−	Yes/No/−
Residual BER	$5*10^{-2}$, 10^{-2} $5*10^{-3}$, 10^{-3} 10^{-4}, 10^{-5}, 10^{-6}	$5*10^{-2}$, 10^{-2} $5*10^{-3}$, 10^{-3} 10^{-4}, 10^{-5}, 10^{-6}	$4*10^{-3}$ 10^{-5} $6*10^{-8}$	$4*10^{-3}$ 10^{-5} $6*10^{-8}$
SDU error ratio	10^{-2}, $7*10^{-3}$ 10^{-3}, 10^{-4} 10^{-5}	10^{-1}, 10^{-2} $7*10^{-3}$, 10^{-3} 10^{-4}, 10^{-5}	10^{-3} 10^{-4} 10^{-6}	10^{-3} 10^{-4} 10^{-6}
Transfer delay (ms)	100 − maximum value	280 − maximum value		
Guaranteed bit rate (Kbps)	<2048	<2048		
Traffic handing priority			1,2,3	
Allocation/ Retention priority	1,2,3	1,2,3	1,2,3	1,2,3
Source statistics descriptor	Speech/ unknown	Speech/ unknown		

TABLE 6.6 QoS attributes in RAB (radio access bearer) service

Traffic class	Conversational class	Streaming class	Interactive class	Background class
Maximum bit rate	x	x	x	x
Delivery order	x	x	x	x
Maximum SDU size	x	x	x	x
SDU format information	x	x		
SDU error ratio	x	x	x	x
Residual bit error ratio	x	x	x	x
Delivery of erroneous SDUs	x	x	x	x
Transfer delay	x	x		
Guaranteed bit rate	x	x		
Traffic handing priority			x	
Allocation/Retention priority	x	x	x	x
Source statistics descriptor	x	x		

TABLE 6.7   Values of RAB (radio access bearer) service attributes

Traffic class	Conversational class	Streaming class	Interactive class	Background class
Maximum bit rate (Kbps)	<2048	<2048	<2048-overhead	<2048-overhead
Delivery order	Yes/No	Yes/No	Yes/No	Yes/No
Maximum SDU size (octets)	$\leq$1500 *or* 1502	$\leq$1500 *or* 1502	$\leq$1500 *or* 1502	$\leq$1500 *or* 1502
Delivery of erroneous SDUs	Yes/No/–	Yes/No/–	Yes/No/–	Yes/No/–
Residual BER	$5*10^{-2}, 10^{-2}$ $5*10^{-3}, 10^{-3}$ $10^{-4}, 10^{-5}, 10^{-6}$	$5*10^{-2}, 10^{-2}$ $5*10^{-3}, 10^{-3}$ $10^{-4}, 10^{-5}, 10^{-6}$	$4*10^{-3}$ 10^{-5} $6*10^{-8}$	$4*10^{-3}$ 10^{-5} $6*10^{-8}$
SDU error ratio	$10^{-2}, 7*10^{-3}$ $10^{-3}, 10^{-4}$ 10^{-5}	$10^{-1}, 10^{-2}$ $7*10^{-3}, 10^{-3}$ $10^{-4}, 10^{-5}$	10^{-3} 10^{-4} 10^{-6}	10^{-3} 10^{-4} 10^{-6}
Transfer delay (ms)	80 – maximum value	250 – maximum value		
Guaranteed bit rate (Kbps)	<2048	<2048		
Traffic handing priority			1,2,3	
Allocation/ Retention priority	1,2,3	1,2,3	1,2,3	1,2,3
Source statistics descriptor	Speech/ unknown	Speech/ unknown		

Unlike UMTS Bearer Service and Radio Access Bearer Service, the QoS attributes for the CN Bearer Service are not explicitly specified in 3GPP. An operator could either choose the QoS capabilities in ATM (Asynchronous Transfer Mode) or adopt the QoS capabilities in IP for its core network. However, if IP is adopted as the backbone, the Diff-Serv defined by the IETF discussed in Section 6.1.2 shall be used. The mapping from UMTS QoS classes to Diff-Serv codepoints is controlled by the operator and is not standardized. The interoperability between operators is based on Diff-Serv architecture as well. It is achieved by the SLA between operators.

6.3.5   Management of End-to-End IP QoS

Sections 6.3.1–6.3.4 discuss the QoS pertinent to UMTS. Assuming the external network is based on IP, this section discusses the management and interaction between the UMTS Bearer Service and the External Bearer Service to provide end-to-end IP QoS.

Figure 6.10 depicts the control plane of the management function to provide end-to-end IP QoS. Compared with Figure 6.7, there are two extra components in

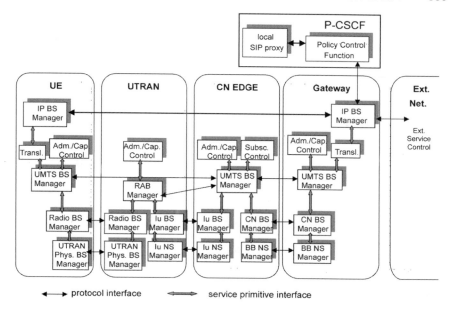

Fig. 6.10 *Control plane for end-to-end IP QoS management*

Figure 6.10: IP BS (Bearer Service) Manager and P-CSCF (Proxy Call State Control Function).

An IP BS Manager controls the external IP bearer service. It utilizes standard IP mechanisms to manage the IP bearer services. The mechanisms may be different than mechanisms used within UMTS and may use different parameters to control services. To interact with UMTS Bearer Service, the IP BS Manager leverages the Translation Function to map the mechanisms and parameters used within the IP bearer service to those used within the UMTS bearer service. In Figure 6.10, there are two IP BS Managers: one in the UE (User Equipment) and one in the Gateway. In realization, the Gateway might be a GGSN. In the rest of this section, we simply use GGSN to represent the Gateway depicted in Figure 6.10. The IP BS Managers in the UE and the GGSN could communicate with each other using relevant signaling protocols. The IP BS Manager may support Int-Serv/RSVP or Diff-Serv edge function. As specified in 3GPP TS 23.207 [13], the Diff-Serv edge function is required for the IP BS Manager in GGSN, and it is optional for the IP BS Manager in UE. For Int-Serv/RSVP, it is optional for both UE and GGSN. Same as Diff-Serv, the PEP function defined in IP policy framework is optional for UE and is mandatory for GGSN. Table 6.8 summarizes the capability of IP BS Manager in UE and GGSN [13].

As discussed in Chapter 3, the P-CSCF is a mobile's first contact point for IP multimedia sessions. It essentially is a local SIP server. Figure 6.10 shows that the P-CSCF also includes a Policy Control Function (PCF). The PCF coordinates the

TABLE 6.8   Capability of IP BS managers in UE and GGSN

Capability	UE	GGSN
Diff-Serv Edge Function	Optional	Required
Int-Serv/RSVP	Optional	Optional
IP Policy Enforcement Point	Optional	Required

applications with the resource management in IP layer. It is a logical entity for policy decision, which conforms to the policy framework defined by the IETF (see Section 6.1.4). By adopting the terminology in IP policy framework, the PCF in Figure 6.10 effectively is a PDP (Policy Decision Point), whereas the IP BS Manager in GGSN is a PEP (Policy Enforcement Point). The interface between PCF and GGSN is identified as G_o interface [10]. It supports the transfer of information and policy decisions between the PCF and the IP BS Manager in the GGSN. The PCF is a logical element that could be implemented separately with the P-CSCF as well. The interface between PCF and P-CSCF, however, is not standardized.

The PCF, utilizes standard IP mechanisms to implement *Service Based Local Policy (SBLP)* in the IP bearer layer. As the name suggests, the SBLP is a local policy utilized by the PCF for making policy decisions. To interwork with external networks, the resource requirements must be explicitly negotiated and provisioned through network management entities. Results then are enforced in edge routers, including GGSN. The GGSN should be able to condition traffic as a Diff-Serv edge router. Resources are allocated along the path based on agreements between the network operators. The edge routers along the path are provisioned with the characteristics of the aggregated traffic that is allowed to pass between different networks.

The policy enforcement of SBLP is defined as a *gate* implemented in GGSN. Functions performed by a gate include packet classifier and traffic meter. When a gate is disabled/closed, all packets in the user plane are dropped. If it is enabled/open, user data are classified by the classifier and measured by the meter. Under the control of PCF, the gate then performs Diff-Serv edge treatment to control and manage media flows based on the defined policy. The Diff-Serv edge function implemented in GGSN should be compliant with the IETF specifications. The PCF effectively authorizes QoS resources and provides final policy decisions to control the allocated QoS resources for the authorized traffic flows. In the middle of a session, the PCF should be able to decide whether new QoS authorization is needed due to changes in media or codec. At the end of a session, PCF should revoke the QoS resource to release the session.

The PCF and the GGSN exchange information over the G_o interface. The functional requirements of the G_o interface specified in [13] include control of Diff-Serv interworking, control of service-based policy *gating* function in GGSN, UMTS bearer authorization, and charging correlation-related function. The G_o interface should conform to the IETF COPS framework. Before demonstrating exemplary signaling flows for end-to-end IP QoS, some messages in G_o interface are itemized as follows:

- **Request (REQ):** A message from PEP to PCF to request SBLP policy information for IP flows.
- **Decision (DEC):** A message containing decision objects sent from PCF to PEP.
- **Report State (RPT):** A message including acknowledgment or error response sent from PEP to PCF.
- **Delete Request State (DRQ):** A message from PEP to PCF to indicate that the request state is no longer available/relevant at the PEP. The corresponding state, therefore, may be removed at the PCF. The DRQ message includes the reason why the request state was deleted.

Figure 6.11 demonstrates the QoS resource authorization for IP bearer service [13], [9]. Because SIP has been adopted by 3GPP as the signaling protocol for packet domain, the QoS authorization process is triggered when receiving a SIP message. As discussed in Chapter 3, the payload of a SIP INVITE usually contains SDP, which specifies the type of media, codec, sampling rate, etc. In addition to IP address and port number, the PCF identifies the connection information such as media and bandwidth requirements for a downlink connection. The PCF then relays the SDP message to the destining UE. Once the SDP from destining UE is received, the PCF identifies the uplink connection information. It also authorizes the requested QoS resources and enforces the IP bearer policy. The SDP message is then forwarded to the originating UE.

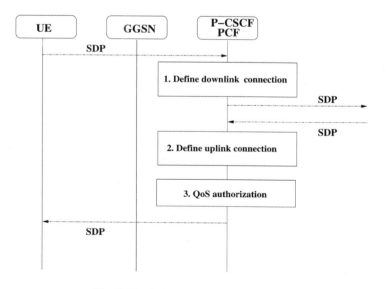

Fig. 6.11   *Authorization of QoS resource*

Figure 6.12 illustrates the resource reservation with SBLP. The SDP parameters are first mapped into UMTS QoS attributes in the UE, which requests a PDP context by sending an *Active PDP Context Request* message to the SGSN. The SGSN sends out a *Create PDP Context Request* to the GGSN as that defined in [14]. The GGSN, the policy enforcement point, sends a *COPS REQ* to the PCF, the policy decision point, through the G_o interface. Once the request is authorized, a *COPS DEC*, which includes the policy information, is sent back to the GGSN. The GGSN then enforces the policy decision and responds with a *COPS RPT* to the PCF to acknowledge the decision or to report errors. A *Create PDP Context Response* is also returned to the SGSN if the results of policy enforcement are positive. Finally, the SGSN returns an *Active PDP Context Accept* to the UE.

Figure 6.13 depicts the approval of committed QoS to enable the allocated QoS resource, which has been authorized and reserved by procedures shown in Figures 6.11 and 6.12. Please refer to Figure 3.15 in Chapter 3 for an exemplary flow combing QoS resource authorization, resource reservation, and approval of committed QoS. As shown in Figure 6.13, once the PCF receives the SIP OK from the destination UE, it approves the committed QoS and initiates a COPS DEC message to the GGSN to enable the gate. The GGSN opens the gate; therefore, the previous authorization resource is enabled. An acknowledgment of COPS RPT is then sent back to the PCF, which further relays the SIP OK to the source UE.

Figure 6.14 shows the authorization of PDP context modification. Figure 6.15 and Figure 6.16 depict the release of PDP context initiated by the SGSN and GGSN, respectively. Although detailed flows are different, these figures exhibit a behavior

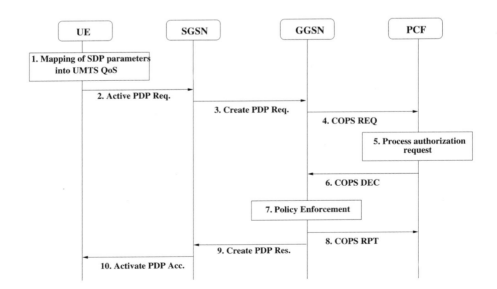

Fig. 6.12 *Resource reservation with SBLP (service based local policy)*

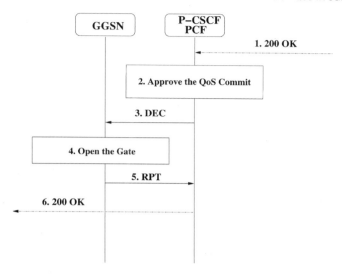

Fig. 6.13   *Approve the QoS Commit*

based on the same framework: The PCF makes a decision, which is enforced by the GGSN.

Figure 6.17 summarizes the end-to-end QoS management by utilizing the policy-based management. As indicated in Figure 6.11, UE uses SIP with SDP to communicate with the P-CSCF. Based on the SBLP, the PCF authorizes the IP QoS and passes the authorized QoS parameters to the IP BS Manager in the GGSN by G_o interface. The GGSN enforces the policy decision received from the PCF. The

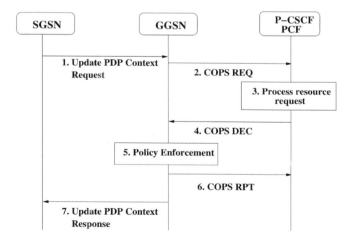

Fig. 6.14   *Authorizations of PDP context modification*

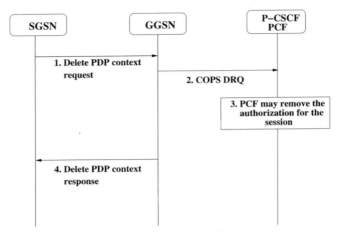

Fig. 6.15 *Release of PDP context initiated by SGSN*

Translation/Mapping functions in the GGSN and the UE convert the IP QoS parameters to the UMTS QoS attributes.

6.4 QoS IN 3GPP2

Many important discussions of QoS mechanisms in 3GPP2 were still going on at the time this book was completed. Similar to 3GPP, the 3GPP2 QoS management for packet data services is based on the IP QoS mechanisms discussed in Section 6.1.

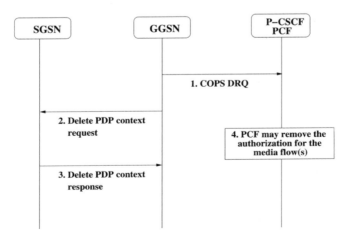

Fig. 6.16 *Release of PDP context initiated by GGSN*

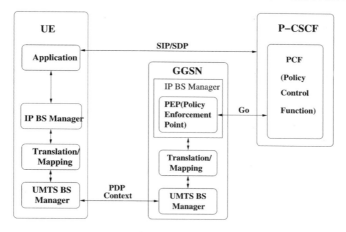

Fig. 6.17   *QoS mapping from application to UMTS bearer*

This section focuses on the system aspects of 3GPP2 QoS based on the specifications defined in [1]–[8].[7]

6.4.1   3GPP2 QoS Architecture

The specification of 3GPP2 S.R0035 [4] defines the requirements for QoS from the perspective of users and network operators. Major QoS principles and practices identified in [4] are itemized as follows:

- The number of user-defined/controlled attributes should be held to a minimum.
- The derivation of QoS attributes from the application requirements should be readily discernible.
- QoS definitions should be unaffected by the introduction of new technologies.
- QoS attributes, or the mapping of them, should not be restricted to one or a few external QoS control mechanisms, but the QoS concept shall be capable of providing different levels of QoS by using specific control mechanisms, not related to QoS mechanisms in the external network.
- All QoS attributes should have unambiguous meaning.
- QoS mechanisms should allow for efficient use of radio capacity.
- QoS should allow for the independent evolution of the core and access networks.
- QoS should allow evolution of the network, eliminating or minimizing the impact of evolution of transport technologies in wireline technologies.

[7]The specification of [5] was not approved yet officially by 3GPP2 when this book was completed.

- The overhead and additional complexity resulting from the QoS scheme should be kept reasonably low, as should the amount of state information transmitted and stored in the network.
- QoS should support efficient resource utilization.
- The number of QoS attributes should be kept to a minimum; increasing the number of attributes will increase system complexity.
- The system should support dynamic QoS behavior (i.e., it shall be possible to modify QoS attributes during an active session.)

Figure 6.18 depicts the 3GPP2 QoS architecture [5]. Similar to 3GPP, 3GPP2 also takes a layered approach to QoS management. In general, application layer QoS requirements between end hosts are identified by SIP/SDP. The QoS requirements in the application layer then are mapped into IP layer QoS parameters. Diff-Serv is utilized to control the Core Network Bearer Service. The External Bearer Service is owned and operated by another service provider. The IP layer QoS parameters are further mapped into link layer. As illustrated in Figure 6.18, the link layer is consisted of cdma2000 Radio Bearer Service and R-P Bearer Service, which utilize the cdma2000 Radio Transport Service and the R-P Transport Service, respectively, to provide physical layer services. CN, BR, HA, PDSN, AGW, RAN, and MS in Figure 6.18 stand for Correspondent Node, Border Router, Home Agent, Packet Data Serving Node, Access Gateway, Radio Access Network, and Mobile Station, respectively.

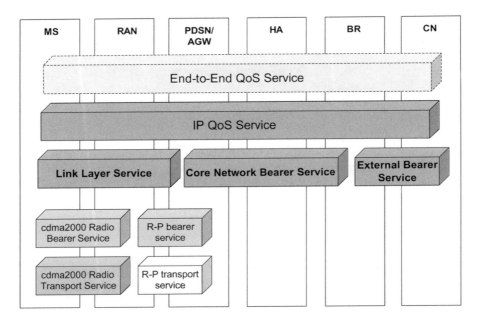

Fig. 6.18 *3GPP2 QoS architecture*

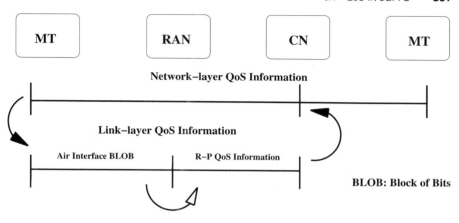

Fig. 6.19   *3GPP2 QoS model*

At the network layer, the policy for IP level Diff-Serv processing is stored in a database accessible by AAA and other QoS components. The characteristics of air link QoS are stored in HLR. Figure 6.19 illustrates the mapping between network (IP) layer QoS information and link layer QoS information [4]. The network layer QoS is maintained from the originating MT (Mobile Terminal) to the target MT. The QoS parameters specified in network layer are mapped into the link layer QoS parameters from the originating MT before traffic enters the RAN. In the edge of the core network (CN),[8] the QoS parameters are converted back to the network-layer QoS parameters.

As depicted in Figure 6.19, the link layer QoS information is further divided into *air interface BLOB* and *R-P QoS information.* The air interface BLOB (Block of Bits) specifics the radio-perspective QoS parameters that are presented by *a block of bits.* The MT sets the field in BLOB to indicate the QoS parameters associated with the user. Details of QoS parameters will be discussed in Section 6.4.4. From RAN to CN, the air interface BLOB is mapped into the R-P QoS information, which specifics the QoS parameters in R-P session. As defined in [2], an R-P interface is the interface between RAN and PDSN. It is the point where the radio-dependent network conjuncts with the radio-independent core network. The PDSN indicates the Diff-Serv class on the R-P session to the RAN for each PPP frame. On a PDSN, the R-P QoS information is then mapped back to the IP level QoS information. The RAN should maintain a QoS profile for each service request. As shown in Figure 6.20, a single PPP session should be associated with the R-P session [8]. A PPP session is maintained between PDSN and MS. The air interface is capable of supporting multiple service instances of the same PPP session. In cdma2000, it specifies that a maximum of six service instances per MS should be supported.

---

[8]In 3GPP2, CN may stand for Correspondent Node or Core Network. It should be apparent based on the context.

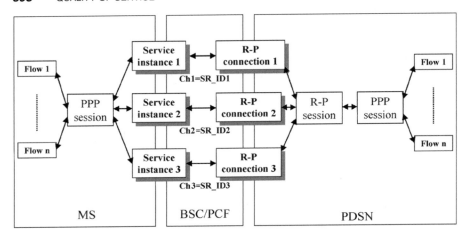

Fig. 6.20 *3GPP2 QoS R-P session*

A specific instance is identified by a unique service number called *Service Reference ID (SR_ID)*.

6.4.2 3GPP2 QoS Management

Figure 6.21 depicts the control plane and user plane of the 3GPP2 QoS architecture in the packet core network [3]. The RAN is based on cdma2000. The core network is based on IP, therefore it is independent of any radio access technologies. The QoS in the packet core network should employ Diff-Serv, although it could also work with Int-Serv. Two major components in user plane are *Access Gateway (AGW)* and *Border Router (BR)*. They are in the boundary of the core network to interface with RAN and external IP networks. In addition to AAA server and various databases, *Core Quality of Service Manager (CQM)* and *Subscription Quality of Service Manager (SQM)* are introduced in the control plane.

The CQM is the primary control entity to manage QoS in the core network. By consulting with the policy database and AAA server, the CQM makes policy decisions for the usage of the core network resource. It makes decisions based on QoS policy and instructs the AGW and the BR to enforce the decision. Effectively, the CQM is the PDP. The AGW and the BR are PEPs. The policy decision made by the CQM is sent to both AGW and BR to condition the traffic accordingly.

The SQM manages QoS resources based on a subscription basis. It communicates with the subscription database and AAA server to authorize the resources for a given subscription. Unlike CQM, the SQM provides the QoS management for each subscriber. The SQM resides in the home network and tracks the resource allocation for each user subscribed to the home network. It accepts or rejects a subscriber's request based on the policy rules and the resources that have been allocated to the

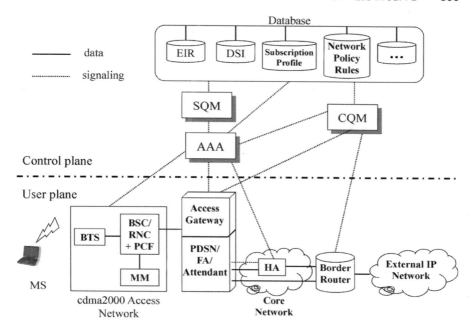

Fig. 6.21   *3GPP2 QoS management in control plane and user plane*

subscriber. The decision is sent to the AAA server to authorize network resources for the subscriber.

The 3GPP2 AGW interfaces the cdma2000 RAN with the core network. It consists of PDSN and other logical functions. The 3GPP2 AGW provides the core network with access to the resources in the cdma2000 RAN. When the AGW receives a QoS request from an MS, it consults with the AAA server for user authentication. The QoS allocation request is forwarded to the CQM, which authorizes the core network QoS resources. Uplink traffic from MS to the core network is aggregated into IP flows for core network services. The AGW conditions user traffic according to the policy decision made by the CQM. Downlink traffic is conditioned as well. Each individual IP flow is mapped into logical link connection in the RAN. The AGW might also propagate appropriate policy decision to the RAN for policy enforcement within the RAN. In addition to QoS mechanisms, AGW supports mobility and handoff as well. It serves as a Foreign Agent (FA) for IPv4 and provides attendant functionality for IPv6. It is the point where NAI (network address identity) maps into MS's identity. Besides, it gathers accounting information in the RAN, which are forwarded to the AAA server. The AGW depicted in Figure 6.21 is a *cdma2000 AGW*. There might be other types of AGW to interface with other types of RANs such as 802.11 and UTRAN. The AGW provides the core network with a common interface to numerous access network technologies. It hides access-specific features thus enables an access-independent core network.

Like AGW, a BR is also in the boundary of core network and enforces policy decision made by the CQM. A BR, however, connects the core network to external IP networks rather than an access network. The functionality of the BR in Figure 6.21 essentially is the same as a Diff-Serv border/edge router discussed in Section 6.1.2. It routes IP packets and conditions incoming and outgoing IP traffic according to the Service Level Specification (SLS).

The databases (DB) in the core network includes EIR (Equipment Identify Register), Dynamic Subscriber Information (DSI), Network Policy Rules, Subscription Profile, and other necessary information that is not shown in Figure 6.21. The EIR contains the equipment-related information. The DSI keeps track of current session registration information. The database of the Network Policy Rules records various rules for QoS-related policies such as subscription resource usage and expected QoS. The Network Policy Rules also provides network-wide policies for AAA and other control entities in the network. As the name implies, the database of the Subscription Profile stores subscriber-specific information.

When an MS initiates a connection, it may mark and classify data traffic if the MS is Diff-Serv aware. The user data are then forwarded to the PDSN by going through RAN and the R-P session. The PDSN may accept the traffic marking and classification executed by the MS. It could also remark packets according to the Service Level Specification or the AAA profile. If the packet is not marked by the MS, the PDSN performs traffic marking to classify packets. The PDSN may also condition traffic based on the service agreement and/or QoS policy. After that, packets are forwarded aggregately inside the core network in accordance with the Diff-Serv requirements. For downlink traffic from core network to MS, the PDSN will process data traffic received from core network in accordance with QoS policy and Diff-Serv requirements. Traffic, again, may be conditioned. User data are forwarded from PDSN to RAN over the R-P session, and then to the MS. Similar to 3GPP, in 3GPP2 supporting Diff-Serv is optional for MS, but it is mandatory for PDSN.

6.4.3   3GPP2 QoS Classes

3GPP2 S.R0035 [4] specifies that at least four traffic classes should be supported: Background Class, Interactive Class, Streaming Class, and Conversational Class. They are defined according to traffic loss, required bandwidth, and tolerance for delay (latency) and jitter. As identified in 3GPP2 S.R0035, *Traffic (Packet) Loss* is the discarding/dropping of packets due to errors or network congestion. *Bandwidth* is the system's ability to provide the capacity necessary to support the throughput requirements for the user's application. *Latency* is the amount of time that it takes to send a packet from a sending node to a receiving node. *Jitter* is a measure of the variation in delay between the arrival of packets at the receiver. It generally occurs where packets are competing for a shared link. Table 6.9 summarizes the traffic classes identified by 3GPP2. Compared with the UMTS QoS classes in Section 6.3.3, they practically are identical.

TABLE 6.9   3GPP2 QoS classes

Traffic Class	Characteristics
Conversational	Two-way, low delay, low data loss rate, sensitive to delay variations.
Streaming	Same as conversational but oneway, less sensitive to delay. May require high bandwidth.
Interactive	Two-way, bursty, variable bandwidth requirements; moderate delay, moderate data loss rate, correctable in part.
Background	Highly tolerant to delay and data loss rate; has variable bandwidth.

In addition to the four classes of traffic described above, 3GPP2 traffic is delivered in two different modes: *Assured Mode* and *Unassured Mode* [7]. In assured mode, a PDU (Packet Data Unit) is guaranteed to be delivered to the destination. On the other hand, it does not guarantee the delivery of a PDU to the destination in unassured mode.

6.4.4   QoS Attributes (QoS Profile)

Similar to that in 3GPP, each subscriber profile contains a QoS profile that maps a contracted traffic class to specific QoS parameters. Each traffic class is defined by a set of attributes. As stated above, bandwidth, delay, jitter, and traffic loss are four attributes that specify the characteristics of an application. The uplink and downlink may have different values to support asymmetric traffic. The IP layer QoS profile should be based on Diff-Serv marking attributes. For RAN QoS profile, it should at least contain *requested QoS class*, *maximum forward/reverse data rate*, and *maximum forward/reverse delay*. Detailed attributes and the allowable values depend on the type of services. Table 6.10 lists the QoS parameters for service option of *cdma2000 High Speed Packet Data* [1]. Each QoS parameter is identified by a bit or few bits. For instance, there is one bit to specify mode of assurance. For the parameters of *Non-assured Priority* and *Forward/Reverse Link Priority*, there are four bits for 16 relative priority levels[9] to prioritize the traffic. The minimum data rate is identified by four bits. As shown in Table 6.10, it could be 8 Kbps, 32 Kbps, 64 Kbps, 144 Kbps, and 384 Kbps. Combining those bits listed in Table 6.10, the BLOB (Block of Bits) is formulated to specify the QoS parameters.

6.4.5   Management of End-to-End IP QoS

The end-to-end QoS management is still under discussion by 3GPP2 TSG-S (Technical Specification Group—Service and System Aspects). Figure 6.22 illustrates

---

[9]Two of the 16 priority levels are reserved.

TABLE 6.10 QoS parameter in 3GPP2 packet data service

QoS Parameter	Length (bits)	Allowable Value(s)
Assured Mode	1	0−Nonassured mode packet data service. This is the default value. 1−Assured mode packet data service.
Non-assured Priority	4	Applies only to nonassured mode. The priority referenced herein is the user's priority associated with nonassured mode packet data service.
Forward Link Priority; Reverse Link Priority	4	Applies only to assured mode. The priority referenced herein is the user's priority associated with assured mode packet data service.
Forward Link Minimum Requested User Data Rate; Forward Link Minimum Acceptable User Data Rate; Reverse Link Minimum Requested User Data Rate; Reverse Link Minimum Acceptable User Data Rate;	4	Applies only to assured mode. 0001−8 Kbps 0010−32 Kbps 0011−64 Kbps 0100−144 Kbps 0101−384 Kbps
Forward Link Requested Data Loss Rate; Forward Link Acceptable Data Loss Rate; Reverse Link Requested Data Loss Rate; Reverse Link Acceptable Data Loss Rate	4	Applies only to assured mode. If RLP does not use its ARQ mechanism, data loss rate is defined as begin numerically equal to the Frame Error Rate. If RLP uses its ARQ mechanism, data loss rate is defined as the ratio of the number of lost data octets to the number of transmitted data octets, measured above RLP. 0001−1% 0010−2% 0011−5% 0100−10%
Forward Link Requested Maximum Delay; Forward Link Acceptable Maximum Delay; Reverse Link Requested Maximum Delay; Reverse Link Acceptable Maximum Delay	4	Applies only to assured mode. Maximum delay is defined as the amount of time user data can be held in the transmit queue (i.e., from the moment it is submitted to RLP for transmission until its actual transmission on a physical channel). The user data may be discarded if the maximum delay restriction is not met. 0001−40 ms 0010−120 ms 0011−360 ms

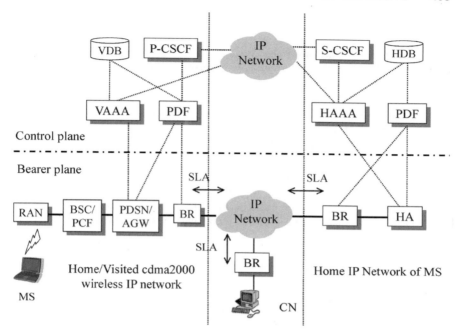

Fig. 6.22   *Reference architecture for end-to-end QoS management*

the reference architecture for end-to-end QoS management specified in the draft version of 3GPP2 S.P0079 [5]. As that discussed in Section 6.4.2, the architecture is independent of radio access technologies although cdma2000 access network is presumed in 3GPP2. Figure 6.22 essentially is extended from Figure 6.21 for end-to-end scenario. In Figure 6.22, home network and visited network are connected by an IP network. The home network maintains a Home AAA (HAAA) server and a Home Database (HDB). In visited network, there is a Visited AAA (VAAA) and a Visited Database (VDB). When the MS initiates a connection to the Correspondent Node (CN), the request is authenticated by the VAAA with the QoS profile stored in VDB. The VAAA may further consult with the HAAA for more information regarding the user. The PDF (Policy Decision Function) in visited network virtually functions like the CQM in Figure 6.21. It makes a QoS decision based on policies stored in related database and instructs the AGW and BR to enforce the policy decision. The AGW classifies and conditions packets to ensure traffic is conformed with service agreement. Each IP network practically is a Diff-Serv domain. The BR conditions traffic from one IP network domain to another IP network domain based on the SLA between each domain. For IP Multimedia Subsystem (IMS), P-CSCF and S-CSCF are served as the SIP servers to establish an end-to-end session. Please refer to Chapter 3 for details of IMS.

REFERENCES

1. 3rd Generation Partnership Project 2 (3GPP2). Data service options for spread spectrum systems addendum 2. 3GPP2 C.S0017-0-2, Version 2.0, August 2000.

2. 3rd Generation Partnership Project 2 (3GPP2). Wireless IP architecture based on IETF protocols. 3GPP2 P.R0001, Version 1.0.0, July 2000.

3. 3rd Generation Partnership Project 2 (3GPP2). IP network architecture model for cdma2000 spread spectrum systems. 3GPP2 S.R0037-0, Version 2.0, May 2002.

4. 3rd Generation Partnership Project 2 (3GPP2). Quality of service, stage 1 requirements. 3GPP2 S.R0035, Version 2.0, September 2002.

5. 3rd Generation Partnership Project 2 (3GPP2). Support for end-to-end QoS, stage 1 requirements. 3GPP2 S.P0079, Version 0.05, December 2002.

6. 3rd Generation Partnership Project 2 (3GPP2). TIA/EIA-41-D based network enhancements for CDMA packet data service (C-PDS), phase 1.3GPP2 N.S0029, Version 1.0.0, Revision 0, June 2002.

7. 3rd Generation Partnership Project 2 (3GPP2). Upper layer (layer 3) signaling standard for cdma2000 spread spectrum systems. 3GPP2 C.S0005-C, Version 1.0, Release C, May 2002.

8. 3rd Generation Partnership Project 2 (3GPP2). Wireless IP network standard. 3GPP2 P.S0001-B, Version 1.0.0, October 2002.

9. 3rd Generation Partnership Project (3GPP), Technical Specification Group Core Network. End to end quality of service (QoS) signalling flows. 3GPP TS 29.208 Version 5.1.0, September 2002.

10. 3rd Generation Partnership Project (3GPP), Technical Specification Group Core Network. Policy control over Go interface, release 5. 3GPP TS 29.207 Version 5.1.0, September 2002.

11. 3rd Generation Partnership Project (3GPP),Technical Specification Group, Radio Access Network. RAB quality of service negotiation over Iu, release 4. 3GPP TR 25.946, Version 4.0.0, March 2001.

12. 3rd Generation Partnership Project (3GPP),Technical Specification Group, Radio Access Network. RAB quality of service renegotiation over Iu, release 4. 3GPP TR 25.851, Version 4.0.0, March 2001.

13. 3rd Generation Partnership Project (3GPP), Technical Specification Group Services and System Aspects. End-to-end quality of service (QoS) concept and architecture, release 5. 3GPP TS 23.207, Version 5.5.0, September 2002.

14. 3rd Generation Partnership Project (3GPP), Technical Specification Group, Services and System Aspects. General packet radio service (GPRS) service description, stage 2, release 5. 3GPP TS 23.060, Version 5.2.0, June 2002.

15. 3rd Generation Partnership Project (3GPP), Technical Specification Group Services and System Aspects. Quality of service (QoS) concept and architecture, release 5. 3GPP TS 23.107, Version 5.6.0, September 2002.

16. B. Awerbuch, I. Cidon, I.S. Gopal, M. Kaplan, and S. Kutten. Distributed control for PARIS. In *Proc. of the Ninth ACM Symposium on Principles of Distributed Computing*, pp. 145–159, Quebec, Canada, August 1990.

17. J.C.R. Bennett, and H. Zhang. WF^{2}Q: worst-case fair weighted fair queueing. In *Proc. of IEEE INFOCOM*, pp. 120–128, San Francisco, CA, March 1996.

18. S. Blake, D. Black, M. Carlson, E. Davies, Z. Wang, and W. Weiss. An architecture for differentiated services. IETF RFC 2475, December 1998.

19. R. Braden, D. Clark, and S. Shenker. Integrated services in the Internet architecture: an overview. IETF RFC 1633, June 1994.

20. R. Braden, L. Zhang, S. Berson, S. Herzog, and S. Jamin. Resource reservation protocol (RSVP). IETF RFC 2205, September 1997.

21. A. Charny, J.C.R. Bennett, K. Benson, J.Y. Le Boudec, A. Chiu, W. Courtney S. Davari, V. Firoiu, C. Kalmanek, and K.K. Ramakrishnan. Supplemental information for the new definition of the EF PHB (expedited forwarding per-hop behavior). IETF RFC 3247, March 2002.

22. J.-C. Chen, A. McAuley, V. Sarangan, S. Baba, and Y. Ohba. Dynamic service negotiation protocol (DSNP) and wireless Diffserv. In *Proc. of IEEE International Conference on Communications (ICC)*, New York, NY, April 2002.

23. W.T. Chen and L.C. Huang. RSVP mobility support: a signaling protocol for integrated services Internet with mobile hosts. In *Proc. of IEEE INFOCOM*, pp.1283–1292, 2000.

24. I. Cidon and I.S. Gopal. PARIS: an approach to integrated high-speed private networks. *International Journal of Digital and Analog Cabled Systems*, pp. 77–86, April 1988.

25. B. Davie, A. Charny, J.C.R. Bennett, K. Benson, J.Y. Le Boudec, W. Courtney, S. Davari, V. Firoiu, and D. Stiliadis. An expedited forwarding PHB (per-hop behavior). IETF RFC 3246, March 2002.

26. D. Durham, J. Boyle, R. Cohen, S. Herzog, R. Rajan, and A. Sastry. The COPS (common open policy service) protocol. IETF RFC 2748, January 2000.

27. S. Floyd and V. Jacobson. Random early detection gateways for congestion avoidance. *IEEE/ACM Transactions on Networking*, pp. 397–413, August 1993.

28. S. Golestani. A self-clocked fair queueing scheme for broadband applications. In *Proc. of IEEE INFOCOM*, pp. 636–646, Toronto, Ont., Canada, June 1994.

29. P.Goyal, H.M. Vin, and H. Cheng. Start-time fair queuing: a scheduling algorithm for integrated services packet switching networks. In *Proc. of ACM SIGCOMM*, pp. 157–168, Stanford, CA, August 1996.

30. D. Grossman. New terminology and clarifications for diffserv. IETF RFC 3260, April 2002.

31. J. Heinanen, F. Baker, W. Weiss, and J. Wroclawski. Assured forwarding PHB group. IETF RFC 2597, June 1999.

32. S. Herzog, J. Boyle, R. Cohen, D. Durham, R. Rajan, and A. Sastry. COPS usage for RSVP. IETF RFC 2749, January 2000.

33. J. Hodges and R. Morgan. Lightweight directory access protocol (v3): technical specification. IETF RFC 3377, September 2002.

34. I. Mahadevan and K.M. Sivalingam. Architecture and experimental results for quality of service in mobile networks using RSVP and CBQ. *ACM Wireless Networks*, pp. 221–234, 2000.

35. S.I. Maniatis, E.G. Nikolouzou, and I.S. Venieris. QoS issues in the converged 3G wireless and wired networks. *IEEE Communications Magazine*, 40(8):44–53, August 2002.

36. B. Moore, E. Ellesson, J. Strassner, and A. Westerinen. Policy core information model—version 1 specification. IETF RFC 3060, February 2001.

37. M.N. Moustafa, I. Habib, M. Naghshineh, and M. Guizani. QoS-enabled broadband mobile access to wireline networks. *IEEE Communications Magazine*, 40(4):50–56, April 2002.

38. K. Nichols, S. Blake, F. Baker, and D. Black. Definition of the differentiated services field (DS field) in the IPv4 and IPv6 headers. IETF RFC 2474, December 1998.

39. T. Robles, A. Kadelka, H. Velayos, A. Lappetelainen, A. Kassler, L. Hui, D. Mandato, J. Ojala, and B.Wegmann. QoS support for an all IP system beyond 3G. *IEEE Communications Magazine*, 39(8):64–72, August 2001.

40. S. Shenker, C. Partridge, and R. Guerin. Specification of guaranteed quality of service. IETF RFC 2212, September 1997.

41. M. Shreedhar and G. Varghese. Efficient fair queuing using deficit round robin. *In Proc. of ACM SIGCOMM*, pp. 231–242, September 1995.

42. A.K. Talukdar, B.R. Badrinath, and A. Acharya. MRSVP: a resource reservation protocol for an integrated services network with mobile hosts. *ACM Wireless Networks*, pp. 5–19, 2001.

43. A. Westerinen, J. Schnizlein, J. Strassner, M. Scherling, B. Quinn, S. Herzog, A. Huynh, M. Carlson, J. Perry, and S.Waldbusser. Terminology for policy-based management. IETF RFC 3198, November 2001.

44. J. Wroclawski. Specification of the controlled-load network element service. IETF RFC 2211, September 1997.

45. J. Wroclawski. The use of RSVP with IETF integrated services. IETF RFC 2210, September 1997.

46. R. Yavatkar, D. Pendarakis, and R. Guerin. A framework for policy-based admission control. IETF RFC 2753, January 2000.

47. T. Zhang, E. van den Berg, J. Chennikara, P. Agrawal, J.-C. Chen, and T. Kodama. Local predictive resource reservation for handoff in multimedia wireless IP networks. *IEEE Journal on Selected Areas in Communications*, pp 1931–1941, August 2001.

Index